P9-DWA-234

BIOTERRORISM and FOOD SAFETY

BIOTERRORISM and FOOD SAFETY

Barbara A. Rasco
Gleyn E. Bledsoe

CRC PRESS

Boca Raton London New York Washington, D.C.

Library of Congress Cataloging-in-Publication Data

Rasco, Barbara.
Bioterrorism and food safety / Barbara A. Rasco, Gleyn E. Bledsoe.
p. cm.
Includes bibliographical references and index.
ISBN 0-8493-2787-3
1. Food—United States—Safety measures. 2. Bioterrorism—United States. I. Bledsoe, Gleyn E. II. Title.

TX531.R37 2005
363.19′2—dc22 2004057049

Direct all inquiries to CRC Press, 2000 N.W. Corporate Blvd., Boca Raton, Florida 33431.

Visit the CRC Press Web site at www.crcpress.com

International Standard Book Number 0-8493-2787-3
Library of Congress Card Number 2004057049
Printed in the United States of America 1 2 3 4 5 6 7 8 9 0
Printed on acid-free paper

Preface

Food security is an extension of food safety programs into a new arena. If only it were possible to return to the time when food security referred to a sufficient and wholesome food, and not to one threatened by intentional contamination. Through this book, we integrate food safety issues, technological developments in traceability, and legal analysis of current and pending regulations with good business practices and tie these to the development of effective and workable food security programs for food businesses. It has been difficult to decide how much attention to give to any particular point, with the intent to provide an overview in an area where we predict there will be rapid advances on the technical, trade, and legal fronts.

Specific sections are provided on biological and chemical hazards *du jour,* with an emphasis on select agents and food-borne pathogens and a synopsis of chemical agents that have been used or could be used to intentionally contaminate food. An analysis of the legal ramifications to food business operations and the international trade in food from new federal food bioterrorism regulations is presented. Development of food security plans based on the extension of current Hazard Analysis Critical Control Point (HACCP) programs or upon Organizational Risk Management (ORM) models is discussed, with suggestions provided on how to develop realistic, effective, and workable food security plans. Advances in traceability for food products and ideas on how to reduce the risk of intentional contamination and improve consumer confidence are also presented.

This work is dedicated to all those who have sacrificed to make the world a safer place, particularly individuals placed in harm's way far from home. In addition, we acknowledge the moral support of our colleagues during this project, particularly Dr. William LaGrange, Scientific Editor, *Food Protection Trends,* for his lifelong service to the profession.

The Authors

Barbara A. Rasco, Ph.D., J.D., is a distinguished lecturer with the Institute of Food Technologists. Dr. Rasco is an attorney and an internationally recognized expert in food laws and food safety regulations. She is currently a professor at Washington State University and formerly a faculty member at the University of Washington. Prior to this, she worked in the private sector as an engineer, research chemist, and quality control manager. Dr. Rasco is licensed to practice law in Washington State and in federal court and advises in the areas of food safety, food security, products liability, and environmental and administrative law. Dr. Rasco conducts research in food processing technology, noninvasive analytical methods, seafood technology, and aquaculture. She has a Ph.D. in food science from the University of Massachusetts, a B.S.E. in biochemical engineering from the University of Pennsylvania, and is a graduate from the Seattle University School of Law. Dr. Rasco has over 90 publications and patents.

Gleyn E. Bledsoe, Ph.D., M.B.A., C.P.A., is an adjunct professor in the College of Agriculture, Human and Natural Resource Sciences at Washington State University. Before coming to Washington State, he was dean of cooperative extension and research at a regional university. As a certified public accountant, Dr. Bledsoe provides management advisory services to fishery, agricultural, and food processing companies and has served on several U.S. Agency for International Development (USAID) projects in Asia, Eastern Europe, and Africa. Dr. Bledsoe has extensive experience in the private sector. He was president and chief financial officer of a multimillion-dollar vertically integrated food processing company. While in industry, Dr. Bledsoe negotiated numerous marketing and processing agreements with European and Asian companies. He fished commercially in Alaska and Washington and has designed and built shore-based and onboard processing facilities, harvest systems, at-sea tendering operations, and automated vessel systems. He has also supervised vessel conversions, negotiated shipyard contracts, and established onboard and in-plant quality control programs. Dr. Bledsoe served in the U.S. Air Force (retired lieutenant colonel) as an intelligence officer with expertise in counterinsurgency and counterintelligence. He has a Ph.D. in fisheries/seafood engineering from the University of Washington, an M.B.A. in management and finance from the University of Idaho, and a B.S.E. from the University of Washington.

Together, Dr. Rasco and Dr. Bledsoe have conducted over 120 food safety and Hazard Analysis Critical Control Point (HACCP) classes for food industry professionals, as well as continuing education programs for attorneys and accountants since 1996. They have been working in the areas of food bioterrorism and industry preparedness for the past 9 years. Their focus is on how to develop realistic, rational, and implementable food security strategies. Of current interest are the new federal public health and food security bills and regulations that have emerged in response to recent terrorist threats and the possible impact these new laws may have on trade and food inspection programs, changes in enforcement strategies, and their impact upon the regulated industry. Drs. Bledsoe and Rasco have estimated the cost of food security programs and possible funding mechanisms, such as investment tax credits, as a means of improving private-sector preparedness. Their research also addresses how environmental issues (e.g., genetically modified (GM) foods, environmental legislation and regulation) may affect liability and financial viability of food businesses.

Contents

Appendix G — Retail Food Stores and Food Service Establishments: Food Security Preventive Measures Guidance 293

Appendix H — Food Producers, Processors, and Transporters: Food Security Preventive Measures Guidance 307

Appendix I — Importers and Filers: Food Security Preventive Measures Guidance 321

Appendix H — Cosmetics Processors and Transporters Cosmetics Security Preventive Measures Guidance ... 335

Appendix I — Traceability in the U.S. Food Supply: Economic Theory and Industry Studies ... 347

Index ... 399

Foods and the Bioterrorist Threat

1

> Terrorism or terror is the enemy of all of us, not the enemy of America. So when we fight terrorism we do that for our own selves.
>
> — **Muammar Kaddafi**[1]

What Is Food Terrorism?

Terrorism is commonly defined as the use of force or violence against persons or property in violation of criminal laws for the purpose of intimidation, coercion, or ransom (FEMA, 1998). The intent of terrorism is to cause property damage, physical injury, or economic damage to people or to an entity, such as a government, corporation or research institute. Terrorists often use threats to generate publicity for their cause while creating fear among the public and convincing citizens that their government is powerless to prevent attacks. An underlying purpose of terrorism is to generate fear and anarchy, hopelessness and hate. A terrorism attack can take a number of different forms — dependent upon the technology available, the nature of the underlying political issue, and the strength of the target. Bombing is the most frequently used tactic in the U.S. and around the world; recall the recent events in Spain, Iraq and Israel, Indonesia, Russia and Western Europe, and, closer to home, at the World Trade Center, Capitol Building and Pentagon, Mobil Oil Headquarters, and the Federal Building in Oklahoma City. More subtle forms of attack include assaults on transportation systems, utilities, public services, and critical infrastructure such as water and food.

On January 30, 2004, President George W. Bush issued the Homeland Security Presidential Directive (HSPD-9) establishing a national policy to defend the agriculture and food system against terrorist attack. It notes that the food system is vulnerable to introduction of disease, pests, or poisonous agents and susceptible to attack due to its extensive, open, interconnected, diverse, and complex structure. Provisions in this directive are outlined in Chapter 4.

Food terrorism has been defined by the World Health Organization (WHO) as "an act or threat of deliberate contamination of food for human consumption with chemical, biological or radionuclear agents for the purpose of causing injury or death to civilian populations and/or disrupting social, economic, or political stability" (WHO, 2003). These agents are inherently terrifying and among the most extreme forms of random violence (Stern, 1999). We would further expand this definition to include deliberate contamination or infection of plants and animals used as food, as the fear associated with contaminated food has the same visceral impact.

Biological terrorism, or *bioterrorism*, involves the use of etiologic or biological toxins in a terrorist act. The term *bioterrorism* has commonly been applied to acts of ecoterrorism as well, since ecoterrorism often involves biological agents and targets (e.g., plots of allegedly genetically modified crops) or ecosystem issues (e.g., forest practices, biodiversity, sustainable agriculture). In response to terrorist threats to the food supply, antiterrorism and counterterrorism strategies have been developed and employed. *Antiterrorism* covers the defensive measures used to reduce the vulnerability of individuals and property to terrorist acts, and *counterterrorism* refers to offensive measures to prevent, deter, and respond to terrorism. The current buzzword *biodefense* is used to encompass both anti- and counterterrorism activities. Unfortunately, the U.S. government has chosen the term *food security* to cover issues associated with intentional contamination of food, causing a great deal of consternation within the public health community. This is because *food security* has been used for decades in the context of *food sufficiency* — having an adequate and reliable supply of wholesome and nutritious food.

Where Does the Threat Come From?

The events of September 11, 2001 focused the world's attention on terrorism and the threat of future terrorist acts. There is an increasing threat from terrorist groups against food research, production, and processing internationally. Until the recent mail attacks involving anthrax in 2001, and ricin in 2003 and 2004, the focus of the public health community regarding bioterrorism involved the potential use of biological weapons (weapons of mass destruction (WMD)) by international terrorist organizations and how the release of biological, chemical, or radionuclear materials could contaminate a food or water supply (WHO, 2003). However, as we are all well aware, the use of anthrax or other pathogenic agents on even a relatively small scale can rapidly overwhelm the response mechanisms in place to deal with the perceived threat.

The deliberate use of these agents to sabotage the food supply has only recently received attention (WHO, 2003). The purpose here is to provide assistance with establishing new food safety systems and strengthening existing food safety programs to address food terrorism, including precautionary measures that organizations can take to prevent or limit the impact of an intentional contamination incident. Recent political events have made food security an area of greater importance.

Gulf War II in Iraq has brought together radical activists of all stripes. The antiwar movement and the radical environmental and animal rights movements have converged around this issue and used it to bolster its support for activities against "the establishment." On March 17, 2003, the first night of the war, demonstrators in Portland, OR, closed off part of the downtown and curtailed public transit during the evening rush hour and into the evening. Following this, they proceeded to the convention center and vandalized a McDonald's restaurant in the vicinity (writing "Meat is McMurder" on the restaurant) and a vehicle dealership. Similar incidents occurred in Seattle, WA, and other large cities prior to and during the early days of the war. As these events show, the greatest fear in the U.S. mind may be of foreign attacks on our food or water supply, but the greatest threat is probably homegrown. Recent history supports this position. The "painted cows" in western Washington in 2004, The Dalles, OR, incident in 1984, contaminated letters in 2001–2004, and intentional incidents of food contamination involving hospital staffs have been instigated and conducted by longtime residents of the U.S.

Unfortunately, the negative impact on communities targeted by terrorists is long lasting, and these impacts are often overlooked after the initial shock and subsequent media coverage subside. Terrorists violate communities and the sanctity of place, injure individuals and indirectly the victims' neighbors and friends, and destroy homes and businesses. Six years later, the residents of Seattle are still dealing with the emotional impact of demonstrations accompanying the World Trade Organization (WTO) summit in November 1999, which resulted in hundreds of arrests and millions of dollars in property damage and lost revenue to city businesses. In these riots, numerous businesses in the downtown area and surrounding neighborhoods were vandalized, disrupting the normal life of the residents and workers there for weeks. How this type of activity can possibly lead to greater support for the cause of the terrorist is beyond comprehension. But then again, these are no longer demonstrations to win the hearts and minds of the citizenry. This is war.

Just about everybody is a potential target of terrorism. In regions of the country where natural resource-based industries are important, there is a heightened level of "activist activity." Agriculture and the associated processing industries have also been popular targets of bio- or ecoterrorists. There seems to be no segment of the private sector immune to attack. Some extremist groups are violently opposed to the development of natural resources. Others consider the "imprisonment and exploitation" of animals and the use of meat and fur anathema. Food and agricultural companies have also been targeted for using or developing genetically modified organisms. Others tied directly or indirectly to natural resource-based industries are also targets.

Within the past 5 years there have been several incidents of animal rights, agricultural or ecoterrorism specifically targeting primary producers, processors, distributors, retailers, shareholders, consumers, vendors/suppliers, investors and investment firms, bankers, insurers, neighboring tenants in leased commercial space, and researchers. Corporations in particular are considered by most terrorist groups to be nonstate or metastate entities and legitimate targets of aggression in their own right (Bascetta, 2000). Universities are deemed culpable through their association with private corporations or corporate foundations. Government research facilities are targeted by groups seeking to make a political statement against an unpopular governmental policy, or for the alleged failure of a governmental agency to take certain types of action that would further the causes of their group. Construction has come to a halt on a research facility in Oxford, England, because of animal rights attacks on the construction firm. Because of these terrorist acts, it has been difficult to find another firm willing to complete the building.

In addition to an actual attack, the use of harmless materials (or even no materials) in a food-tampering incident can be very damaging. A credible hoax can be very effective in precipitating a recall and creating significant economic loss for the company involved. A hoax can negatively impact trade relations as evidenced by the August 2004 "Lemongate" debacle involving imported citrus fruit. (See Chapter 3 for a detailed discussion.) Simply claiming that a product has been purposely contaminated with dangerous material is sufficient to precipitate an extensive product recall with the associated adverse publicity, short-term economic loss, longer-term loss of market share, and resultant economic impact (Bledsoe and Rasco, 2001a, 2001b). The regulatory requirements for a Class I recall are a reasonable probability that the use of or exposure to a violative product will cause adverse health consequences or death (21 Code of Federal Regulations (CFR) § 7.3(m)(1)). However, the indirect costs of a recall can be 10 or more times higher than the costs of recovery and replacement of the goods and the lost value of the recalled food.

Rhodesia had a policy of using food as a weapon during the Liberation War. Initially this campaign involved food rationing, restriction of bulk food purchases, and restrictions on the transfer of food from black workers on white farms to others. Later, dubious tactics were employed. During the last years of the war, special psychological units of the Rhodesian government in conjunction with forces from South Africa injected canned meat with thallium and then provided this meat to insurgents through channels that led the guerillas to believe they were being resupplied by other friendly insurgents (Martinez, 2003). The guerillas gave this thallium-laced meat to innocent villagers, killing them. The insurgents, fearing that the villagers would attempt to poison them, murdered them in a preemptive strike. In related food contamination incidents tied to political insurrections in Africa, holes have been drilled into bottles of whiskey and the contents laced with cyanide, paraoxon and possibly the pesticide parathion. Various other chemical and biological agents were evaluated on civilians and captures during human trials. For example, Warfarin was used to contaminate food, water, and clothing, creating scares that hemorrhagic fever was prevalent in guerilla camps inside Mozambique. Other incidents involved contamination of cigarettes with anthrax spores.

Thousands of food, supplement, and pharmaceutical products each year are subject to malicious tampering and accidental contamination that precipitate a product recall or market withdrawal. Stringent process control and quality assurance programs in the food industry are designed to hopefully prevent, but at least to better contain, a contamination incident. These crisis management programs, preparedness, and response planning will take on a new twist as food becomes more political, as international markets grow, and as price sensitivity increases.

Nonintentional food scares have a major impact, one that governmental incompetence can further magnify. Finding a single bovine spongiform encephalopathy (BSE)-infected "mad cow" in Mabton, WA, caused the beef industry to grind to a halt right before Christmas 2003. The cow had been tested as required by the U.S. Department of Agriculture (USDA); however, the agency permitted release of the animal for processing before the test results had come back.

The infected cow was born on April 9, 1997 on a dairy farm in Calmar, Alberta, Canada and moved in September 2001 along with 80 other cattle from that dairy to operations in Oregon, Washington, and Idaho. About 189 investigations were initiated causing complete herd inventories to be conducted on 51 premises involving an identification determination on roughly 75,000 animals (www.usda.gov/newsroom.0074.04 10 Feb 04). This resulted in locating 255 "animals of interest," defined as animals that could have been on the same farm with the "mad cow," on ten farms in the tri-state region. These animals were "depopulated" although none tested positive. An additional 701 cows were killed (a.k.a. "selectively depopulated") at facilities located primarily in eastern Washington. In addition, 2,000 tons of meat and bone meal were removed from the marketplace because of the possibility that these materials could have been contaminated by protein from the BSE-positive cow. This material was landfilled.

As a result of this snafu, the value of beef and beef products dropped dramatically, with an immediate reduction of more than 20% for both feeder and live cattle. As a precaution, beef was pulled from the market in key West Coast markets, generating a small degree of hysteria, even though most of the product from the contaminated cow would likely have been long gone. The incident was greeted with resignation and cynicism in the eastern U.S., where the impact on beef sales was not as great. As a result of this incident, roughly 4% of the U.S. public stopped eating beef and an additional 16 million consumers reduced beef consumption. Early loss estimates of nearly $10 billion were projected, with an estimated impact on the U.S. retail sector of $3 billion per year from those who had stopped eating beef and roughly $6+ billion per year from those reducing consumption. It was predicted that the U.S. domestic beef market would recover quickly, and there were signs of this by the end of January 2004. Unfortunately, the same cannot be said for export markets, where over 90% of the U.S. exported beef market was placed in jeopardy, with a December 2004 loss of $3 billion in trade.

As bad as the impact was on beef sales, the greatest long-term impact of this single mad cow, however, was in related industries, particularly animal feed production and rendering. Governmental bans on the sale of beef-containing animal feed products have had a major impact on these sectors. Bans have included the sale of products containing no nerve tissues (such as blood) and provide further restrictions on the sale of beef by-products for nonfood uses. The rendering industry has lost sources of raw material, and any remaining suppliers/customers have seen their costs double. In an earlier incident occurring in the summer of 2003, a mad cow was found in Canada, with the U.S. response effectively freezing commerce in animal feed products for more than 6 months. Inappropriate governmental actions tied to both of these mad cow incidents provide little confidence to either the food industry or to consumers that an actual attack against our food supply would be handled in a rational and responsible manner.

Terrorist Strategies and Tactics

Actions by terrorist groups are often well organized and orchestrated and will increase. Terrorists commonly employ both overt and covert intelligence methodologies to damage or destroy property or commerce, threaten public health and safety, and injure people, physically or emotionally tormenting them. A common goal is to put a company or industry out of business by affecting stock value or equity or product availability or marketability in a malicious way. Such a program is directed toward elimination of a specific food, ingredient, or agricultural product or practice. Another strategy is to hamper the importation of competing crops, research, or development in a particular area and impose pressure to erect trade barriers.

Attacks against a country's crops and livestock still remain a viable aggressive weapon in the strategic planning of many governments, particularly those with reduced conventional weaponry. The use of both biological and chemical weapons was a strategy of the Rhodesian Army's psychological operations unit to eliminate "terrorist" attacks on the white minority farmers while retaining what little support remained for the apartheid government by reducing the food and water available for rural villages to provide to guerillas (Martinez, 2003). These well orchestrated attacks over a three- or four-year period included the use of biological and chemical weapons against civilians and agricultural operations. Intentionally contaminated water supplies spread cholera, and livestock were killed with anthrax.

Many of the tools of the food terrorist are cheap and simple. These may include flooding a company's website or e-mail, mail services, or communications systems with harassing correspondence or repeated requests for information; the filing of false consumer complaints with state consumer protection, environmental, food safety, or revenue agencies; targeting a company at business functions and through shareholder, professional, or trade groups; and making false claims of tampering. Other tactics include spurious complaints to regulatory agencies, media "tips," filing frivolous lawsuits or administrative actions, creating potentially huge legal bills for the defendant company, boycotts, lockouts, and publicity stunts. Unfortunately, bombings, arson, product tampering, including poisonings, crop destruction, and facilities vandalism, or the threat of all of these, are also common. Finally, food and animal rights terrorists have harassed and physically attacked employees, suppliers, customers, financiers, and insurance providers.

The terrorist groups perpetrating these acts tend to be organized in small cells with no central organization (or one that is difficult to trace), making it difficult to break such an organization. Most members of these groups blend in well with the local community, may lead otherwise respectable lives, and have no arrest record. For example, Al-Qaeda provides specific instructions to its members on how to blend into Western society to remain below the radar and avoid detection.

Property Destruction: Arson and Vandalism

Property destruction is probably the most common visible tactic employed by ecoterrorists and against food companies, particularly distributors. Incidents have involved sabotaging construction vehicles (e.g., "large yellow machines of death") by sifting salt into dry cement on vehicles involved in a controversial highway project in Minneapolis, MN, in 2000, and burning tract homes in Indiana purportedly threatening a local water supply. The notorious and very sophisticated $12 million arson fire in Vail, CO, in October 1998 remains unsolved despite a thorough investigation. This arson protested the expansion of a resort into prime lynx habitat. There were no arrests from this incident even though a federal grand jury exhaustively questioned everyone known to be on the mountain that night. Several hundred gallons of jellied gasoline were employed, ignited by a series of firebombs set off simultaneously across a half-mile swath. The fire was designed so that alarms would not go off until the buildings were fully engulfed. In a similar vein, Boise Cascade regional headquarters suffered a $1 million fire on Christmas Eve 1999 as a "lesson to all greedy multinational corporations who don't respect ecosystems. The elves are watching …"

Highly detailed bomb construction manuals are available on the Internet. These manuals include simple strategies, such as making cash purchases for bomb components. The purchases may be traceable back to a retailer, but not to an individual consumer. For vandalism activities, instructions provide for the use of gasoline mixtures that are difficult to identify and trace, and how to leave no fingerprints, identifiable tire tracks, or shoe prints (e.g., wear a larger size), and how to handle and dispose of gloves, masks, hoods, and other clothing used in raids (Murphy, 2000).

"Nighttime gardening" is a popular form of terrorism directed at production agriculture of both animals and plants. This is a tactic that involves tampering with or destroying a crop during off hours. There are dozens of these incidents reported each year, with a spike in activity between 1998 and 2001, before new state laws specifically criminalizing these activities were proposed or went into effect. Increased security at both private farms and research institutions (university or otherwise) in the regions where there have been attacks, or for producers with the commodities that have experienced the greatest impact, have been somewhat effective. As a result, in 2002–2004 a shift in activity was observed into regions of the country that have had less experience with these forms of attack. Much of the information promoting or describing attacks was shunted through third parties or "spokesmen." One such group, GenetiXAlert, was active around 2000 promoting an antiagricultural biotechnology "reporting service" issuing "Green Sheets" in a move to educate people about the antiecological, corporate-dominated motives of "so-called public institutions." Many research facilities were targeted numerous times in the late 1990s and early 2000s (http://www.ainsfos.ca/ainfos0872.html). In one case the "elves" motivated by the group destroyed a 0.5 acre test plot of genetically modified (GM) corn at the University of California–Berkeley Gil Tract Research facility. Actions included mixing and removing crop labels, as well as collecting crops in isolation bags, including 5 bushels of cobs, which they mixed together and trampled into the mud. Much of the time nighttime gardening activities include destruction of both conventional and modified crops. Increasingly sophisticated tactics are being employed to discover the exact nature of research activities. This is, at least by some groups, allegedly to avoid destruction of conventional crops. However, this work is more likely conducted to collect intelligence to direct future activities. Sometimes leaves are collected from plants in test plots and sent off for laboratory analysis to confirm if and how they have been genetically modified.

One Green Sheets statement protested what the group calls a second wave of GM research ("pharming") in which crop or horticultural plants are genetically modified to produce drugs or vaccines. The group also claims that multinational corporations are bankrolling university public relations programs (in addition to research programs) and that agricultural colleges have sold out.

For example, Novartis has provided the University of California–Berkeley with tens of millions of dollars in research funding. Research at other colleges with large plant biotechnology programs is funded in a similar manner. As a direct result, experimental plots at corporate sites and other institutions with sponsored research by the same corporation are targeted (for example, the uprooting of crops at a Novartis facility in Hawaii in 2000 and later activities in 2003 and 2004). Even a small amount of corporate-sponsored research has made individuals a target. This is probably one of the most effective psychological tactics employed as it garners significant publicity. Over Christmas break in 1999, the Environmental Liberation Front (ELF) claimed responsibility for the destruction of Michigan State University research laboratories. Years of work were destroyed, with over $1 million in property damage. One researcher was targeted and as a result of this attack lost all slides, papers, notes, books, and teaching materials simply because she administered a mere $2000 project from Monsanto that was to be used to sponsor five African scientists to attend a conference on biotechnology. Much of her department's work was funded by the U.S. Agency for International Development (USAID) and involved training foreign scientists to develop biotechnology research promoting insect and viral resistance in crops.

Products other than food crops are also targeted. For example, the Anarchist Golfing Association (AGA) attacked the Pure Seed Testing facility in Canby, OR, in June 1999 by holding a "nocturnal golfing tournament," sabotaging experimental grass plots by destroying plants, removing or rearranging markers, and leaving graffiti, golf balls, and figurines with their moniker AGA. This incident caused $300,000 to $500,000 in damages and jeopardized

5 or more years of research. The group's statement on its Web site during that period was:

> The biotech industry usually hides behind the racist aura of feeding the third world, but as you can see, it is quite obvious that these crops are grown for the profit and pleasure of the rich and have no social value, better weed-free putting greens for our local corporate execs ... Grass, like industrial culture, is invasive and permeates every aspect of our lives ... While the golf trade journals claim that golf courses provide suitable habitat for wildlife, we see them as a destroyer of all things wild.

Sometimes nighttime gardening activities are tied to other protest activities. During the World Trade Meeting in Seattle, November 30 to December 2, 1999, numerous protests were scheduled on everything from globalization and sustainable agriculture to biodiversity to workers rights in the Third World. As part of the protest, computers at the poultry diagnostic laboratory at the Washington State University experiment station in Puyallup, WA (about 40 miles south of Seattle), were vandalized along with experimental plants.

Tactics not accepted by more moderate groups in the environmental movement, including Earthfirst! and SHAC (Stop Huntington Animal Cruelty), involve death threats, personal physical attacks, threats and the use of physical violence to family members and business associates. Other anarchist groups, like Reclaim the Seed, claim to engage in "self-defensive measures on behalf of all living things," which run the gamut from nonviolent protests to arson. Regardless, the activities of these groups have not lessened since 9/11. There was only a short period with reduced media coverage. This sort of homegrown terrorism is becoming increasingly sophisticated and must be taken seriously by food producers.

Extortion

Extortion is commonly employed by terrorist groups and is generally coupled with other forms of activity to induce a victim to modify his or her behavior. There are a number of motivations for extortion:

1. Political and consumer terrorism, publicity causes
2. Malicious purposes such as revenge on a former employer by a disgruntled employee
3. Actions with no obvious motivation
4. Copycat incidents
5. Criminal (extortion for profit, publicity threatened but not actually sought)

Other tactics incorporate extortion and bomb threats. The infamous Mardi Gras bomber, Edgar Pierce, was recently sentenced to 21 years in the U.K. for using homemade bombs to extort money from Barclay's and from Sainsbury, a retail food chain that also owns Shaw's supermarkets in the U.S.

Legal extortion is a tactic that uses spurious litigation and the associated bad publicity to target victims. For example, the vegetarian activist group Physicians Committee for Responsible Nutrition sought an injunction to block the issuance of USDA dietary guidelines until nondairy sources of calcium were listed in the document. The group called the USDA plan racist, claiming that the agency was not sensitive to the needs of blacks, Hispanics, and Asians, who tend to experience a high incidence of lactose intolerance.

Ecotage, Environmental Extortion, and Fraud

A new face of radical environmentalism has moved beyond the simple monkey wrenching and tree spiking of the 1990s. Several groups now resort to extortion and violence for political purposes. Many conduct powerful public relations campaigns and generate a great deal of public sympathy. Extortion is the underlying theme and can be politically motivated, to generate consumer fear or publicity for a cause, or simply to generate as much economic damage as possible for the target (which may take the form of profit or increased contributions for the perpetrator). Malicious acts may also be committed for revenge by disgruntled employees. Copycat crimes are relatively common.

Targets can be small, such as a retail food shop or university experiment stations. The Internet is often passively instrumental in aiding and abetting this form of domestic terrorism by permitting groups to initiate unregulated propaganda campaigns no matter how inaccurate or damaging the information is. The ability of activists to communicate globally, almost instantly via Internet, fax, or phone, provides new tools for targeting, fund-raising, propaganda dissemination, and operational communication (Anon., 1998). These tools provide for the emergence of very fluid, nonhierarchical international terrorist organizations, and diffuse focus from a single cell or individual.

Environmental extremist groups have adopted these tactics and those of the Animal Liberation Front (ALF), an international animal rights organization that often employs violence. This is one of the 350 to 400 animal protection groups active in Canada and one of over 7000 animal rights groups in the U.S. Public sympathy for the underlying, allegedly legitimate cause of this and similar organizations has obscured an astonishing amount of physical damage to retail outlets, laboratories, and clinics (Smith, 1992). Animal rights activism is promoted as both fashionable and progressive. Take note of the plots of recent movies targeting teenage females, such as *Legally Blonde II*, where the principal theme is to have the main character's dog's mother released from an animal testing lab so she (the dog) can attend the wedding of the main character. Animal rights and now the environmental movement provide a chic atmosphere and an exciting outlet for trendy, young, and middle-class individuals bored with their tame lifestyles. Unfortunately, militant extremists within these organizations with violent agendas use environmental and animal rights groups as a means to secure popular support for radical causes (Smith, 1992). A recent example from May 2000 involves a jet skier who disrupted the whale[2] harvest by the Makah Indian Nation in northwestern Washington State. Harvesting marine mammals has been a traditional part of the culture of these people from time immemorial. The whale harvest was halted voluntarily by the Makah tribe until stocks were replenished and remained in effect for over 70 years. The U.S. government restricted whale harvest as well as the harvest of other mammals under the Marine Mammal Protection Act, passed in the 1970s even though adequate resources existed for a limited harvest by indigenous peoples. The jet skier intentionally raced in front of a moving vessel and was hit. Fortunately, none of the Native American hunters, floating in an open canoe in the Pacific Ocean, were injured.

From 1982 to the present, ALF has been involved in extensive urban terrorism activities, including vandalism, breaking and entering, and arson of furriers and mink farms, meat and fish shops, livestock auction facilities, pharmacies, veterinary schools and clinics, and medical or scientific research labs.[3] More recently the trend of these organizations has been to become more violent in their attacks on people not just facilities. Individuals have been injured in car bomb attacks and firebomb attacks to protect animal rights.[4] Criminal activity does not directly track the day-to-day activities of the target. For example, sellers of animal products, including meat, may be attacked in protests against animal research.[5] Ironically, animals are often killed in these attacks.[6] Tactics include leafleting schools where the children of scientists involved in animal research attend, death threats, hate mail, obscene phone calls, and harassment of public officials and their families (Smith, 1992). People have come perilously close to being killed by poisonings.[7] These hoaxes have proven to be very effective in causing economic damage.[8]

ELF is one group with the intent to inflict as much financial harm as possible on corporations whose interests are deemed to be odds with the environment. And in that they have been very successful (Tubbs, 2000). ELF picks times and places where no one expects them to be (Tubbs, 2000). A key ELF spokesman, Craig Rosebraugh, a vegan baker from Portland, OR, operated an ELF press office. Federal agents raided Rosebraugh's home and offices in February 2000, seizing his computer, Internet passwords, phone bills, tapes, and books. The government found e-mail regarding Michigan State and Washington State Universities, both targets of anti-GM vandalism (Tubbs, 2000). In front of a grand jury and risking 18 months in prison on contempt charges for failing to provide contacts, Rosebraugh claimed that he did not know who ELF members were, but only publicized their activities because he agreed with them. He was offered immunity from prosecution if he cooperated. Rosebraugh stated: "I want people to understand that these are not random acts of lawlessness but actions that have a definite purpose and that is the end of abuse and exploitation. People are tired of spending an incredible amount of time and energy to try and have campaigns legally that basically get nowhere at all. Individuals in the ELF want to see results. They want to pick up where the law is leaving off." ELF has a long history of setting fires to protest logging, developments, and genetically modified organisms (GMOs). The immediate goal of the organization is to cause economic damage. As Rosebraugh also stated: "I want to see these actions increase, not just at Michigan State University, its not just universities, it each and every entity involved in genetically modifying organisms that are going to be targeted."

Misrepresentation is another interesting tactic. At least one antibiotech group has misappropriated the name of respected and credible agriculture organizations as a front for vandalism. "Future Farmers of America" (FFA) vandalized vegetable seed research at the Seminis Vegetable Seeds Research Center in Woodland, CA, on May 24–25, 2000. Besides being an illegal use of a trademarked name, there is the unfortunate potential for creating public confusion, damaging the FFA's agriculture educational programs and negatively impacting student members.

Cheery reports of vandalism are reported on many different websites, one of the most successful by a group called BAN. In an attempt to reduce association with criminal activity, webmeisters post disclaimers like the following with their reports:

> GenetiXAlert is an independent news center which works with other above-ground, anti-genetic engineering organizations. GA has no knowledge of the persons(s) who carry out any underground actions. GA does not advocate illegal acts, but seeks to explain why people destroy genetically engineered crops and undertake other nonviolent actions aimed at resisting genetic engineering and increasing the difficulty for entities which seek to advance genetic engineering or its products. GA spokespeople are available for media interviews. [Note — "call cell phone" first.]

Because most of the perpetrators are judgment-proof, and the supporting terrorist organizations so diffuse, collecting civil damages is not a viable option in most cases. Criminal prosecution is difficult, because the ecoterrorists are well versed in the legal tactics necessary to tie up local district attorneys and counties, making prosecution excessively costly. For example, in the 1999 WTO riots, several hundred arrests were made, but most cases were not brought or were dropped because the state could not afford to prosecute them. New laws to make it easier to maximize damages and criminalize certain activities will only be effective if the laws are enforced.

Information Warfare

Information warfare employs "weapons of mass corruption" (Ramthell, 1997). It is a modern variant on command-and-control warfare; however, practioners recognize the dependency of national and multinational organizations (including the military, government, and national economies) upon rapid, reliable information processing.

There are three basic types of information warfare:

1. Attacks on military reconnaissance, surveillance, dedicated communications, command and control, fire control, and intelligence assets
2. Attacks on the basic communications links in society (e.g., voice, video, data transfer, electric power, or telephone systems)
3. Using television, radio, or print media to attack or influence the attitudes of the military or civilian population, political or economic leaders

Information warfare signifies the damage that can be done across a society without directly causing any first-level physical damage. Cyberwarfare or civilian "cyberattacks" are intrinsically different from conventional warfare (Devost, 1995; Wardlaw, 1998) and, in general, from terrorist activities in the sense that:

1. It is waged relatively cheaply and anonymously (determining accountability is difficult).
2. It is not well defined and therefore not yet entirely taboo or clearly illegal.
3. There is low cost to human health or safety (so far).
4. Boundaries are blurred in cyberspace, and ordinary distinctions between public and private interests, such as between war and crime, are less pronounced.
5. Attacks are time and not location specific.
6. There are no effective early warning systems, permitting a first-strike advantage. Opportunities abound in cyberspace to manipulate perception of an attack (magnitude, impact, etc.).

7. Information warfare has no front line, and the potential "battlefield" (criminal or military) is anywhere a network system permits access. Convergence of points of attack and other nodes represent vulnerable points.
8. Defensive measures against it are difficult and costly. Firewalls and other preventive measures can be erected, but these are not necessarily effective. The most effective defensive strategy, at least at low-budget operations, is to shut the targeted system down to prevent further damage to the system.
9. It is possible to reap highly visible payoffs by making low-risk attacks at computer systems.

Information warfare can be employed at different levels, each with the potential for far-reaching and catastrophic results:

1. Net warfare is the disruption of or damage to opposition infrastructure.
2. Net warfare affects public opinion, reducing confidence in governments or others in charge of protecting communications infrastructure. With a reduction in military spending, more technical information functions are in the hands of civilians. Now 95% of U.S. military communications activity is conducted over the same networks that civilians use for telecommunications. In the private sector, technical information functions have been handed off to outside contractors, many not in the same time zone (and sometimes in different countries).
3. Operations targeting information infrastructure would be conducted according to information-related principles, moving to network structures that require some decentralization of command and control. For example, the Mongols were able to easily defeat the hierarchical defenses put up by feudal systems from China throughout Central Europe to protect a centralized command-and-control network. Similarly, guerilla tactics were used successfully by the Viet Cong against U.S. forces during the Vietnam War to undermine communications.

Cyberterrorism will increase in importance in intrastate and transnational conflicts by substate actors and will increase markedly with the increased availability of information warfare tools and the increasing vulnerability of networks now used in industrialized countries. Private enterprise is susceptible to attacks that involve data destruction, penetration of a system to modify its output, and system penetration with the goal of stealing information or sensitive data. Systems can also be "bombed" with identical or repeated messages and attached files or "spammed" with numerous e-mails to a large number of users. This can overload a system quickly. Limited foci for communication and electronic product distribution make systems highly vulnerable.

Entire societal functions are now in cyberspace. This creates a greater necessity for proper functioning of inherently unstable information transfer networks. In fact, the physical and functional infrastructure of our society is becoming increasingly electronic. This includes informational activities of education, research, engineering, design, mass information and entertainment, and private and public records (there is now a strong preference for electronic records, and there may be no redundant system or backup paper system in case data files are lost). Transactional activities are widely conducted in cyberspace, including any commercial, business, or financial transaction and government activities. Substate actors will instigate information warfare by:

1. Intelligence gathering, communications, money laundering, and propaganda
2. Physical violence against information activities of target entities
3. Digital attack techniques against information activities of the target entity

Specific forms of attacks and electronic monkey wrenching come with their own special jargon. Reported corporate cyberattacks have involved these forms (GAO, 1996):

1. Installation of malicious code in e-mail sent over networks. As a send mail program scans the message for an address, it will execute the attacker's code. Send mail operates at a system's root level and has all the privileges to alter passwords or grant access to an attacker.
2. Password cracking and theft. This is relatively easy with a powerful computer searching program that can match number or alphanumeric passwords to a program in a limited amount of time. Success depends upon the power of the attacking computer.
3. Packet sniffing. This is an attack that inserts a software program at a remote network or host computer that monitors information packets sent through the system and reconstructs the first 125 keystrokes in the connection. This would normally include a password and any logon and user ID. By packet sniffing, a hacker can obtain a password of a legitimate user and gain access to the system.
4. Stealing information. Attackers who have gained access to a system can damage it by stealing information from authorized users. van Eck

emissions enable hackers to capture contents of computer screens up to 200 m away using low-cost devices (Devost, 1995).

5. Denying service. Threats to deny service were popular 3 or 4 years ago and were used by ecoterrorists or electrohippies. Numerous companies, including Microsoft, Yahoo, eBay, Amazon.com, CNN, ZD-Net, and E*Trade, have been hit. Others have been attacked on the day of the initial public offerings of stock (e.g., Buy.com). Targeted sites would receive hits on their servers at incredible rates, up to 1 gigabyte of data per second, making the server inaccessible for legitimate business purposes for hours.
6. Virus. Copies of destructive software are propagated and sent electronically to other users.
7. Trojan horse. A Trojan horse is an independent program that when called by an authorized user, performs a useful function but also performs an unauthorized function, which may usurp the user's privilege.
8. Worm. A worm is a program that pretends to be benign but is destructive. It tricks the user into running it by claiming to perform a useful function. A worm is like a virus, but it can travel along a network on its own.[9]
9. Logic bomb. A logic bomb is an unauthorized code that creates havoc when a particular event occurs, like Millennium Day (a major fear at the turn of the century), the anniversary of an important political event, or, more mundane, the dismissal of an employee on a certain date.

Each of our critical industries — telecommunications, electric power and energy distribution systems,[10] oil and gas, water, food, transportation, banking and finance,[11] emergency and public health services — are susceptible to attacks tied to loss of communications or ability to transfer information effectively. Normal accidents cause enough havoc. An inadvertent cable cut in Eagan, MN, delayed two thirds of the Northwest Airlines flights out of their Minneapolis hub for 1 day (Anon., 1998). Failure of one ATT switch in New York City put the New York Stock Exchange (NYSE) out of business for a whole day, causing it to lose billions of dollars in trading values, 4.5 million blocked domestic long-distance calls, 500,000 international calls, and 80% of the FAA circuits. A similar failure in 1991 in Boston resulted in a loss of 60% of all normal phone traffic.

Can you imagine what would happen if terrorists manipulated phone communications to divert calls or to eavesdrop? A small group of individuals could target several large phone networks at once and electronically paralyze large parts of the country. Loss of function of a portion of a single communications satellite in 2002 due to sunspots affected cellular phone operation in the western U.S. for several weeks. Corporate systems have been under increasing attack, with a cost of increased computer security in 1998 alone of $100 to $300 million (Adams, 1999), and with recent figures more than twice that.

Undermining computer systems involves hackers (amateurs) and professionals in the employ of business, government, business intelligence firms, criminal groups, or professionally motivated groups of substate actors, and political activists (Rathmell, 1997). Individuals from any of these groups would be able to disrupt communications in a cheap and undetectable manner. Political activists and terrorists have already damaged or threatened to disrupt information transfer as a way of making political statements.

Hacking causes physical destruction and deception within a network, or provides a basis for psychological operations through its attack on information flow. An international subculture supports hacking and involves the disenfranchised individuals who will resort to a virtual space to commit acts of terrorism far more effective than what could be done by other means. Hackers today are much more likely than those in the mid- or late 1990s to be employed by organized crime or a (pseudo)governmental entity. When hackers crashed the White House site in 1999, they posted a message: "Why did we hack this domain? Simple, we (expletive) could" (Adams, 1999). As an example of the type of damage that can be done, a band of Russian hackers siphoned tens of thousands, possibly millions, of dollars of research and development secrets from the U.S. government and private companies and then sold these to the highest bidder over the Internet (Adams, 2000). Other examples include a Portuguese hacker crashing the FBI website to protest the NATO bombing of the Chinese Embassy in Belgrade, and Serbian hackers knocking out the NATO site during the Kosovo conflict. Much of the planning and implementation for the 9/11 attack was conducted electronically and via satellite technologies. Recent viruses and worms propagated through Microsoft operating systems may have had terrorist motivations.

These systems are vulnerable to cyberattack by relatively unsophisticated hackers from anywhere in the world. Many government systems are vulnerable and have poor computer security (*C&EN News*, 2000). For the second time in 2000, the General Accounting Office (GAO) chastised the Environmental Protection Agency (EPA) for poor security, noting that EPA systems are still "highly vulnerable to tampering, disruption, and misuse from both internal and external sources."[12] This lack of security jeopardizes huge volumes of confidential data submitted by companies as required by the agency for registrations and other permitting. The American Chemistry Council is concerned that lax security would enable economic sabotage, noting that industrial spies are increasingly using the Internet to gain access to confidential business information.

No longer can we say that Yugoslavia or Portugal is a world away (Adams, 1999). Hackers can use scripting features in off-the-shelf software programs to add malicious programs and spread them by e-mail.[13] Copycat hackers simply modify the program and resend it. One variant of the "love bug" showed up as a virus alert claiming to fix the love bug, but instead it destroyed crucial systems files, one as a joke and another confirming the purchase of a Mother's Day present (Stone, 2000). Other tactics include replacing the front page of a corporate website, reading e-mail and files, stealing information (intellectual property, customer lists, financial information), and causing havoc for a site in general, for example, by denial of service attacks.

The Melissa virus from 1999 caused over $80 million in damages (Stone, 2000). The May 4, 2000, "love bug" worm wreaked techno-havoc globally by targeting Microsoft Outlook users, infecting millions of computers worldwide. It originated in the Philippines, which at that time had no laws for prosecuting cyberterrorism. The worm love bug, when activated, sent itself in an "I love you" message to all of the victims' electronic contacts. An attachment with it ran a program that searches out and destroys digital photographs and music files on the user's hard drive. Estimates are that 80% of the businesses in Sweden and 70% in Germany were hit. Most of the high-tech companies in Silicon Valley took their e-mail servers offline for a day to clean up the mess. Zurich's Kloten Airport was plagued by flight delays, and several newpapers lost photo archives. The British House of Commons, the U.S. Defense Department, and George W. Bush's 2000 presidential campaign were hit.

New provisions in U.S. law to control hacking and crimes using the Internet (e.g., money laundering, conspiracy) will permit law enforcement to gather intelligence in ways and using methods (e-mail privacy, information architecture systems) never before used in a developed democracy. We have seen these provisions coupled with those of the U.S. Patriot Act of 2001, regarding searches and information access, chip away at civil liberties and will have to decide as a society whether this is how we want to continue to live.

Better consumer awareness coupled with a concerted effort to design antivirus software has meant that recent attacks from Novarg and SoBig in 2003 and 2004 had less of an impact than they would have had a couple of years earlier. Unfortunately, the apparent frequency of attacks is increasing, and it has been difficult to trace their origins.

The very nature of the Internet means that information storage and transfer merges the public and private sectors into one network. So there is no longer any relevant distinction between matters that have been regarded as public and those that are private (Blyth, 1998). Blurring the distinction between public and private sectors is leading in part to the decline of the relevance of the nation-state in its traditional form. This may be one of the reasons that sectors of the environmental movement have become increasingly hostile and militant, because many in these groups feel that governmental action, rightly or wrongly, will be ineffective in either promoting their "legitimate" cause or impeding their illegal or antisocial ones. Threats to multinational corporations are one manifestation; these are nonstate and metastate entities and terrorist targets in their own right. As the importance of the nation-state declines, supranational organizations such as the United Nations, subnational groups, including private enterprise and criminal and ethnic factions, and nonnational groups, including terrorist groups not aligned with a state and multinational businesses, become increasingly significant as sources of political and economic power (Steele, 1998). The porosity of corporations and their entrenchment in a newly configured military–industrial complex give corporate entities the appearance in the eyes of terrorist groups to be legitimate targets in an old-fashioned guerilla war employing arson and vandalism, as well as targets using weapons created to target the new information age. Regardless, the distinction between civilian and state (and military) targets is becoming blurred (Molander et al., 1998). This increases the vulnerability of the private sector and nonmilitary government targets to attack.

Furthermore, the Internet is a powerful public relations tool that has led to the creation of an international unregulated public relations network that can raise havoc almost instantaneously. Unregulated campaigns, no matter how inaccurate, are widely promulgated, distributing damaging information across the world from sources that are difficult to track. Campaigns to regulate the content of Internet sites have been largely unsuccessful. Broad constitutional protections transfer to Internet speech, if recent experience with Internet pornography is any guide. Many groups have latched on to these First Amendment protections and taken advantage of the opportunities it provides. Antibiotech groups have launched numerous masterful websites, including the commercial www.cropchoice.com, which will accept advertising, but not from agricultural, chemical, seed, or biotech companies. As part of the new net war, a consortium of biotech companies launched a $50 million public relations campaign to promote their industry in the ether and in print media, including the websites www.betterfoods.org and www.whybiotech.com. All of this makes one somewhat nostalgic for the distribution of propaganda leaflets of by-gone years, with the military variant of airborne leafleting affectionately referred to among old pilots as b___s___ bombing.

In short, terrorism today can be quite different from the good old days of physical sabotage (Adams, 1998). Use of the computer and associated information transfer systems leverages the capacity to reach people everywhere. It also leverages the amount of damage a terrorist can cause. Computer-generated attacks, in addition to or in place of more conventional car, truck, or suicide bombers, vandalism, or cable cutting, could unleash a cascade of events, collapsing a service grid, pipeline, or air traffic control system. The long-range power outage hitting the northeastern U.S. in the summer of 2003 shows the impact of failures at a single power station within the grid and the personal and economic costs of a single and relatively small infrastructure failure.

Terrorist Motivation

The most likely perpetrators of terrorist activity targeting food have a variety of different motivations. The motivation can be economic (targeted to financially impact a specific commercial entity or industry segment), political (making a statement, influencing the outcome of an election, forcing a particular political outcome), or malicious mischief (the infamous copycatter). Trends are for terrorists to come from disenfranchised groups. Motivation is generally tied to the elimination of real or imagined injustices directly associated with the affected food product or the food industry. However, there could just as likely be some concrete or nebulous political or economic connection. Facts are irrelevant and normally do not inhibit the activities of these extremist factions. The focus of these groups are directed against perceived injustices, and while their actions are not necessarily encompassed within the realm of conventional terrorist activities, the results often are.

The most probable perpetrators of food terrorism are groups promoting causes with a certain degree of public support. Many individuals engaged in food terrorism may initially have been well-intentioned activists gone wrong from the animal rights, consumer protection, and environmental conservation movements. The largest group, besides political anarchists, is from the environmental movement, coalescing around issues of pesticide use and animal drugs, sustainable agriculture, and, most recently, genetically modified plants and animals as food. Paul Watson, former director of Sea Shepherd, famous for harassing the Makah whalers in 1999 and Canadian seal harvesters, actively promoted terrorism as an appropriate tactic for environmental groups in his takeover attempt of the Sierra Club. Other activists fear technology, innovation, or social progress, while other individuals attracted to such activities are anarchists seeking a new venue for criminal activity.

Food terrorists may even emerge from religious movements gone awry and from groups threatened by innovation. Commonly, bio- or ecoterrorists are anarchist factions tied directly or indirectly to mainstream groups that reasonably and peaceably strive to promote their political causes. These spin-off terrorist factions typically form loosely organized, fluid networks or cells with anonymous memberships.

The issue with any of these groups is money. Funds are commonly diverted from what appear to be legitimate charitable, religious, or nonprofit organizations to extremist activities. Although recent laws, such as the Patriot Act, have made it easier to trace money laundering activities and short-circuit funding of some terrorist groups, tactics used pose some risk to our civil liberties. Furthermore, U.S. laws will have limited long-term impact, since within a short period, funding activities will simply move offshore.

A clash in civilizations provides the primary incentive for individuals to join groups like Al-Qaeda (Stern, 1999). Unfortunately, we are naïve in assuming that all rational people share beliefs regarding the relative importance of rights and responsibilities between citizens and the state, the extent of governmental liberty and scope of governmental authority, and the equality of individuals regardless of gender, race, ethnic origin, or religious belief. Terrorist groups motivated by religion are becoming more common, and fanatical terrorists of all stripes are posing an unprecedented threat to Western society as these groups become increasingly sophisticated and gain access to chemical and biological agents (Stern, 1999). As potential recruits, Al-Qaeda specifically targets smugglers, political asylum seekers, adventurers, the unemployed or needy, and employees at borders, airports, ports, coffee shops, restaurants, and hotels for recruitment into the organization (*Al-Qaeda Manual*, no date).

New Laws Tied to Food Terrorism

The threat from terrorists and terrorist groups against food research, production, and processing is increasing. Even bottled water has been a target. A recent incident in November 2003 led to the serious illness and death of several children in Italy from the injection of bleach or polar solvents into bottled water.

Although food terrorists are motivated by perceived injustices, these actions are not necessarily encompassed within the realm of conventional terrorist activities in the minds of authorities charged with maintaining public safety. Until quite recently, ecological and bioterrorist acts directed against the food or agriculture sectors were not taken as seriously as they should have been by law enforcement. This is not entirely the fault of the police and district attorneys, as criminal law in many jurisdictions did not have adequate provisions to discourage this sort of antisocial behavior. Beginning in 2001, many states have passed specific agroterrorism laws, making it easier to charge and prosecute cases of food and agro terrorism. Depending upon the jurisdiction, these laws have increased the civil penalties and liability for damages tied to food terrorism acts. For example, under earlier statutes, the only damages permissible by law may have been the replacement value of a corn plant. Under the new laws, the damages could also include the research and development costs that went into developing that specific genetically engineered corn plant, the costs of maintaining an affected test plot, and, in addition, a reasonable estimate of the value of the plant products, if the crop was grown for nonfood purposes. Business disparagement laws have also gone into effect; however, since there are no criminal penalties associated with these, there is less likelihood that these laws will have much effect on inhibiting food terrorism.

Anti-Terrorism Laws

Threats from terrorism come in different forms[14] and both civil and criminal laws to prevent or to limit the impact of terrorism are in effect or are being developed. The most prominent law is the "Effective Counterterrorism Act of 1996." (Title 1. Sec. 101. 18 USC Sec 2239A). This law makes it an offense to influence or affect the conduct of a government by intimidation, coercion or to retaliate against government conduct.[15] It is a federal crime[16] to provide material support or resources for, or to conceal or disguise the nature of materials provided to, a terrorist. It is also a federal crime to knowingly provide material support or resources to any organization that a person should know is a terrorist organization or which the government has designated to be a terrorist organization (18 USC Sec. 2239b). Additional penalties apply if a child is targeted.[17]

Tampering with consumer items including food is also a federal crime. Under 18 USC Part 1 Chapter 65 Sec. 1365 it is a federal crime to attempt, conspire, threaten to, or tamper with the contents, container or labeling of a consumer product in a manner that could cause death or bodily injury.[18] Under this Antitampering Act, it is also a crime to intentionally cause injury to a business or to communicate false information about the tainting of a consumer product.

States have also introduced legislation to impose civil or criminal penalties for attacks against crops or disparagement against food businesses. In 2000 a new law in California imposed civil liability for crop vandalism. This bill was introduced because California law[19] did not provide sufficient deterrents and did not adequately address the costs of crop vandalism including the cost of research and development. Chapter 1 of Division 8 of the Food and Agricultural Code: Article 6 Destruction Sec 52100(a) (b) provides that any person or entity who willfully and maliciously damages or destroys any field crop products (Sec. 52001), that part of a testing or product development program conducted by, or in conjunction with, a program recognized by a state institution of higher education or other state or local governmental agency is liable for damages. Financial liability is for twice the value of the crop damaged or destroyed. When considering an award of damages, the court is instructed to consider research, testing, and crop development costs directly related to the crop that has been damaged or destroyed. These laws appear to have provided some deterrents to food and agricultural terrorism activities in recent years; however, risks remain.

In summary, among the objectives of food terrorism are:

1. The use of food as a means of instilling fear and causing injury or death in a civilian population
2. The use of food as a weapon to affect political change
3. Localized acts of sabotage or individualized attacks for personal revenge or no apparent purpose
4. Threats to food animal or food plant health to reduce the availability of food or the quality of the food supply
5. The desire to undermine research and development of food or to eliminate a specific food, ingredient, or agricultural practice
6. Severely impacting a company and putting it out of business by affecting the stock price, product availability, or marketability in a malicious way
7. Prohibition of the importation or trade of competing crops, or of research or development in a particular area
8. Pressure to erect trade barriers

Contaminating Food

Contaminating food has been a popular method of terrorizing civilian populations during wartime throughout recorded history. Although there is much talk about weapons of mass destruction, and these remain a potential threat, they are not the primary risk to food systems or to the public at large, as these agents, for the most part, are relatively difficult to stabilize, transport, and effectively disseminate on a large scale. But as mentioned earlier, a small-scale or even botched operation involving any agent would be sufficiently alarming to have a significant negative economic and possibly public health effect. The deliberate contamination of food or water with a chemical or biological WMD agent would probably be easier to control than contamination from a targeting and public health perspective. However, this does not mean that a deliberate contamination incident would not be catastrophic.

We know that incidents of *unintentional* contamination have wide-ranging effects on public health, on the economic viability of food businesses, and upon consumer food choices following an incident. The largest recorded food-borne illness outbreak was an incident in 1991 involving hepatitis A-contaminated clams in Shanghai, China, affecting 300,000 people. Other large incidents include the 1994 *Salmonella enteritidis* contamination of ice cream, infecting 224,000 people in 41 states (WHO, 2003). A community outbreak of waterborne *Salmonella typhimurium* occurred in Gideon, MO, in which 1100 people were supplied with unchlorinated water, leading to 650 people becoming ill, 15 being hospitalized, and 7 dying. About one third of the individuals in the community did not heed instructions to boil the water with about half of those failing to follow instructions becoming sick. Reasons for noncompliance included not remembering (44%) and disbelieving (25%) instructions from public health authorities (Angulo et al., 1997).

These examples illustrate how a well-designed intentional contamination incident could have a devastating impact on public health. Various models have been developed to predict the scope and impact of an intentional contamination incident. Although these do not use food as a target, the models would still apply (Stern, 1999).

The economic impact of an incident of unintentional contamination can be huge. Recovery of market share, or a position in the market at all, is greatly jeopardized with any incident of widely publicized product contamination. In short, it is hard to get your good name back. Some countries capitalize on consumer fears regarding safe food. For example, the Korean government has been conducting a misinformation campaign against imported foods for the past 20 years. Market bias from Japan has been more subtle. However, these disinformation campaigns have been successful in creating a general market perception in both of these countries that imported foods are less safe than domestic products. Therefore, any contamination incident in these markets from an imported product would confirm existing biases and jeopardize trade. Because of this mind-set, it is likely that any food-borne outbreak in these two countries will target imported products first, at risk to public health. Recent history has borne this out. For example, the *Escherichia coli* outbreak in Hokkaido in 1998 first implicated imported salmon caviar (ikura) from the U.S. It was not until later that the outbreak was traced to locally produced daikon (white radish) sprouts.

Loss of market confidence in Europe over the safety of Israeli citrus followed a 1978 product contamination incident with mercury, and aldicarb contaminated watermelon from the U.S. in 1985. There have been a number of incidents of contamination and contamination threats from Palestinian militants in the intervening years implicating Israeli produce as a means of destroying confidence in the safety of food exported from Israel. Cyanide contamination of Spanish cooking oil in 1981 led to 800 deaths and 20,000 illnesses, with serious market disruption across three continents (WHO, 2003). A recall of 14 million kg of suspected, unintentionally contaminated ready-to-eat meats in the U.S. resulted in plant closure and direct costs of product recalls ranging from $50 million to $70 million. The overall costs of these incidents have not been firmly established but reach into the billions of dollars.

Dioxin contamination of animal feeds in Europe in the late 1990s followed on the heels of health scares tied to the prion-based disease BSE, which is linked to Cruetzfeldt–Jakob syndrome. These incidents together caused multi-billion-dollar impacts to the markets for muscle foods across the world. Locating a single affected cow in Canada during the summer of 2003 led to the closure of U.S. borders to Canadian cattle and to any feed product containing meat by-products. This closure caused much consternation in the U.S. beef industry, but also had wide-ranging effects on aquaculture and poultry operations, as sources of feed for their animals had been cut off. It is a common practice to ship feeder cattle from the northern U.S. to Canada, have them grown out, and then shipped back. This practice was halted for several months and resulted in the highest spot market prices for beef in decades for U.S. farmers. Continuing restrictions on the importation of Canadian cattle remained in effect as of the fourth quarter of 2004 and are the basis for a multimillion dollar legal action for unfair trade practices by Canadian farmers against the U.S. government. Although trade restriction actions must be taken to protect public health, the implications are incredible and much wider reaching than initially anticipated.

Reported incidents of intentional contamination have been less common, but their impact on international trade has been substantial. In 1989, the *intentional* contamination of Chilean grapes resulted in several hundred million dollars in losses and caused more than 100 growers and shippers to go bankrupt. Federal inspectors found cyanide in two grapes out of the numerous shipments of produce passing through Philadelphia on that day. There is no rational explanation for this fortuitous detection except that the grapes were either planted and found on a tip, or the whole incident was a hoax. The Chilean industry sought $300 million in compensation for economic damages resulting from this incident and suspect complicity of U.S. Customs, possibly through an employee with sympathies toward the United Farm Workers Union. The United Farmworkers were promoting a boycott against U.S. grapes to protest what they considered to be unfair labor practices. As a result, market demand for grapes dropped and remained low for several months.

Similar staged contamination incidents occur routinely in foreign markets. In 1995, the first shipment ever of Golden Delicious Apples from Washington State to Japan were contaminated with the chemical additive DPA, an unapproved food additive in Japan. The news of this hit the press just as the product was to be launched into the Tokyo market. The timing was so perfect that the only rational explanation was deliberate product contamination intended to protect Japanese producers from foreign competition. Suspicions were that a city worker in Tokyo contaminated this shipment. The short-term impact was roughly $45 million in lost markets in 1995. However, the long-term impact was the loss of the Japanese market for U.S. apple imports, even 8 years later.

Questionable lab test results for imported products also cause market havoc, along with huge costs for embargoing perishable food shipments and product testing. How many of these problems arise from monkey-wrenching and how many are from human error are difficult to determine. For example, in the early 1990s, following the Alar scare in the U.S., Alar was found in samples of citrus juice imported into Korea. Alar is never used on citrus products, but how many consumers know that? Whether this was a "false positive" lab result or intentional contamination will never be known. But by the time an investigation is complete, the market damage would already have been done.

Probably the most notorious case of intentional food contamination occurred in Oregon in 1984. What many individuals consider to be the only real recent case of intentional mass food poisoning in the U.S. occurred in September 1984 in the city of The Dalles. Members of a religious cult purchased *Salmonella typhimurium* from the American Type Culture Collection (ATCC), grew it up, and conducted a series of trial runs contaminating water before selecting food as the delivery system of choice. Their purpose was to throw a local election and take over local government — the current zoning board did not take too kindly to the rapid expansion of the cult into Antelope, OR, and the surrounding community and local residents were not particularly fond of a movement to rename the town Rajneeshpuram. Members of the Rajneeshee cult misted pathogens onto salad bars in 10 restaurants in The Dalles. This town is significant because it is a regional hub situated on a major east–west interstate highway tracking the Columbia River. Over 1000 individuals reported symptoms, with 751 confirmed cases.

This case was further complicated by refusal of federal and state authorities to consider the possibility that criminal activity was involved, despite strong supporting evidence. Despite several laboratory confirmations of the same pathogenic strain, two confirmed outbreaks (September 9 and 25), reported illnesses from individuals who had eaten at 10 separate restaurants, and suspicions advanced by a local authority (Judge William Hulse; see Miller et al., 2001), the deputy state epidemiologist concluded in his November 1984 report that there was no evidence to support the hypothesis that the outbreak was the result of deliberate contamination. Instead, the epidemiologist stated that the contamination "could have occurred where food handlers failed to wash their hands adequately after bowel movements and then touched raw foods." This misconception received further support from the Epidemic Intelligence Service of the U.S. Centers for Disease Control and Prevention in its report issued in January 1985, which stated that it too "was unable to find the source of the outbreaks and that food handlers were probably to blame." The Centers for Disease Control and Prevention (CDC) report reasoned that because workers preparing the food at the affected restaurants had fallen ill before most patrons had and because some minor violations of sanitary practices at a few restaurants had been detected, food handlers "may have contaminated" the salad bars. Again, the CDC asserted that there was " 'no epidemiologic evidence' to suggest that the contamination had been deliberate" (Miller et al., 2001). This report put a kibosh on the local criminal investigation despite a strong belief on the part of local authorities that criminal activity may have been involved. It was not until September 16, 1985, a year after the outbreaks, that law enforcement officials were able to reopen the investigation of the incident, but only after the leader of the Rajneeshees (in conjunction with his deportation) alerted officials that rogue members of his group had deliberately perpetrated this act of bioterrorism.

Some of the most heinous intentional contamination incidents in recent history are associated with the military use of both biological and chemical agents against civilians by the Rhodesian government during the late 1970s. This included the use of ricin, anthrax, cholera, organophophate pesticides and heavy metals to poison wells, intentionally spread disease and kill livestock (Martinez, 2003). Poisoning a well which was the sole source of drinking water with an unknown agent killed at least 200 people. Other incidents involved poisoning livestock watering holes, stagnant pools of water and slow moving streams near guerilla camps. Cholera was spread by Rhodesian operatives inside Mozambique to disrupt supply lines of guerilla forces. The Ruya River was contaminated with cholera causing an epidemic with numerous fatalities. Efforts involving the use of cholera were stopped after it was determined that cholera dissipated too quickly to provide any lasting tactical advantage. In an interesting strategy, uniforms were treated with organophosphates. This caused guerilla recruits to die a slow and miserable death in the African bush on their way to join insurgent camps. Many ill recruits were shot by their compatriots to end the suffering.

Food scares, real or imaginary, cause long-term damage to food markets. Overzealous celebrity involvement in a food scare only makes matters worse. One of the most widely publicized and notorious food scares involved Alar in apples in 1989. At that time, Alar, a trade name for daminozide, was one agricultural chemical in a long list that consumer activists wanted off the market. Alar can improve the aesthetic qualities of apples. For example, growers could ensure that Red Delicious Apples had an elongated shape and the classic five distinct nubs at the base of the apple, good color, and firm texture by using daminozide. The safety of this compound was under review by the Food and Drug Administration (FDA) when the scare arose. Alar was suspected of being carcinogenic in 1985, but test results did not clearly substantiate this with additional testing in process by the federal government when the "media investigation" began. As a result of the controversy over the use of this chemical, and because of a conservative approach to product safety, major retailers and buyers were requiring assurance from growers that Alar was not being used before the news story broke. Unfortunately, monitoring of fruit for Alar residues by retailers was sporatic and contaminated product was found in retail markets. When the scare hit, Alar was immediately removed from the market by the manufacturer.

The media coverage of this incident was irresponsible. Media campaigns specifically targeted children. Instead of focusing on fresh fruit, the target became apple juice as a means to increase the apparent threat to public safety and the scare value of the story. Children are the largest consumers of apple juice, and through a series of televised news magazine shows and news reports spanning a period of 3 months, this point was emphasized again and again. The direct cost of this scare to Washington State apple growers alone was $130 million that year, with a direct loss of $100 million to growers in other states. This figure does not include losses to producers of processed apple products, primarily juice, but also sauces, fillings, and dried or frozen products. As is usually the case with a media frenzy, similar products are impacted as consumers become concerned with the safety of produce.

This scare precipitated wide-scale testing programs by retailers for agricultural chemicals in fresh fruits and vegetables. Programs were in full force in the U.S. through the mid-1990s. Pesticide residue-free marketing campaigns have waned nationwide, but are still important in regions with high proportions of environmentally conscious consumers. There was little public health justification for these wide-ranging testing programs. The costs of these marketing and testing programs were passed on to consumers, even though there is probably little benefit to the public from them. The primary beneficiary of these pesticide testing programs was organic production. These scares clearly moved organic farming off dead center.

The Alar scare and associated fallout led to the passage of the Food Quality Protection Act (1996), which requires a complete safety review of all 10,000+ food pesticide chemicals and their established tolerances within an unrealistic time frame. Under this law, the EPA is to establish binding upper limits or tolerances for all pesticide residues in foods under new health-based standards for cumulative exposure and considering risk to susceptible individuals, particularly children. In addition, risks from dietary exposure and exposure through drinking water and other sources (such as a personal residence) must be taken into consideration.

The economic losses from Alar were a direct result of a perceived, not a real, health risk. The risk was sensationalized by the media, and by formerly prominent celebrities with failing careers who hopped on the bandwagon to promote "safe" food, protecting children, and themselves. The lack of understanding of these individuals of the underlying issues, their failure to present an unbiased perspective, and the flagrant self-serving promotion had a far-reaching negative impact. The opportunity to increase ratings and to ensure heightened publicity cannot be overlooked; however, the same scenario does not play out for food producers. Any producer or seller of a consumer good, particularly food, would be unlikely to effectively defend itself under such a media attack without risking a loss of credibility and appearing biased. Furthermore, the longer a producer fights, the longer the story is in the media and the greater the resulting economic damage.

Food and ecoterrorists have taken the incidents described above to heart and have used them to develop specific tactics for targeting food business. The types of attacks terrorists have directed against the food industry to date range from false statements or accusations to overt acts designed to destroy property, information and communications systems, crops, animals, and people (Bledsoe and Rasco, 2003).

Product tampering (real or hoaxes) and vandalism have proven to be particularly productive in terms of perpetrator notoriety and economic damage to targets. Thousands of products each year are subject to malicious tampering and accidental contamination, which precipitate product recalls or market withdrawals (Hollingsworth, 2001). Food, beverages, pharmaceuticals, agricultural chemicals, fertilizers, pest control media, and genetically modified crops are among the products more commonly affected. Products developed using biotechnology are facing the greatest threat recently, and numerous organizations (private or public) supporting biotechnology have been targeted (Bledsoe and Rasco, 2001a, 2001b, 2002, 2003). Food contamination cases and precautionary recalls are looming possibilities and are a major motivating force behind the stringent process controls and quality assurance procedures in the food industry. However, crisis management planning will take on different twists as food becomes more political, as international markets grow, and as price sensitivity increases (Hollingsworth, 2001). Unfortunately, the management of food safety functions within government is seriously flawed (Merrill and Francer, 2000) and has recently become more complicated as food safety-related functions from several federal agencies have been transferred to the Department of Homeland Security. Suffice it to say, both the FDA and USDA will retain jurisdiction over the presence of adulterants in any of the foods they regulate. Although there may be changes in the scope of FDA jurisdiction, specifically as a result of the recent bioterrorism legislation (see Chapter 3), the mission and effectiveness of these agencies will remain relatively unchanged. Unfortunately, the efforts of these agencies are primarily directed at reacting to an incident and not to the development of preventive measures. In short, any improvement in food safety or security will be a result of private-sector endeavors. The regulatory presumption is that it is the role of business to prevent overt or covert contamination of the products they sell or distribute. As far as liability for contamination incidents is concerned, food companies are on their own. The government will not be there to help you, although agents may show up to seize your product, shut you down, impose civil penalties, or throw you in jail.

References

Adams, J. 2000. Testimony of James Adams, Chief Executive Officer, Infrastructure Defense, Inc. Committee on Governmental Affairs. United States Senate. March 2, 2000.

Adams, J. 1999. Opinion: Hacker pranks are not laughing matter. An analytical look at a contemporary issue. The Bridge News Forum. www.bridge.com. June 8, 1999.

Adams, J. 1998. The enemy within: A new paradigm for managing disaster. Disaster Forum '98 Conference. June 29, 1998.

Adams, J. 1998b. Big problem — bad solution. The crisis in critical infrastructure and the federal solution. Online news summit. May 18, 1998.

Al Qaeda Manual. No date. Located by Manchester Metropolitan Police, Manchester, UK as a computer file in the apartment of an Al Qaeda operative describing the military series related to the Declaration of Jihad. Section UK/BM-93.

Angulo, FJ, Tippen, S, Sharp, DJ, Payne, BJ, Collier, C, Hill, JE, Barrett, TJ, Clark, RM, Geldreich, EE, Donnell, HD, Jr., Swerdlow DL.1997. A community waterborne outbreak of salmonellosis and the effectiveness of a boil water order. *Am. J. Public Health.* 87(4):580–584.

Bascetta, Cynthia A. 2000. Combatting terrorism. Chemical and biological medical supplies are poorly managed. GAO/T-HEHS/AIMD-00-59 March 8,2000; GAO-HEHS/AIMD-00-36.Oct. 29,1999.

Bledsoe, GE and Rasco, BA. 2003. Effective food security plans for production agriculture and food processing. *Food Protection Trends.* Feb. pp. 130–141.

Bledsoe, GE and Rasco, BA. 2002. The threat of bioterrorism to the food supply. *Business Briefing. Food Tech.* World Market Series. World Markets Research Centre Ltd. Leatherhead Food RA. Surrey, UK. pp. 16–20.

Bledsoe, GE and Rasco, BA. 2002. Addressing the risk of bioterrorism in food production. *Food Technology.* February, 56(2):43–47.

Bledsoe, GE and Rasco, BA. 2001. What is the risk of bioterrorism in food production? *Food Quality.* November/December, 33–37.

Blyth, T.1998. Cyberterrorism and private corporations: new threat models and risk management implications. Solicitor, Supreme Court of New South Wales, Australia.

C & EN News, 78:8, August 21, 2000; 78:13, February 28, 2000.

Devost, M. 1995.National Security in the Information Age. University of Vermont. MS. Thesis.

FEMA. 1998. Terrorism in the United States. Fact sheet. Federal Emergency Management Agency, Jan. 10, 1998. Washington DC.

Galdi, T. 1995. CRS Report for Congress – Revolution in Military Affairs? Competing Concepts, Organizational Models, Outstanding Issues. Pp 1,4.

GAO. 1996. Information security: computer attacks at the Department of Defense pose increasing risk. Chapter report. 05/22/96. GAO/AIMD-96-84. www.gas.org/irp/gao.

Hollingsworth, P. 2000. Know a crisis when you see one. *Food Technology.* 54(3):24.

Martinez, I. 2003. Rhodesian anthrax: the use of bacteriological and chemical agents during the Liberation War of 1965–1980. *Indiana International and Comparative Law Review.* 13:447–475.

Merrill, RA and Francer, JK. 2000. Organizing federal food safety regulation. *Seton Hall Law Review,* 31:61–170.

Miller, J, Engelberg, S and Broad, W. 2001. Chapter 1 – The Attack. In: *Germs – Biological Weapons and America's Secret War.* Simon and Schuster, New York. pp. 15–24.

Molander, R, Wilson, P, and Mussington, D. 1998. Strategic information warfare rising. Rand Ord. MR-964-osd. Pp3.

Murphy, K. 2000. Disruption is activist business. *Los Angeles Times.* April 2, 2000.

O'Rourke, D. 1990. Anatomy of a disaster. *Agribusiness.* 6(5):417–424.

Rathmell, A. 1997. Cyberterrorism: The Shape of Future Conflict. Royal United Service Institute J. 40–46. Pp 1–5.

Smith, GD. 1992. Militant activism and the issue of animal rights. Commentary No. 21. Canadian Security Intelligence Publication, CSIS Ottawa, CN.

Steele, R. 1998. Information peacekeeping: the purest form of war. In: Matthews, J (ed) *Challenging the United States Symetrically and Asymetrically: Can America be Defeated.* US Army War College. Strategic Studies Institute, pp. 144.

Stern, J. 1999. *The Ultimate Terrorist.* Harvard University Press, London.

Stone, B. 2000. Bitten by Love. *Newsweek.* May 15, 2000. Pg 42.

Tubbs, D. 2000. FBI's former counter-terrorism chief. *Los Angeles Times,* April 25, 2000.

Tubbs, D. 2000. 4/13/00: The face of ecoterrorism. Elf defines the use of violence.

Wardlaw, G. 1989. *Political Terrorism.* Second Ed. Cambridge University Press, Cambridge, UK. Pp 26.

WHO, 2002. Food Safety Issues. Terrorist Threats to Food Guidance for Establishing and Strengthening Prevention and Response Systems. World Health Organization, Food Safety Department, Geneva, Switzerland.

Notes

1. Quoting Libyan president Kaddafi in an interview commenting on President G. W. Bush's praise for his decision to end Libya's weapons of mass destruction and on his cooperation in the war with Al-Qaeda. In Klaidman, D. 2004. Kaddafi, Reformed? *Newsweek*, March 15, p. 42.
2. The Makah harvested their first whale after a seventy year ban in May, 1999. This single event had a profound impact, pulling the young people together and reenergizing the community. Unfortunately, other whale hunts have been held up in court proceedings, but may resume after futher population studies are conducted to ascertain which animals are resident in Puget Sound and which are migratory and can be hunted. Other restrictions on the harvest of marine mammals are being lifted because of the negative impact uncontrolled growth of the marine mammal populations have had on the ecosystem. The end of seal hunting by native and non-native fishers has been responsible in part to the decimation of the cod stocks on the East Coast and salmon on the West Coast. There is growing support among the community at large for a controlled harvest of marine mammals as a way of restoring important fishery runs.
3. Activities include smashing windows of buildings or cars, graffiti to buildings and cars and paint stripping vehicles, pouring glue into locks, releasing animals. This included an attack in April 15, 1987 causing multimillion dollars of damage to a veterinary diagnostic laboratory at the University of California, Davis, CA and 17 nearby vehicles.
4. One incident in Britain in 1990 involved a car bombing directed at two scientists. These individuals were almost killed. In a second car bombing, a 13-month-old infant passing along the sidewalk in a carriage was injured. (Smith 1992)
5. On April 24, 1989 two meat markets in Vancouver, BC were destroyed by arson on "World Laboratory Animal Day."
6. In one incident on December 17, 1991, several live lobsters and crabs were killed in an attack on a fish shop in Edmonton, Alberta. (Smith, 1992)
7. In November, 1984, ALF allegedly placed rat poison in Mars chocolate bars in the United Kingdom. Millions of bars were recalled after notes were found inside candy wrappers in 6 UK towns.
8. Animal rights activists allegedly injected 87 candy bars in Edmonton and Calgary with oven cleaner, accusing the developer of the candy bar with abusing animals during product development. Tens of thousands of bars were pulled off the market in 250 retail outlets in British Columbia, Alberta, Saskatchewan, and Manitoba. Candy production was halted by the company, which resulted in the lay off of 22 employees. This was a hoax, although one bar was found to contain an alkali which could cause burns if consumed.

   In another incident, ALF threatened to contaminate a popular drink, Lucozade. The manufacturer ordered 5 million bottles withdrawn at a cost of several hundred thousand dollars. (Smith, 1992).

9. The famous worm of Robert Morris in 1998 slowed internet sites and infected 6200 computers within 12 hours. It took days to clean up over 1 million damages. Morris was convicted and sentenced to a $10,000 fine, 3-year probation and 400 hours of community service.
10. For example, there are 4 electrical grids in the U.S. (Texas, Eastern U.S., Midwest and Northwest) connected in Nebraska. Computer security is designed to anticipate no more than two disruptions at a time. A massive power grid failure would be difficult to repair and take a long time. This would put national security at risk and leave permanent economic, personal and political scars (Devost, 1995). Malfunction at a single power station in Ohio in the summer of 2003 caused a power outage in parts of Eastern Canada, New England and the mid-Atlantic states to Philadelphia. Although much of the region was back on-line within 24 hours, other areas remained down for several days. Communications were disrupted for weeks and the impact to commerce was in the billions of dollars. Similarly, a fire in an underground tunnel in Baltimore in July of 2001 destroyed communications transmission cables, took weeks to repair and led to unreliable and sporatic phone and internet service for much of the mid-Atlantic region for the remainder of the summer.
11. There was a one billion dollar loss of revenue from loss of communications as a result of the World Trade Center bombing in New York in 1993. Destruction of the Trade Center in September, 2001 caused hundreds of billions of dollars in increased operating expenses and lost revenues for businesses. It solidified the political will caused in part by the invasion of Afghanistan in October, 2001 and Iraq in March, 2003. Both of these military efforts to date have cost in excess of 300 billion dollars, generating large budget deficits. This has put a crimp in the availability of capital for business financing and governmental activities.
12. The GAO tested EPA systems by breaking into them. The GAO found that EPA firewalls were defective and password protections were poorly encrypted. Also, the Agency's operating system was inadequate. (*C&EN News,* 78(34):8, August 21, 2000)
13. Hacker tools are readily available on the internet: 88% are effective at penetrating computer systems, 86% of all system penetrations are undetected, 95% of instances when penetration is detected nothing is done to prevent it (Devost, 1995).
14. Societal threats follow one of five different models: (1) Phenomenological threats — environmental disasters, health epidemics, famine and illegal immigration, (2) Non-national threats of force against political, racial, religious, and ethical conflicts which challenge a nation state, (3) National threats such as organized crime, piracy and terrorism operating outside the authority of the host nation state, (4) Meta-state national threats such as religious movements and international criminal movements that operated beyond the nation state, and (5) Recognizable internal security forces, infantry-based

armies and armor mechanized-based armies (in other words, the usual mechanisms for conventional war) (Galdi, 1995).

15. Sec. 2332B. DEFINITIONS (1) federal crime of terrorism is an offense that (A) is calculated to influence or affect the conduct of government by intimidation or coercion, or to retaliate against government conduct and (B) is a violation of special provisions designed to protect certain individuals, businesses or infrastructure or to prevent certain types of activities. These include the following: Sec 32 (aircraft/facilities) 27 (violence at international airports) 81 (arson within special maritime or territorial jurisdictions) 175 (biological weapons) 351 (assassination, kidnap or assault of congressional, cabinet, supreme court members), 831 (nuclear attacks), 842 (use of plastic explosives), 844 (arson and bombing of property), 956 (conspiracy in foreign country) 1114 (protection of officers or employees of the United States), 1116 (manslaughter of foreign officials, official guests, internationally protected persons), 1203 (hostage taking), 1361 (destruction of government property), 1362 (destruction of communication lines), 1363 (injury to property or buildings within special maritime and territorial jurisdiction of the United States), 2155 (destruction of national defense material, premises, utilities), 2156 (production of defective national defense materials, premises or utilities), 2280 (violence against maritime navigation), 2281 (attacks on maritime fixed platforms), 2332 (certain homicides and types of violence outside the United States), 2332a (provision or use of weapons of mass destruction), 2332b (terrorism transcending national boundaries), 2339a (providing material support to terrorist organizations), 2340a (torture), 46502 (aircraft piracy), 60123(b) (destruction of interstate gas or hazardous liquid pipeline facility), 1366 (destruction of an energy facility) 1751 (assassination, kidnap or assault of the president or presidential staff), 2152 (injury of harbor defenses).

    Further offenses include Title 18 Sec. 106, a conspiracy to harm people and property overseas and under Sec. 956 of Ch 45 of Title 18: a conspiracy to kill, kidnap, maim or injure persons or damage property in a foreign country.

    Sentences are severe. For persons within the jurisdiction of the United States, imprisonment for any term of years or for life is possible if the offense is conspiracy or murder. The sentence can be 35 years for conspiracy to maim. A sentence of up to 25 years in prison can be imposed for damage or destruction of specific property situated within a foreign country and belonging to a subdivision of a foreign government such as a railroad, canal, bridge, airport, airfield, or public utility, public conveyance, or public structure, or any religious, educational, or cultural property so situated (18 USC Ch 45 Sec. 956 (2)(b)).

16. The federal crime of terrorism under 18 USC Sec. 2332B is an offense that is calculated to influence or affect the conduct of the government by intimidation or coercion, or to retaliate against government conduct and includes: attacks against aircraft or airlines: violence at international airports or maritime navigation; aircraft piracy; attacks on harbor defenses; arson or injury

to property or buildings within special maritime or territorial jurisdictions; biological or chemical weapons; assassination; kidnap or assault of congressional, cabinet level or supreme court officials, the president or presidential staff; nuclear attacks; attacks with plastic explosives; arson or bombing of property; certain violent crimes and homocides outside the US; conspiracy in a foreign country; conspiracy to kill, kidnap, maim or injure persons or property overseas; manslaughter of foreign officials, official guests, or internationally protected people; hostage taking; destruction of government property including national defense material, premises and utilities; destruction of communication lines or an energy facility, interstate gas or hazardous liquid pipeline facility, or maritime fixed platform; international terrorism; providing materials to terrorists or terrorist organizations; etc.

Under the "Effective Counterterrorism Act of 1996" (Title I. Sec. 101. Sec. 2239A or Title 18 USC), it is a crime for anyone within the United States to knowingly provide material support or resources for a terrorist or a terrorist organization or to conceal or disguise the nature, location, source or ownership of material support or resources, knowing or intending that they are to be used in preparation for or in carrying out a terrorist act, or in preparing for or in concealing a terrorist act or in aiding in an escape. Punishment for violations includes fines and imprisonment of not more than 10 years or both. A person "knowingly provides" material support or resources if he or she knows or should have known that these activities were supporting a terrorist organization (Sec. 2339B).

Types of material support or resources include: currency or other financial securities, financial service, lodging, training, safehouses, false documentation or identification, communications equipment, facilities, weapons, lethal substances, explosives, personnel, transportation, and other physical assets, except medicine or religious materials.

Sec. 2339B of Title 18 USC amended further stipulates: "Whoever within the United States knowingly provides material support or resources in or *affecting interstate or foreign commerce* (emphasis added), to any organization which the person knows or should have known is a terrorist organization that has been designated under this section as a terrorist organization shall be fined under this title or imprisoned not more than 10 years or both."

17. In addition to the above provisions, under 18 USC Sec. 2332B, whoever intentionally commits a federal crime of terrorism against a child, shall be fined or imprisoned for any term of years or for life or both. This law does not prevent imposition of a more severe penalty for the same conduct under another federal law.
18. Federal AntiTampering Act. USC Title 18 Part I Crimes. Chapter 65 Malicious Mischief. Section 1365 Tampering with consumer products.
    (a) Whoever, with reckless disregard for the risk that another person will be placed in danger of death or bodily injury and under circumstances manifesting extreme indifference to such risk, tampers with any consumer product that affects interstate or foreign com-

merce, or the labeling of, or container for, any such product, or attempts to do so, shall

in the case of an attempt, be fined under this title, or imprisoned not more than 10 years, or both;
if death of an individual results, be fined under this title or imprisoned for any term of years, or life or both;
if serious bodily injury to any individual results, be fined under this title or imprisoned not more than 20 years, or both; and
in any other case, be fined under this title or imprisoned not more than 10 years, or both.

(b) Whoever within intent to cause serious injury to a business of any person, taints any consumer product or renders materially false or misleading the labeling of, or container for, a consumer product, if such consumer product affects interstate or foreign commerce, shall be fined under this title or imprisoned not more than three years, or both.

(c) (1) Whoever knowingly communicates false information that a consumer product has been tainted, if such product or the results of such communication affects interstate or foreign commerce, and if such tainting, had it occurred, would create a risk of death, or bodily injury to another person, shall be fined under this title or imprisoned not more than five years or both.
(2) As used in paragraph (1) of this subsection, the term "communicates false information" means communicates information that is false and that the communicator knows is false, under circumstances in which the information may reasonably be expected to be believed.

(d) Whoever knowingly threatens, under circumstances in which the threat may reasonably be expected to be believed, that conduct that, if it occurred , would violate subsection (a) of this section will occur, shall be fined under this title or imprisoned not more than five years, or both.

(e) Whoever, is a party to a conspiracy of two or more persons to commit an offense under subsection (a) of this section, if any of the parties intentionally engages in any conduct in furtherance of such offense, shall be fine under this title or imprisoned not more than ten years or both.

(f) In addition to any other agency which has authority to investigate violations of this Section, the Food and Drug Administration and the Department of Agriculture, respectively, have authority to investigate violations of this section involving a consumer product that is regulated by a provision of law such Administration or Department, as the case may be, administers.

(g) As used in this section —

the term "consumer product" means —

any “food”, “drug”, “device”, or “cosmetic”, as those terms are defined in section 201 of the Federal Food Drug and Cosmetic Act (21 USC 321); or

any article, product, or commodity which is customarily produced or distributed for consumption by individuals, or use by individuals for purposes of personal care or in the performances of services ordinarily rendered within the household, and which is designed to be consumed or expended in the course of such consumption or use;

the term “labeling” has the meaning given such term in section 201(m) of the Federal Food Drug and Cosmetic Act 21 USC 321(m));

the term “serious bodily injury” means bodily injury which involves –

substantial risk of death,

extreme physical pain,

protracted and obvious disfigurement or protracted loss or

impairment of the function of a bodily member, organ, or mental faculty; and

the term “bodily injury”

a cut, abrasion, bruise, burn, or disfigurement,

physical pain,

illness

impairment of the function of a bodily member, organ or mental faculty; or

any other injury to the body, no matter how temporary.

19. This bill was introduced because California law did not adequately address the cost of vandalized crops that that are products of testing and research. Existing law provides that any owner of livestock or poultry which is injured or killed by a dog, may recover as liquidated damages from the dog's owner, twice the actual failure of the animal killed or twice the value of the damages sustained. In 1999, several acts of ecoterrorism occurred at UC Davis and Berkeley, and a private site in Woodland including destruction of corn, sugar beets, walnut tees, melons, tomatoes, sunflowers and related equipment for research. Proponents of this and similar legislation stated that such acts of violence did not have substantial penalties to deter agroterrorism. These new laws include damage for the intrinsic value of crop from the research costs necessary to develop it.

    Text: Section 1/ Article 6 (commencing with Section 52100) is added Chapter 1 of Division 8 of the Food and Agricultural Code to read: Article 6 Destruction. 52100 (a) Any person or entity who willfully and maliciously damages or destroys any field crop product, as defined in Section 52001, that is the subject of testing or a product development program being conducted by, or in conjunction with , or in a program recognized by the University of California, California State University, or any other state or local government agency shall be liable for twice the value of the crop damaged or destroyed.

(b) in awarding damages under this section, the courts shall consider research, testing, and crop development costs directly related to the crop that has been damaged or destroyed. http://democrats.assembly.ca.gov/members/a08

2 Potential Biological and Toxic Chemical Agents

> Chemicals continue to be the weapons of choice for terrorist attacks.[1]

The FBI issued a warning in April 2003 that terrorists may try to improvise weapons from common household items with the intent of contaminating food or introducing biological or chemical contaminants into the environment (Anderson, 2003). Producing cyanide or ricin (or abrin, a closely related toxin) in a home laboratory is not difficult. Likewise, growing crude cultures of *Salmonella* sp. or *Clostridium botulinum* (botulism toxin) is possible by an individual with minimal training or skill. Capture of Al-Qaeda's chief of operations, Khalid Shaikh Mohammed, provided fresh evidence that this terrorist group was experimenting with such weapons. Raids on a mosque in Manchester, England, and in neighboring apartments in February, 2003 uncovered ricin production in home labs. The Department of Defense is also of the belief that terrorists may use biological agents to contaminate food or water because they are extremely difficult to detect (Anon., 1998). Again, the purpose of these weapons would be primarily to terrorize unprotected civilians and not as weapons of war.

The effective use of these agents to contaminate food would depend upon:

1. The potential impact to human, animal, or plant health
2. The type of food material contaminated
3. How easy it would be to detect contamination through discernible changes in appearance, odor, or flavor of the food
4. The point at which the contamination is introduced into the food supply

5. The potential for widespread contamination
6. The fear factor people associate with either the food or the toxic agent

Biological Agents

A biological agent can be living infectious microbes or toxins produced by microorganisms that cause illness or death in people, animals, or plants. Microbes have the insidious capability to grow and multiply within the target organism, causing disease. Biological agents can be dispersed in liquid or solid media or as aerosols or airborne particles. Biological agents tend to be highly *specific* for target victims (human, animal, or plant). Biological agents pose difficult issues with *controllability.* First of all, environmental conditions necessary to transmit the biological material may be difficult to manage, such as air movement or, in the case of food, temperature or pH. Factors affecting the viability of the strain used may not be well understood and may also be difficult to control. Furthermore, limiting the impact to a specific target may be difficult due to secondary transmission and other biological vectors, for example, further transmission of an infective microbe from person to person, or transmission to people outside the targeted area by biological vectors such as insects, rodents, or birds.

Toxins are similar to chemical agents and, if delivered as such, are, like a chemical agent, effective upon delivery. These can be dispersed as an aerosol, liquid, or component of a solid. Biological toxins are highly *toxic* and can kill or debilitate at levels as low as one to ten parts in 1,000,000,000,000,000. Concentrations are usually measured in parts per million (1 part per 1,000,000) to parts per trillion (1 part in 1,000,000,000,000). In general, the residual effects of biological toxins are less than those for chemical agents, and certainly lower than those remaining after a radiological incident. However, there are important exceptions to this: the spores of potential agents such as *Bacillus anthracis* or *Clostridium botulinum* persist in soil and in the estuarine environment for years and could similarly persist inside insulation, building materials, or structures for a long period.

Live agents, most commonly microbes, generally act more slowly than chemical agents or biological toxins. Microbes act through the processes of either *infection* or *intoxication*. Both processes are relatively slow because of the need of the microbe to grow inside the tissue of the target organism or in the food prior to ingestion. If the microbe directly produces disease, this is an infection. However, if the microbe must grow inside the victim and then produce a toxin that has the debilitating effect on the victim, this is intoxication. Another form of intoxication occurs when the microbe grows in a food and produces a toxin. Illness occurs when an individual consumes the contaminated food. Both the greatest potential and the greatest downside of using a live biological agent is the long time it may take before detrimental effects are noted. The effect of bacterial toxins, such as the protein toxin from *Staphylococcus aureus*, takes at least 30 minutes, and usually 8 hours or longer to cause illness.[2] Other bacterial or fungal toxins take much longer to produce a notable effect, and oftentimes symptoms of intoxication can be misdiagnosed (for example, the neurological effects of botulinum and ciguatera toxins). Furthermore, for infection, the bacteria must survive digestion (or possibly inhalation), reach the target tissue, grow, and then cause illness symptoms or produce the toxin that then produces an illness. The onset of a food-borne infection takes at least 24 hours and, in cases, such as with *Listeria*, may be as long as a number of weeks. The onset of illness from anthrax, salmonella, and *Escherichia coli* takes 1 to 6 days. Symptoms of food-borne illness are often misdiagnosed as gastroenteritis. In the case of an intentional food contamination incident, the community may or may not be notified that it has been targeted, or may be falsely informed regarding the nature of an incident, complicating treatment of affected individuals. Unfortunately, for treatment to be most effective for a bacterial food-borne illness, it should start before symptoms appear.

Most agents used in biological warfare weapons or considered agents for contaminating food are diverse and widely distributed in the environment (Atlas, 1998). Almost any pathogen can be used to intentionally spread disease. The Centers for Disease Control and Prevention (CDC) along with other sources have developed a list of high-risk biological agents. A list of these agents is presented in Table 2.1, along with symptoms, lethality, and modes of transmission.

Proposed biological agents fall within the following classifications: viruses, bacteria, fungi, and rickettsia. Viruses are commonly defined as genetic material (most commonly DNA (deoxyribonucleic acid) but also RNA (ribonucleic acid)) surrounded by a protein coat. Viruses are too primitive to reproduce on their own and can only reproduce by invading a host cell and pirating its nucleic acid and protein-replicating systems. *Retroviruses* are even more primitive and are composed of unenveloped particles of nucleic acid; these are essentially viruses with no protein coat. Most viruses do not survive well outside the host. Exceptions are the rhinoviruses that cause the common cold, which survive well in humid environments, making them highly contagious. Encephalitis, psittacosis, yellow fever, and dengue fever are agents most commonly cited as being potential biological warfare agents. The likelihood of these being used to contaminate food is not high. When you consider the havoc that can be caused by a small and relatively well contained outbreak with common hepatoviruses, for example, the November 2003 outbreak of hepatitis A in western Pennsylvania tied to green

Table 2.1 Potential Biological Agents for Food Contamination

Organism	Disease Infective Dose	Symptoms	Onset	Transmission	% Lethality if Not Treated	Stability
Bacteria						
Bacillus anthracis	Anthrax 8,000–50,000 spores	Inhalation: Mild fever, muscle ache, coughing, difficulty breathing, exhaustion, toxemia, cyanosis, shock, respiratory failure, meningitis Cutaneous: Usually painless skin lesion develops into depressed black eschar; may also be fever, malaise, headache, regional lymphadenopathy Gastrointestinal: Severe abdominal pain, fever, septicemia; oropharyngeal and abdominal forms of disease possible Oral: Lesions on pharynx and at base of tongue, with dysphagia, fever, and lymph swelling Abdominal: Lower bowel inflammation, nausea, loss of appetite, fever, abdominal pain hematemesis, bloody diarrhea	Inhale: 1–7 days; can be up to 60 days Cutaneous: 1–12 days Skin: 1–7 days	Air/respiratory Ingestion or through skin Limited contagion Except cutaneous	Inhale: >85% Cutaneous: 20% GI: 25–60%	Very high
Yersinia pestis	Pneumonic plague Bubonic plague Septicemic plague 100–500 org	Pneumonic plague: High fever, headache, chills, toxemia, cyanosis, respiratory failure, circulatory collapse, cough blood Bubonic plague: Malaise, high fever, tender lymph nodes Septicemic plague: Intravascular congestion, necrosis, skin lesions, "black death" gangrene in fingers and nose	1–6 days pn 2–8 days bu	Air/respiratory or direct skin contact Highly contagious	100% pneumonic 50% bubonic	Somewhat

Vibrio cholerae	Cholera 1000–10,000,000 org	Vomiting, abdominal distension, pain, diarrhea, severe dehydration, shock	2–4 days	Oral Not contagious	25–50%	Low
Escherichia coli	Hemolytic uremic syndrome (HUS) >10 org	Acute bloody diarrhea, abdominal cramps, kidney failure, seizures, stroke	12–36 hours	Oral Secondary transmission possible	3–5%	Low
Salmonella enteritidis	Salmonellosis 1,000,000 org	Vomiting, abdominal cramps, diarrhea	6–48 hours	Oral Secondary transmission Possible	<5%	Low
Salmonella typhi	Typhoid fever 100–1,000,000 org	Septicemia, runny stools, fever, hemorrahages, peritonitis, headache, constipation, malaise, chills, myalgia, confusion, delirium	3–56 days	Oral Chronic carrier state common	12–30%	Low
Shigella	Shigellosis			Oral Secondary transmission possible		Low
Listeria monocytogenes	Listeriosis >100 org	Flu-like symptoms, headache, meningitis, encephalitis, endocarditis, spontaneous abortion, septicimia	1–90 days	Oral Not contagious	13–34%	Somewhat
Clostridium botulinum toxin	Botulism µg/kg (type A) 0.7–0.9 µg, inhale 70 µg, oral 0.09–0.15 µg, IV	Nausea, weakness, vomiting, respiratory paralysis, dizziness, dry throat, blurred vision, difficulty swallowing and speaking	12–72 hours typical 2 hours–8 days	Oral, IV, or inhalation	5–90%	High
Clostridium botulinum organism	Botulism	Same as above (organism colonizes host)	Several days	Oral, wound Not contagious	Same	Very high

Table 2.1 Potential Biological Agents for Food Contamination (Continued)

Organism	Disease Infective Dose	Symptoms	Onset	Transmission	% Lethality if Not Treated	Stability
Brucella sp.	Brucellosis Undulant fever	Acute form (<8 weeks from illness onset): Symptomatic, nonspecific, and flu-like, including fever, sweats, malaise, anorexia, headache, myalgia, and back pain Undulant form (<1 year from illness onset): Undulant fevers, arthritis, and orchiepididymitis in males Neurologic symptoms may occur acutely in up to 5% of cases, the chronic form (>1 year from onset): Chronic fatigue syndrome-like, depressive episodes, and arthritis	Weeks to 1 year	Air/respiratory Oral, skin wounds Not contagious	Low	Somewhat
Francisella tularensis	Tularemia 10–50 org	Cough, chills, muscle ache, swollen glands, pleuropneuomonitis, respiratory distress, exhaustion, can be confused with Q fever	2–4 days	Air/respiratory tract Not contagious	30–40%	Good
Staphlyococcus aureus, B enterotoxin	Staph infection 30 ng/person (incapaciting) 1.7 µg/person (lethal)	Oral: Fever, headache, chills, nausea vomiting, diarrhea Inhaled: Nonproductive cough	3–12 hours	Oral and inhalation	<5%	High
Rickettsia						
Coxiella burnetii	Q-fever 1–10 org	Fever, aches, headache, fatigue, hepatitis, endocarditis	10–40 days	Air/respiratory tract Rarely contagious	<5%	Good
Coccidioides immitis	Valley fever	Fever, pain, weakness		Air/respiratory tract	<5%	High

Virus						
Variola major	Smallpox 10–100 org	Fever, characteristic blister like rash, Back pain, headache	7–17 days	Air/respiratory tract Direct contact with skin Contagious	35%	High
Flaviviridae	Flavirus Yellow fever Omsk hemorrhagic fever Kyasanur forest disease	Yellow fever: Fever, myalgias, facial flushing, conunctial injection, bradycardia, jaundice, renal failure, hemorrhagic complications Omsk: Fever, cough, conjunctivitis, papulovesicular eruption on soft palate, facial flushing, swelling of glands, pneumonia or CNS failure Kyasanur: Similar to Omsk but biphasic illness with 50% developing meningoencaphalitis	3–15 days, general Yellow fever: 3–6 days Omsk: 2–9 days	Air/respiratory tract Insect vector Contagious	<5% generally Yellow fever: 20% Omsk: <10% Kyasanur: 3–10%	Somewhat
Filoviridae	Filovirus Ebola Marburg 1–100 org	Ebola: High fever and severe prostration, diffuse maculopapular rash within 5 days, bleeding, intravascular coagulation Marburg: High fever, myalgia, skin rash, bleeding, intravascular coagulation	Ebola: 2–21 days Marburg: 2–14 days	Air/respiratory Person to person Highly contagious	Ebola: 50–90% Marburg: 23–70%	Low
Arenaviridae	Arenavirus Lassa New world arenaviridae	Lassa: Gradual onset of fever, nausea, abdominal pain, severe sore throat, cough, conjunctivitis, oral ulcers, severe swelling in head and neck, pleural and pericardial effusions New world: Gradual onset of fever, myalgia, nausea, abdominal pain, conjunctivitis, flushing in face and trunk, bleeding, CNS dysfunction, seizures	Lassa: 5–16 days New world: 7–14 days	Air/respiratory Person to person Highly contagious	Lassa: 15–20% New world: 15–30%	Low

Table 2.1 Potential Biological Agents for Food Contamination (Continued)

Organism	Disease Infective Dose	Symptoms	Onset	Transmission	% Lethality if Not Treated	Stability
Bunyaviridae	Nairovirus (Crimean–Congo hemorrhagic fever) Phlebovirus (Rift Valley fever) Hantavirus (hemorrhagic fever with renal syndrome)	Rift Valley: Fever, headache, retro-orbital pain, photophobia, jaundice	2–6 days	Air/respiratory Person to person Ingestion of fecal-contaminated food for Hantavirus Rift Valley: Mosquito vector Hantavirus: Rodent vector	<1%	Somewhat
Hemorrhagic viruses generally	Hemorrhagic fevers 1–100 org	Headache, fever, vomiting, diarrhea, chest pain, cough, easy bleeding, prostration, shock	Varies, 4–21 days	Air/respiratory Contagiousness varies	Varies	Somewhat
Venezuelan equine encephalitis (VEE)	VEE encephalomyelitis 10–100 org	Joint pain, chills, headache, nausea, vomiting with diarrhea, sore throat	2–5 days	Air/respiratory tract Direct contact with skin Low contagion	70%	Low
Toxin						
Ricin (abrin is related toxin)	Ricin intoxication	Inhale: Fever, cough, nausea, chest tightness, heavy sweating, pulmonary edema, weakness, fever, cough, severe respiratory distress Ingest: Vomiting, diarrhea that may become bloody, severe dehydration, high blood pressure, hallucinations, seizures, bloody urine, liver, spleen, and kidney failure Inject: Weakness, fever, vomiting, shock, multiorgan failure	6–24 hours GI: 1-4 hours	Oral, injection Air or respiratory Not contagious	Presumed high Death can occur in 36-72 hours	Good

Abrin (similar to ricin)	Abrin intoxication	Ingest: Muscle weakness, tremors, spasms, low blood pressure, fast heart rate, irregular heart rhythms, cardiovascular shock from severe dehydration, central nervous system disorder, drowsiness, disorientation, hallucination, seizure, coma GI: Burning pain in mouth, abdominal pain, nausea, severe vomiting, diarrhea, bleeding and swelling of GI tract lining, liver damage, pancreas damage, blood in urine, low urine output, kidney cell damage Eyes: Dilated pupils, retinal hemorrhage, tearing, swelling, pain, redness, corneal injury Skin: May be absorbed, redness, blister, pain, cyanosis; may cause severe allergic reactions Inhale: Irritation, sensitization	<24 hours	Oral Skin or eye contact Respiratory	Symptoms can be delayed 1–3 days Lethal dose is 0.005–0.007 mg/kg	Good
Aflatoxin		Headache, jaundice, GI distress, liver disease	>24 hours	Oral Not contagious	Low	High

Note: pn = pneumonic; bu = bubonic; org = organism.

Source: From Adams and Moss, 2000; Anon., 2003; Arnon et al., 2001; Atlas, 1998; Borio et al., 2002; Inglesby et al., 2001, 2002; MMWR, 2001a, 2001b, 2003; NIOSH, 2003; Stern, 1999.

onions, the use of simpler and more common agents to intentionally contaminate food is more realistic.

Bacteria are simple single-cell organisms that can reproduce either inside or outside of a host organism. They have a nucleus containing genetic material, but they do not have a complex intracellular structure like animal and plant cells do. Certain bacteria can enter into a dormant state as a spore when challenged by heat, low-moisture conditions, or chemical agents such as chlorine-sanitizing agents. Bacteria act by causing an infection or by producing a toxin (intoxication) that causes illness.

Fungi are a more complex microorganism than bacteria. These organisms can spread by producing hyphae, which resemble small branches, outward from a central core. Sexual reproduction through spores also occurs. The spores can be spread by air and are often quite resistant to heat or chemical treatment. The food contamination risk from a fungus is not the organism itself, but the toxin it produces. However, the intentional spread of a disease-producing mold within the agriculture community could be devastating. For example, the easy-to-grow rice blast fungus *Pyricularia oryzae* has been proposed as a biological warfare agent.

Parasites have also been proposed as biological weapons. Most foodborne parasites (such as *Anisakis* sp. in seafood products and Trichinae in meats, or Giardia sp. from contaminated water) are difficult to grow outside appropriate hosts and would not be suitable agents to target humans directly although a couple of notable exceptions mentioned here are highly infectious and can be transmitted by air; however, the risk of parasitic disease to plants and animals poses a significant threat. Rickettsia is one example of a proposed biological agent that causes animal and human disease. *Coxiella burnetii*, the cause of Q-fever, is a highly infectious disease that rarely kills but is incapacitating.

Biological agents have the unfortunate characteristics of low visibility, high potency, substantial accessibility, and relative ease of delivery (Atlas, 1998). Transmission through the air is technically somewhat difficult, but is the dissemination method most likely to affect the largest number of people. Infection of water or food and the use of insect vectors and animal hosts are other possible methods of transmission.

Small quantities of a biological agent can cause a serious public health emergency. One millionth of a gram of pure anthrax is a lethal inhalation dose. Small quantities of agent are easy to conceal and transport. Many of the potential agents are present in the environment and are often used for legitimate purposes. Most biological agents are alive, can adapt to an environment once released, and grow rapidly with victims of an attack not being aware of exposure to an infective agent until days or weeks later (Atlas, 1998). After an attack, an infection could spread to the individuals providing aid to the victim — family members and health care workers. Before diagnosis, an infection could be widely spread through the air and by direct contact with the victim (Atlas, 1998). The illnesses that many of these agents cause, at least in the initial stages, have symptoms that resemble the flu.

Biological (and chemical) agents are the poor man's nuclear weapon with specific strategic and tactical military objectives. At least 10 governments have programs for biological weapons[3] (Anon., 1998; Carus, 1998). There are huge amounts of these materials dispersed across the planet, and many are not under any form of reasonable control. The former director of the Soviet civilian biological weapons program claims that the USSR stockpiled 20 tons of powdered smallpox virus and smallpox (Henderson et al., 2001) and large amounts of anthrax (possibly some recombinant forms), plague, and tularemia for use as biological weapons (Atlas, 1998). The Soviets had the industrial capacity to produce smallpox (possibly recombinant strains or those with enhanced virulence) and had weaponized its use in bombs and intercontinental ballistic missiles (Henderson et al., 2001). Similarly, North Korea may have a stockpile of biological agents, including *Clostridium botulinum*, *Vibrio cholera*, hemmorhagic fever-causing viruses such as yellow fever, plague bacteria, and typhus. Countries have also developed programs to create economic havoc anticrop agents. Allegedly accidental releases of biological agents, likely tied to weapons programs, have occurred. One incident, the 1979 outbreak involving a highly lethal strain of anthrax from a Soviet weapons facility in Sverdlovsk, was initially treated by the public health community as a natural occurrence (Berns et al., 1998; Inglesby et al., 2002). Anthrax is particularly troublesome because airborne dispersions could travel for several kilometers before dissipating. Large-scale tests with anthrax have been conducted by Iraq and the USSR and, in 1960, by the U.S. near the Johnson Atoll in the South Pacific. In this particular study, a 32-mile-long line of anthrax traveled more than 60 miles before it lost its infectiousness (Inglesby et al., 2002). An estimated 50 kg of anthrax spores released over an urban center of 5 million would sicken 250,000 individuals and kill roughly 100,000. Between 130,000 and 3 million deaths would occur from a release of 100 kg, making the lethality of an anthrax release of this type similar to that of a hydrogen bomb (Inglesby et al., 2002).

South Africa maintained a stockpile of biological weapons for "defensive use" even after signing the Biological Weapons Convention in 1972 (Martinez, 2003). Countries in support of apartheid developed programs to kill people and animals, and to wreak economic havoc using biological agents. Anthrax was intentionally used to kill cattle during the 15-year civil war (1965–1980) in Zimbabwe (formerly Rhodesia), possibly through an aerosol spraying program (Martinez, 2003). This rise in incidence was attributed to bioterrorism because the risk of anthrax in Rhodesia was low. Prior to 1978, there were an average of 13 anthrax cases per year with a total number of 355 cases from 1950–1978. However, in 1979, there were 182 deaths from cutaneous anthrax, and 10,783 individuals infected. The impact to livestock in the region was devastating. Zimbabwe provided a perfect environment for anthrax dissemination. It had proper soil conditions and the people were dependent upon livestock to make a living. In rural areas of southern Africa, wealth was measured by the number of cattle a family had. If the cattle died, the family lost its source of income. Because of the political instability in Rhodesia and the surrounding countries from the 1960s onward, immunization programs for livestock had fallen into disarray making herds more susceptible to infection. The collapse of veterinary services contributed to the anthrax deaths and to soaring rates of malaria, biharzias, tick borne diseases and sleeping sickness during the period of the anthrax spread. Between 1975 and 1979, 250,000 cattle died. The propaganda campaign directed to the affected tribes regarding anthrax was that the disease was being introduced into Rhodesia from Mozambique by the guerilla insurgents.

Iraq had active chemical and biological weapons programs. Both the Iraqi Salam Pak and the Al Hakam biological production center were under control of the military and within 100 km of Baghdad. These facilities had the capacity to produce huge quantities of microbes (Barton, 1998; Zilinskas, 1998). Long lists of biological material and equipment were imported into Iraq in the late 1980s, including 150 liters of fermenters, freeze dryers, biological cabinets, continuous-flow centrifuges, large shaking incubators, and over 40 tons of bacterial growth media sufficient to produce over 1 million liters of culture broth. The total cost of the Iraqi biological weapons program was projected at $200 million to $300 million, employing approximately 200 to 300 scientists and support staff. This effort was small compared to Iraq's multi-billion-dollar chemical weapons program, which employed over 1000 individuals. It was also small compared to the huge Soviet Union biological weapons effort, which employed over 30,000 people at 50 research and production facilities (Barton, 1998; Zilinskas, 1998). The objective of these weapons programs was to produce anthrax (*Bacillus anthracis*), botulinum toxin (produced by the bacteria *Clostridium botulinum*), and *Clostridium perfringens*, the organism causing gangrene (Zilinskas, 1998).

The March 2003 invasion of Iraq was justified in part on Saddam Hussein's illicit weapons programs. Iraq had more botulinum weaponized than any other agent in the program. In 1991, Iraq admitted to having 19,000 liters of concentrated botulinum toxin — three times the amount of toxin necessary to kill everyone on the planet. Earlier, in 1990, Iraq had deployed at least 180 field-ready warheads with biological agents (Barton, 1998). This includes specially designed missiles with a 600-km range: 13 with botulinum toxin, 10 with aflatoxin, and 2 with anthrax, possibly for targets in Kurdistan,

Iran, and Israel. Saddam had also deployed 180-kg bombs, 100 with botulinum, 50 with anthrax, and 7 with aflatoxin (Arnon et al., 2001). Iraq may also have possessed roughly one hundred 100-km FROGs (free rockets over ground) with chemical or biological capability. There remains a strong likelihood that numerous biological weapons are buried in the sands of Iraq along with confirmed airplanes and munitions. Hopefully these can be recovered before they cause harm to the civilian population. Besides the typical agents, there were programs for targeting various ethnic groups not politically supportive of the regime with camel pox, plus programs to produce economic weapons using fungal and viral agents to spread plant disease.

Numerous nonstate actors and terrorists have evaluated the use of biological agents. The Aum Shinrikyo cult in Japan attempted to acquire anthrax bacilli, botulinum toxin, and the Ebola virus during the 1990s (Berns et al., 1998). The largest known U.S. incident of intentional contamination to date was with *Salmonella typhimurium* in The Dalles, OR, in the fall of 1984. This incident created a national impact, but was initially dismissed as not being a terrorist incident. Members of a religious cult were intent upon taking over control of the county government and decided to incapacitate residents so they would not be able to vote. Led by a renegade nurse and her minions, home-grown *Salmonella typhimurium* was misted onto 10 salad bars. Although there were 751 reported illnesses, it is likely that there were many more, as travelers to points all across the U.S. could have been affected. This incident is described in greater detail in Chapter 1.

The U.S. Congress appropriated over $500 million for smallpox vaccination programs. Individuals with the U.S. government suspected that smallpox would be an agent of choice in a biological attack (although many experts considered anthrax and botulinum to be more likely choices). Smallpox had been produced in great quantities under the Soviet weapons program and could pose a serious threat (Atlas, 1998), as loss of control of this material became evident in the early 1990s. Routine vaccination for smallpox stopped in 1980 after the disease had been completely eradicated worldwide (1977). As a result, there is no natural immunity to the disease in humans, making the impact of an outbreak devastating, with at least a 30% mortality rate and permanent disfigurement for the survivors. Until vaccination became common, almost everyone eventually contracted one of the two forms of the disease, variola major (smallpox) or variola minor (alastrim), the milder form (Henderson et al., 2001).

Other forms of biological agents can serve as low-tech weapons (Berns et al., 1998). The Tartars catapulted bodies of bubonic plague victims over the city walls of 14th-century Kaffa. During World War I, the Germans used foot-and-mouth disease and other agents against livestock in South America, targeting horses being shipped to the U.S. cavalry (Atlas, 1998). Japanese forces during World War II attempted to float hot air balloons containing plague-infected ticks and fleas for release over the western U.S. (Berns et al., 1998). Low-tech methods of aerosol dissemination such as crop dusters, backpack sprayers, and handheld atomizers could be used for limited attacks and have been suggested as possible delivery systems for terrorists of biological agents. Handheld misters were the delivery system for salmonella in The Dalles incident in 1984. Canadian researchers confirmed the effectiveness of dispersing powdered anthrax spores via common envelope (primary aerolization) in an indoor environment, showing rapid delivery of a high localized concentration of spores. Indoor airflows, activity patterns, and heating and cooling systems further disperse spores throughout other parts of the building (Inglesby et al., 2002). The porosity of the paper is greater than the diameter of the spores. Spores can settle on surfaces or inside porous materials and then be dispersed (secondary aerolization) at a later date, making further recontamination possible. Secondary aerolization poses a limited risk as long as spore density remains below 1 million spores per m^2 (Inglesby et al., 2002).

Tularemia was one of the agents studied by the Japanese during World War II as a biological agent and may have been disseminated in Manchuria during the war (Dennis et al., 2001). Similarly, tularemia outbreaks that affected tens of thousands of Russian and German soldiers on the Eastern European front may have been from deliberate releases of this organism. In the 1950s and 1960s, the U.S. developed aerosol delivery systems for this microbe and stockpiled it. Estimates for the economic impact of an aerosol dispersement incident with tularemia would cost $5.4 billion for every 100,000 people exposed. Reported terrorist attempts with this microbe are lacking, but cannot be ruled out.

Plague, one of the most dreaded diseases of antiquity, is considered to be a likely choice for weaponization by terrorists and has been a component of weapons arsenals. In World War II, the Japanese dropped plague-infested fleas over population centers in China, causing numerous sporatic outbreaks of the disease (Inglesby et al., 2001). Following the war, the U.S. and Soviet Union developed technologies to aerosolize plague to increase the reliability of dispersion. The Soviets were capable of weaponizing large quantities of plague, with thousands of scientists at 10 institutes reported to have experience with this agent. Through the 1990s, cultures of plague bacteria and other potential biological weapons agents were commercially shipped to suspicious purchasers before stricter federal controls were instituted. In one incident, a microbiologist with suspicious motives was arrested after fraudulently obtaining *Yersinia pestis* cultures through the mail (Inglesby et al., 2001). In a World Health Organization (WHO) study, 50 kg of plague released as an aerosol over a city of 5 million could cause 150,000 cases of pneumonic plague and result in 36,000 deaths (Inglesby et al., 2001). The agent would remain viable for an hour and be dispersed over an area of up to 10 km.

Hemorrhagic viruses pose a new threat. These are agents with high morbidity and mortality at very low infective doses. Hemorrhagic viruses are highly infectious, easily transmitted through aerosols and from person to person, making widespread outbreaks a possibility. In addition, there is no reliable treatment for most of these agents. Fortunately, large-scale use of hemorrhagic viruses is not yet technically feasible, but of course, that could change. Both the U.S. and Soviet Union were able to weaponize hemorrhagic fever viruses. In the early 1990s the USSR had stocks of Marburg, Ebola, Lassa, Junin, and Machupo. The U.S. weaponized both yellow fever and Rift Valley fever. North Korea may currently have yellow fever-based weapons (Borio et al., 2002). Recently, the Japanese terrorist group Aum Shinrikyo unsuccessfully attempted to obtain Ebola as a terrorism agent (Borio et al., 2002). One of the most likely scenarios is widespread infection of humans with Rift Valley fever initiated by infecting livestock via mosquito vector transmission.

Terrorists have attempted to use botulinum toxin as a bioweapon, with the most notorious being the Aum Shinrikyo cult on at least three occasions between 1990 and 1995. They had isolated the toxin from soil cultures collected in northern Japan (Arnon et al., 2001). During the Japanese occupation of China beginning in the 1930s, prisoners in Manchuria were killed by feeding them cultures of *Clostridium botulinum*. Both Germany and the U.S. had botulinum bioweapons programs in place during World War II. The USSR continued its botulinum bioweapons program until its fall, with several tests conducted at the Aralsk-7 site on Vozrozhdeniye Island in the Aral Sea. Soviet scientists report attempts to splice botulinum toxin genes into other bacteria, and this expertise has been hired out following the collapse of the Soviet Union to develop botulinum weapons programs in Iran, North Korea, and Syria.

Because the U.S. had concerns that the Germans had weaponized botulinum during World War II, over 1 million doses of botulinum toxoid were prepared for the Allied troops in anticipation of the Normandy invasion on June 6, 1944 (Arnon et al., 2001). Recently, California adopted a similar strategy and is promoting the development of botulinum antitoxin in case this toxin is used to poison food. California requested an expedited approval process from the Food and Drug Administration (FDA) for this application (Anon., 2003).

Botulinum toxin is produced by a common bacterium that can be easily recovered from the soil.[4] Although the microbe is somewhat difficult to culture, because oxygen-free conditions are required for its growth, this is far from an impossible task for a marginally competent scientist. The toxin has become much more widely available as of late. Recently, botulinum toxin has been licensed for medical uses (at concentrations of about 0.3% of the estimated human lethal inhalation dose), for the treatment of headache, pain, stroke, traumatic brain injury, cerebral palsy, achalasia, and various dystonias (Arnon et al., 2001). Botulinum toxin is widely used in cosmetic surgery to temporarily remove wrinkles, crow's-feet, and laugh lines.

The risk remains of botulinum as a biological agent, and although there are skeptics regarding the utility of botulinum as an effective agent against military targets, any deliberate release against a civilian population would cause substantial damage. A point source aerosol release could kill or incapacitate 10% of the people within a 0.5-km range (Arnon et al., 2001).

In recent terrorist incidents anthrax was sent through the mail to prominent politicians and members of the media in September 2001, with 22 confirmed or suspected cases, 11 of which were inhalation cases and 11 of which were cutaneous, resulting in 5 deaths (Inglesby et al., 2002). Contaminated letters were sent from Trenton, NJ, to New York City, Florida, and Washington, D.C. In addition, disease investigations involving confirmed or suspected cases were conducted in New Jersey, Pennsylvania, and Virginia. Of the seven inhalation cases, five affected individuals were postal workers. Of the total individuals made ill, seven were associated with the media (including one infant) and eight were postal workers (MMWR, 2001b). A number of false reports of contaminated mail surfaced across the western U.S., targeting major corporations such as Microsoft and various government officials. The letter sent to Senator Daschle contained 2 g, with between 100 billion and 1 trillion spores per g (Inglesby et al., 2002). The letter contained a weapons-grade strain (the Ames strain) traced to a lab in College Station, TX. As of this writing (September 2004), mail delivery to Washington, D.C., is still somewhat unreliable. At least $23 million has been spent on environmental remediation of the Hart Senate Office Building alone. The costs for decontaminating other government buildings in D.C., the expensive changes in mail-handling practices, and the impact of these changes on business and governmental operations, have yet to be calculated. However, the overall cost of this incident to the public is estimated in the many billions of dollars.

Another series of terrorist acts with anthrax involved Aum Shinrikyo. It released aerosolized anthrax though the Tokyo subway system on at least eight occasions in the mid-1990s. Fortunately, the cult used a strain commonly used for animal vaccination that had a low virulence in humans. This is most likely the reason there were no illnesses from these attacks (Inglesby et al., 2002).

It is our opinion that a likely form of biological attack on food (including water) would involve either the limited use of one or more pathogens developed specifically for biological warfare purposes or a common bacterial food-borne or zoonotic agent. Zoonotics are animal disease causing agents (e.g., anthrax (*Bacillus anthracis*), plague (*Yersinia pestis*), and rabbit fever (*Franciscella tularensis*)) that can be transmitted to humans. Contamination of food would not require that a highly purified form of a biological agent be used. Other possible biological attacks involve economic terrorism targeted at a specific commercial entity or industry segment that involve the real or threatened introduction of an animal or plant pathogen (or its genetic material) at a production or agricultural facility or at a point of sale or distribution. This would also include the actual or threatened introduction of genetic material(s) into a food or agricultural product.

Specific Diagnosis Issues with Biological Agents

Botulism has incredible potential for use as a biological agent because it is heat stable and lethal in low quantities. However, the onset of the illness varies widely, and it is commonly misdiagnosed as an acute flaccid paralysis, polyradiculoneuropathy (Guillain–Barré or Miller–Fisher syndrome), myasthenia gravis, or a central nervous disease (Arnon et al., 2001).

Plague is not common and could initially be misdiagnosed. Initial symptoms of the most lethal form, pneumonic plague, would be fever and cough with dyspnea and possibly gastrointestinal symptoms. At a later state, the disease would resemble severe respiratory illness or pneumonia (Inglesby et al., 2001). An individual can have multiple forms of the disease, for example, bubonic plague with secondary pneumonic infection, or pneumonic plague with secondary septicemia.

Similarly, early diagnosis of inhalation anthrax is difficult and requires a high degree of suspicion (Inglesby et al., 2002). Initial symptoms are nonspecific, including fever, cough, dyspnea, headache, vomiting, chills, weakness, abdominal pain, and chest pain. Furthermore, laboratory studies may also be nonspecific. Sometimes a brief recovery period follows within a few days, further complicating diagnosis (Jernigan et al., 2001).

Reporting

Public health agencies have very effective established and integrated reporting systems for disease outbreaks, although full compliance with reporting requirements is difficult to achieve. Over 100 nations have agreed to contact WHO in Geneva, Switzerland, within 24 hours of suspected cases of cholera, plague, and yellow fever (Woodall, 1998). WHO reporting procedures are also in place for influenza, arthropod-borne viruses, hemorrhagic fevers, and antibiotic resistance. In addition, livestock-transmitted diseases, including anthrax, Rift Valley fever, brucellosis, and shiga-like toxin producing *Escherichia coli*, are also reported.

However, the reporting of animal and plant diseases is much less developed (Woodall, 1998). At least 148 nations report List A outbreaks to the Office International des Epiozootis (OIE). These are quarantinable diseases appearing for the first time or reappearing after control in a member country. Unfortunately, compliance with these voluntary reporting requirements may be suppressed by a government for political or trade reasons. This poses a risk that an incident of intentional introduction of an animal or plant disease may not be reported, causing widespread harm that could otherwise have been successfully contained. Voluntary reporting is not necessarily in anyone's best interest if government hysteria blows an incident out of proportion. Shipments of animal feeds into the U.S. were prohibited from Canada for months during 2003 when a single incident of bovine spongiform encephalopathy (BSE) was detected. This prohibition covered not only feeds containing beef components, but any item produced in a facility were meat by-products were processed, shutting down an important source of feed for U.S. agriculture and aquaculture operations unnecessarily. Similarly, widespread slaughter of hogs and chickens has occurred in China, Hong Kong, and Europe as means of controlling animal disease outbreaks; sometimes these are completely legitimate, but sometimes they are not.

Illness Investigations

"Physicians need to be open minded about the unexpected," according to Marcelle Layton, MD, MPH, Director of the Office of Communicable Disease at the New York City Department of Health (Vastag, 2001). Incidents that could be a public health trigger for a bioterrorism event could be a suspicious disease outbreak, unusual or unexplained illness clusters, and unusual symptoms or disease manifestations, such as a healthy young person dying of the flu or pneumonia. Other triggers would include manifestation of an illness in an odd location or person who would not otherwise be exposed, for example, anthrax, glanders, or in an individual from an urban setting who does not hunt or have contact with farm animals. Similarly, an unusual or exotic disease, or an isolated case with unusual features, should raise suspicions (Woodall, 1998). Unfortunately, public health officials are not used to thinking about intentional contamination, and for those that do, the focus is on the possibility of a large-scale localized incident more than on a diffuse one.

Since 9/11 there have been several tests of preparedness systems across the country. Some have been real-life enactments, such as the multi-million-dollar national disaster training session in spring 2003 involving fire, bombing, and aircraft accidents simultaneously across the U.S. The purpose was to evaluate integration of emergency response both locally and regionally and the impact of multiple regional disasters on a national scale. Simulations or exercises involving various disaster and bioterrorism scenarios have also been run recently, again with the effect of determining how good the response is. All these systems lack a grassroots approach to preparedness that incorporates the needs of the private sector. Although local environmental health professionals recognize the need for these programs, funding for their development has been limited and spotty (Berg, 2004). Hopefully some of these deficiencies will be addressed through the presidential directive on food and agriculture biosecurity, but full implementation is still years away.

A better model for preparedness is that developed jointly by the Oregon Department of Human Services and Oregon Department of Agriculture for food security (Shibley and McKay, 2003). Here the state developed a network of different food safety and law enforcement agencies with businesses to respond to food safety incidents. The advantages of the Oregon system is that it puts all the affected parties at the table together, lets them get to know each other, and then allows participants to develop a system that will work for their specific area. In addition, the Oregon system provides the needed capability to notify industry promptly of a risk or threat and provides companies that may be affected with an understanding of what the possible risk might be, what the level or severity of the risk is, and how they might respond to it. Unfortunately, simple efforts such as these that have a high probability of real success are underfunded.

Organizational Readiness: Generally

Most organizations are ill prepared to deal with tampering, let alone other manifestations of bioterrorism. Issues of product liability, insurance coverage, crisis management, and maintaining business viability are of critical concern. Recommendations include analysis of an organization's risk prior to an incident, utilizing best practices to avoid a tampering or contamination event, formulating and instituting a crisis management and communication plan, conducting a cost–benefit analysis for transferring the risk through insurance coverage, conducting product recalls, litigating a tampering or recall case, and conducting forensic accounting to quantify losses and analyze claims (ACI, 2000). High-profile consumer product tampering instances from the 1980s made companies aware of new risks; however, we have unfortunately entered a brave new world of well-organized, internationally based targeting of organizations and products in and related to the food industry. Recent conferences have addressed techniques for monitoring open-space research, covert sensor technology, and crime prevention training. According to the FBI, domestic crime-targeting biotechnology is the emerging antitechnology crime of the new millennium (FBI, 2000). However, techniques and tools for protecting facilities are lacking, and new measures under the Public Health Security and Bioterrorism Preparedness and Response Act of 2002 may increase the ability of government to identify and then respond to a bioterrorist incident, but they do little to improve preparedness in the private sector.

Chemical Agents

The use of poison is a treacherous, unacceptably cruel sign of cowardice and an atrocity dating to Cicero in Ancient Rome, a tool "worthy of brigands and not of princes" (Stern, 1999). In the general public, there is a visceral dread of poisons coupled with a recently reignited chemophobia that makes the use of chemical agents particularly attractive for terrorists targeting foods. Poisoning is a much greater fear to most of us than being shot, because the option to run away or to physically remove a small tangible object from one's body following an injury is not possible. Poisons are invisible and, in the surrounding environment, become inhaled or ingested and cannot be easily removed once inside the body. The fact that many potential chemical food contamination agents are readily available and do not have to be grown or extracted, like biological toxins, is a major advantage. Chemical agents kill or incapacitate people, destroy livestock, or ravage crops. Many chemical agents are odorless and tasteless and are difficult to detect. The most likely chemical agents commonly available include toxic organic compounds, pesticides, industrial chemicals, heavy metal-containing compounds, and microbial or plant toxins. Hoaxes and scares after the anthrax incidents in 2001 led to submission of numerous white powders for analysis including: aspartame, Gold Bond medical powders, collagen, sugar, baking powder (Ferguson, 2004), and starch. Emergency response to chemical contamination incidents has improved significantly in recent years. Cooperation of local public health officials and law enforcement is key. As an example, on June 5, 2004, 10 dairy cows in Enumclaw, Washington were painted with a toxic, sticky, red chromium-containing material killing three of the aminals. The incident was immediately reported by the farmer to the State Veterinarian and the county sheriff. No potentially affected milk entered commerce and animals were properly quarantined. The perpetrator, who had a history of similar activity, was arrested on short order. There were complaints that the federal government should have become involved in the investigation early on (Weise, 2004), but a rapid integrated response on the local level suggests that this is not always necessary. A partial list of hazardous chemicals proposed as food contamination agents is presented in Table 2.2.

Chemical agents have been used in legitimate warfare and by terrorists or terrorist regimes. For example, phosgene and chlorine were used extensively during WWI as choking agents. Phosgene was responsible for the largest number of deaths from exposure to chemical agents during this war

Table 2.2 Possible Chemical Agents

Toxins			
Abrin	Ricin	Cannabinoids	Fentanyls and other opioids
LSD	Quinuclidinyl Benzilate (Bx) Psychodelic agent 3		
Metals			
Arsenic	Arsine (SA)	Methyldichloroarsine (MD)	Diphenylchloroarsine (DA)
Diphenylcyanoarsine (DC)	Ethyldichloroarsine (ED)	Phenodichloroarsine (PD)	Cadmium
Chromium	Mercury	Red phosphorus (RP)	White phosphorus
Thallium	Titanium tetrachloride	Zinc oxide	
Cyanide			
Cyanogen chloride (CK)	Bromobenzylcyanide (CA)	Hydrogen cyanide (AC)	Potassium cyanide
Sodium cyanide (NaCN)			
Nerve Agents			
Sarin (GB)	Cyclohexyl sarin (GF)	Tabun (GA)	VX
Soman (GD)			
Chemical Agents			
Distilled mustard	Mustard/lewisite	Mustard (T)	Nitrogen mustard (HN-1, HN-2, HN-3)
Sesqui mustard	Sulfur mustard or mustard gas (H)		
Phosgene	Diphosgene	Phosgene oxime (CX)	
Adamite (DM)	Lewisite (L, L-1, L-2, L-3)		
Industrial Chemicals and Pesticides			
Ammonia	Benzene	Chlorine	Hydrogen chloride
Hydrofluoric acid	Ethylene glycol	Paraquat	Nitrogen oxide
Phosphine	Perfluoroisobutylene (PHIB)	Sulfur-trioxide-chlorosulfonic acid (FS)	
Other			
BZ	Agent 15	CR	CS
Chloroacetophenone (CN)	CN in chloroform (CNC)	CN in benzene and carbontetrachloride (CNB)	
Chloropicrin (PS)	CN and PS in chloroform (CNS)		
Phenothiazines			

Source: Data from CDC, 2002; DTS 2004.

(CDC, 2002). A variant, phosgene oxime, has been produced as a chemical warfare agent, but has yet to be used.

During and following the Gulf War of the early 1990s, Saddam Hussein used chemical weapons (possibly tabun, sarin, or VX and mustard agents) and blister agents against civilians and to suppress uprisings of the Kurds in northern Iraq. Saddam also employed chemical weapons during the disasterous 8-year Iran–Iraq War beginning in 1980, targeting both military personnel and civilians (Barton, 1998). Evidence of his continued development of chemical weapons in violation of United Nations Security Council Resolutions (Section C, UNSCR 687 and UNSCR 715) was found in April 2003 by U.S. forces during the second Gulf War and again in June 2004 during an attack on U.S. forces (Hosenball, 2004).

Cults have also targeted innocent people with chemical weapons. On March 20, 1995, the Aum Shinrikyo cult placed packages of sarin in subway cars in Tokyo, killing and injuring numerous people (Woodall, 1998). To date, this is the only successful large-scale open-air dissemination of a chemical agent tied to a terrorist group.

However, limited documented successes have not kept terrorists from trying to use chemical agents in an attack. One toxin, ricin, extracted from castor beans, has an interesting history as an agent in a series of assassination attempts in London in the late 1970s and early 1980s (Atlas, 1998). One of these attempts killed Georgi Markov, a Bulgarian writer. He died after being stabbed with an umbrella rigged to inject a pellet of ricin underneath his skin (CDC, 2003). Ricin has also been found at Al-Qaeda sites in Afghanistan (CDC, 2003) and is making its way around the world. Al-Qaeda-sponsored ricin production labs have been found in Europe. Ricin was found in March 2003 in baggage lockers in the Paris Metro and in the apartments of Al-Qaeda operatives in Manchester, England. In November 2003, President George W. Bush received ricin-contaminated mail, and in February 2004, Senator Frist (R-TN) was similarly targeted.

Although there have not been any confirmed cases of ricin intoxication from these recent incidents, indications are that terrorist activities were being planned for it. On October 15, 2003, a post office in Greenville, SC, received an envelope with a sealed container of ricin plus a note threatening to contaminate water supplies if demands were not met (MMWR, 2003). No illnesses were reported from this incident, and no detectable environmental release of ricin within the facility was found.

Chemical agents tend to be faster acting than biological agents. They can have an immediate effect (in a few seconds to few minutes) or a delayed effect (several hours to several days). In general, acute effects include respiratory and neurological impacts and often vomiting if materials are ingested. Longer-term effects include neurological, organ, and tissue damage, increased rates of cancer, and fetal abnormalities. Chemical agents can be lethal or incapacitating at very low doses, in the parts per thousand to parts per million range. Unlike biological agents, the selectivity of a chemical agent is generally not high and can affect a large range of living things. The possible exceptions to this are specific neurological agents, including some pesticides.

Controllability of a chemical agent is problematic and dependent upon environmental conditions. Because transmission through biological vectors is not likely, the rate of secondary transmission of the adverse affects of a chemical agent is lower than that for a biological agent. The residual effects of a chemical agent vary widely, with highly volatile compounds like sarin or chlorine dissipating within minutes. However, less volatile agents like VX and pesticides can persist for hours or days, and mustard agents can persist for several months if contained underground.

Chemical agents are generally classified as follows: biological toxins, blood agents (e.g., hydrogen cyanide and cyanogen cloride), nerve agents (tabun, sarin, soman, V agents, and new novichok agents) (Stern, 1999), blister agents (e.g., mustard and Lewisite), urticant or nettle agents (phosgene oxime), and choking and incapacitating agents (e.g., chlorine and phosgene). Nerve or blood agents are the most likely chemical warfare agent candidates for intentional contamination of food or water. However, from a practical food safety perspective, contamination of food with agricultural or industrial chemicals is more likely, and many possibilities exist, including organochlorine and organophosphate pesticides and carbamates. Industrial chemicals that have seen use as warfare and terrorist agents and could be used to contaminate food or water. These include cyanide-containing compounds, carbonyl chloride, and arsine. Heavy metals compounds (e.g., arsenic, cadmium, lead, and mercury) are widely available, as these have many agricultural and industrial uses. Heavy metals have a long history of use as food poisons and have been implicated in recent unpublished contamination incidents in Israel. Heavy metal-containing compounds are stable, are effective at relatively low levels, and can be applied in such a manner as to not greatly affect the sensory properties of the food. A compilation of some possible agents and symptoms of exposure is presented in Table 2.3. It should be noted once again that the contaminant need not be present in any significant quantity. The mere detection of a contaminant in a food product can do serious economic damage.

Biological Toxins

Ricin is recovered using a rather simple procedure from the castor bean (*Ricinius communis*). Ricin is a water-soluble, heat-sensitive globular protein

Table 2.3 Possible Chemical Agents for Terrorist Attacks

Agent	Features	Symptoms	Exposure	Time of onset	Stability	
Nerve						
Sarin or GB	Water soluble, clear, colorless, tasteless liquid. Highly volatile.	Runny nose, watery eyes, small pinpoint pupils, eye pain, blurred vision, drooling, excessive sweating, cough, chest tightness, rapid breathing, diarrhea, increased urination, confusion, drowsiness, weakness, headache, nausea, vomiting, abdominal pain, slow or fast heart rate, low or high blood pressure, loss of consciousness, convulsions, paralysis, respiratory failure	Inhale, ingest, skin contact	A few seconds (vapor) Minutes –18 hr (ingestion) Antidote must be taken quickly.	Short-lived — highly volatile.	Fatal in 1–10 min.
VX	Water miscible. Oily amber colored liquid, odorless and tasteless. Volatility similar to motor oil.	Constriction of pupils, visual effects, headache, pressure sensation, runny nose, nasal congestion, salivation, chest tightness, nausea, vomiting, giddiness, anxiety, difficulty thinking, difficulty sleeping, nightmares, muscle twitches, tremors, weakness, abdominal cramps, diarrhea, involuntary urination and defecation. Convulsion and respiratory failure.	Ingest, eye, skin contact, inhale	Minutes or hours depending upon dose.	High — persists in environment for days or months. Surface contamination is very persistent.	Fatal in 4–18 hr.
Tabun or GA	Water soluble, clear, colorless, tasteless liquid with faint fruity odor. Vaporizes with heating. More volatile than VX, less volatile than sarin.	Similar to sarin	Inhale, skin contact, ingest	Same as sarin	Short-lived, low persistence in environment, will remain on surfaces longer than sarin.	Fatal in 1–10 min.
Soman or (GD)	Water soluble, clear, colorless, tasteless liquid with a slight camphor or rotting fruit aroma. It can be vaporized by heating.	Similar to sarin	Inhale, skin contact, ingest[1]	Same as sarin	Similar to tabun.	Fatal in 1–10 min.

Choking agents						
Phosgene or CG	Colorless, corrosive, nonflammable gas with limited miscibility in water. Can form a white to pale yellow cloud, odor of freshly cut hay or green corn, at high concentrations odor can be unpleasant. Fog-like in initial concentration, colorless as it spreads. Liquid under pressure or refrigeration. Can cause flammable substances to burn. Decomposes immediately in contact with water.	Coughing, burning sensation in throat and eyes, watery eyes, blurred vision, difficulty breathing, nausea, vomiting, pulmonary edema (may be immediate or take up to 48 hr hr) "dryland drowning," low blood pressure, heart failure. Chronic bronchitis and emphysema in survivors Skin: lesions similar to frostbite	Skin or eye contact, ingest, inhale	Effects can be immediate or delayed (48 h). No antidote.	Degrades slowly.	Fatal in minutes to days.
Vesicants or blister agents						
Mustard gas, mustard agent or sulfur mustard	Low volatility, colorless and odorless liquid. When mixed with other compounds, becomes brown with a garlic, onion or mustard-like smell. Solid and gaseous forms.	Inhale: irritation and blistering of mucosal membranes, runny nose, coughing, sinus pain, shortness of breath, bronchitis, respiratory damage. May increase lung cancer risk. Skin: redness, itching, irritation and yellow blistering. Liquid is more likely to cause second and third degree burns than vapor. Eye: irritation, pain, swelling, mild to moderate sensitivity to light, can cause temporary blindness (up to 10 days). Can cause permanent blindness if exposure is extensive. GI: abdominal pains, diarrhea, fever, nausea, vomiting Reproductive: Reduced sperm counts and DNA damage. Can suppress immune system.	Inhale, skin or eye contact, ingestions less likely	Effects occur within 2–24 hours. Repeated exposure has cumulative effects. No antidote.	Degrades in air, soil, or water in minutes to days. Persists longer in cold climates.	Fatal at high levels from respiratory failure or burns.

Table 2.3 Possible Chemical Agents for Terrorist Attacks (Continued)

Agent	Features	Symptoms	Exposure	Time of onset	Stability	
Vesicants or blister agents						
Nitrogen mustards (HN-1, HN-2, HN-3)	Nitrogen mustards are colorless to yellow oily liquids with a low volatility. Vapors are heavier than air. HN-1 has a faint, fishy or musty odor. Originally it was designed to remove warts. HN-2 has a soapy odor at low concentrations and a fruity odor at high concentrations. Used temporarily as a cancer treatment. HN-3 smells like bitter almond.	Inhale: irritation of mucosa, coughing, bronchitis and respiratory damage. Skin: irritation, blister and burning, damage greatest to moist tissue within 6–12 hr. Eye: causes irritation and burns, can cause blindess. GI: abdominal pain, diarrhea, nausea and vomiting. CNS: tremors, loss of coordination, seizures. Immune: damage to bone marrow in 3–5 days causing anemia, bleeding, increased risk of infection. Can be fatal. Other: Can cause leukemia in humans and cancer in animals.	Inhalation, skin, or eye contact	Effects are generally within a few hours.	Break down quickly in water or moist soil. Can persist in air for several days.	Fatal at high levels from respiratory failure, burns or suppression of the immune system
Lewisite or L mustard, Lewisite mixture (HL or HD)	Oily, colorless liquid in pure form, amber to black in an impure form with odor of geraniums. Vapor is heavier than air. Contains arsenic.	Skin: pain and irritation occur immediately, redness in 15-30 minutes, blisters form in several hours. Small centralized blister expansion. Some discoloration of skin. Eyes: irritation, pain, swelling, tearing on contact. Extensive exposure can cause permanent damage to the cornea and cause blindness. GI: diarrhea, nausa, vomiting CV: low blood pressure, shock, damage to blood vessels and bone marrow. Reproductive: may have reproductive effects.	Inhale, skin or eye contact, ingestion	Skin: within seconds or minutes. Other symptoms within minutes or hours. Antidote is arsenic chelating agent.	Persists in the environment for days. Breaks down in water and moist soil. Arsenic moiety has long environmental persistence.	Fatal at high levels.

Urticant or nettle agents						
Phosgene oxime or CX	Solid: colorless. Liquid: water soluble, yellow brown. Corrosive, disagreeable, irritating odor.	Skin: Intense itching, rash similar to hives, corrosive, instant unbearable pain upon skin. Eye: severe pain, irritation, tearing, temporary blindness. Respiratory: immediate irritation with runny nose, hoarseness and sinus pain, pulmonary edema, shortness of breath, coughing. GI: no information on symptoms.	Skin or eye contact, ingest, inhale	Immediate to several hours. No antidote.	Degrades in soil within 2 hr at normal temperatures, breaks down in water in a few days.	Fatal at high levels within a short period.
Biological						
Abrin (similar to ricin)	White or yellowish powder. Toxic protein (63–67KD) resistant to digestive enzymes.	Ingest: muscle weakness, tremors, spasms; low blood pressure, fast heart rate, irregular heart rhythms, cardiovascular shock from severe dehydration, central nervous system disorder, drowsiness, disorientation, hallucination, seizure, coma. GI: burning pain in mouth, abdominal pain, nausea, severe vomiting, diarrhea, bleeding and swelling of GI tract lining, liver damage, pancreas damage, blood in urine, low urine output, kidney damage. Eyes: dilated pupils, retinal hemorrhage, tearing, swelling, pain, redness, corneal injury. Skin: may be absorbed, redness, blister, pain, cyanosis. May cause severe allergenic reactions. Inhale: irritation, sensitization.	Ingest Skin or eye contact, inhale	Within one to several hours. No antidote.	Heat stable under pasteurization conditions (140°F (60°C) for 30 min), toxin inactivated at 176°F(80°C) for 30 min.	Symptoms can be delayed 1–3 days. Lethal dose is 0.005–0.007 mg/Kg.

Table 2.3 Possible Chemical Agents for Terrorist Attacks (Continued)

Agent	Features	Symptoms	Exposure	Time of onset	Stability	
Biological						
Ricin (Abrin is a related toxin)	Whitish powder.	Ingest: vomiting, diarrhea that may become bloody, severe dehydration, low blood pressure, hallucinations, seizures, bloody urine, liver, spleen, and kidney failure. Inject: weakness, fever, vomiting, shock, multi-organ failure.	Ingest, inject, inhale	Within one to several hours. No antidote.	Similar to abrin	Presumed high Death can occur in 36–72 hr.
Blood Agents						
Cyanide-Hydrogen cyanide (AN) Zyklon B, Cyanogen chloride (CK)	Colorless gas (HCN, CNCI) or whitish crystal (NaCN, KCN). Gas is less dense than air. May have a bitter almond odor. CK is a colorless gas with sharp pepper odor similar to most tear gas. Odor can go unnoticed because it is so irritating. Slightly soluble in water. Liquid at <55°F.	Ingest: within minutes-rapid breathing, restlessness, dizziness, weakness, nausea, headache, heart pain, rapid heart rate, changes in blood chemistry, enlargement of thyroid gland and loss of thyroid function, vomiting. Other symptoms: weakness in toes and fingers, difficulty walking, loss of visual acuity, deafness, convulsions, low blood pressure slow heart rate, loss of conciousness, lung injury. Respiratory failure can be fatal. Permanent heart and brain damage possible. Inhale: similar symptoms, more toxic than ingestion. Skin: irritation and sores.	Ingest, inhale, hand to mouth, skin	Symptoms occur within minutes.	HCN persists in the environment and has a half life of 1–3 years. HCN in water dissipates as a gas.	Can be fatal. Death from respiratory failure.

Toxic Metals						
Arsenic	Present in soil and as oxygen, chlorine and sulfur complexes.	Inhale: sore throat and irritated lungs. Ingest: nausea and vomiting. Skin: redness and swelling. General: decreased production of red and white blood cells, blood vessel damage, abnormal heart rhythm, "pins and needles" sensation in hands and feet. Chronic ingestion or inhalation of low levels can cause darkening of the skin and appearance of small corns or warts on the palms, soles of feet and on the torso.	Inhale, ingest, skin and eye contact	Onset of toxicity symptoms can take weeks. Organic forms are less toxic than inorganic forms. Antidotes are chelating drugs such as BAL (British anti-Lewisite dimercaprol).	Stable	Ingesting high levels is fatal.
Arsine or arsenic hydride	Colorless, flammable, water soluble gas. Heavier than air. Garlic like or fishy odor (at around 0.5 ppm).	Inhale: nonirritating; there may be no immediate symptoms. Initial symptoms include malaise, dizziness, headache, thirst, shivering, nausea, abdominal pain (liver) and dyspnea. Hemglobinuria within hours. Jaundice within 1–2 days. Hypotension with severe exposure. Death is from hemolysis causing renal failure. CNS: restlessness, memory loss, disorientation and agitation within several days. Peripheral nerve damage can occur 1–2 weeks after exposure. Polyneuropathy within 1–6 months after exposure. Skin and eye: contact with liquid (pressurized gas) resembles frost bite. Changes to skin from inhalation include a bronze tint induced by hemolkysis. There may be red staining of the conjuctiva in early stages. Ingest: Metal arsenide are solids that can react with gastric contents releasing arsine in the stomach.	Inhale, skin, ingest	Initial symptoms within 30–60 minutes, but can be delayed for several hours at lower exposures (<3 ppm). No antidote.	Stable	Highly toxic and fatal if inhaled in high quantities. Irreversible health effects at > 0.5 ppm for 1 hour.

Table 2.3 Possible Chemical Agents for Terrorist Attacks (Continued)

Agent	Features	Symptoms	Exposure	Time of onset	Stability	
Toxic Metals						
Mercury	Metallic mecury is a shiny silver white, odorless liquid. Gas is colorless and odorless. Mercury salts (chlorine, sulfur or oxygen) are whitish powders or crystals. Organic complexes are formed by microbes in soil and water.	Inhalation: lung damage. Ingestion: nausea, CNS damage, vomiting, diarrhea, increase in blood pressure or heart rate Skin and eye: skin rash and eye irritation.	Ingest, inhale, skin and eye contact	Onset of toxicity symptoms can take weeks or months.	Stable with exposure from various environmental and food sources. Bioaccumulation of organic complexes poses serious health risk.	Fatal at high ppm levels from chronic or acute exposure.
Thallium	Grayish powder.	Ingestion: vomiting, diarrhea, temporary hair loss, loss of nerve function, damage to heart, liver and kidney.	Ingest	Onset of symptoms can be prolonged.	Stable	Fatal at high levels.
Pesticides						
Chlorine	Yellow-greenish gas. Liquid under pressure.	Skin and eye: corrosive hypochlorous and hydrochloric acids burn and scar. Skin can develop a bluish tinge. Skin contact causes redness, burning pain and blistering. Respiratory: burns and scars lung tissue. In sensitive individuals, reactive airways dysfunction syndrome (RADS), a type of asthma, is common. GI: irritates, burns and scars digestive tract lining. Long-term health complications from chlorine exposure include eye and skin burns and bronchitis from individuals who may have developed pneumonia.	Skin contact, inhalation, ingestion	Immediate exposure at levels from 1–10 ppm causes coughing, sore throat, eye and skin irritation. At higher levels rapid breathing, bronchial constriction and difficulty breathing can occur. Other symptoms of intoxication are pulmonary edema (2–4 hr); a burning sensation in the nose, throat, and eyes; watery eyes, blurred vision; nausea, vomiting.	Highly volatile gas. Compounds containing chlorine (such as hypochlorite are very stable).	Can be fatal at high ppm levels. Respiratory failure.

Paraquat and similar compounds	Colorless, odorless liquid.	Ingestion: pain and swelling of the mouth occurs, followed by nausea, vomiting, abdominal pain and bloody diarrhea. Dehydration, an electrolyte imbalance and low blood pressure may also result. Within a few hours to a few days, a victim may experience pulmonary edema, confusion, coma, seizures, muscle weakness, heart injury or organ scarring. Skin: absorption through the skin causes symptoms similar to ingestion.	Ingestion, open skin lesions, skin contact. Inhalation not likely.	Immediate pain and swelling of the mouth and throat. First signs of illness are GI in nature. Other symptoms appear within a few hours or days and include: pulmonary edema, confusion, coma, seizures, muscle weakness and respiratory problems. Ingestion of small amounts can cause liver, kidney or heart failure or lung scarring over several days.	Stable. To avoid accidental consumption, liquid is marketed with a dye (paraquat is blue) to prevent confusion with beverages. Pungent odor and agents to induce vomiting are added.	Heart, kidney, or respiratory failure from acute or cumulative exposure can be fatal.
Phosphine	Colorless, flammable, explosive gas at ambient temperature. Has odor of garlic or decaying fish. Metal phosphide pesticides release free phosphine upon contact with stomach acid.	Ingest: abdominal pain, nausea, vomiting, At high levels weakness, bronchitis, pulmonary edema, enlarged spleen, shortness of breath, convulsions and death can occur. Heart damage can occur in children. Inhale: bronchitis, pulmonary edema and other symptoms listed above. Long term exposure to low levels causes anemia, bronchitis, gastrointestinal illness, depression of visual, speech and motor skills	Ingestion, inhalation, skin contact	Onset of symptoms usually within hours.	Dissipates quickly having a half life of one day in air.	Fatal at low levels.

Source: ATSDR. 1995–2002; CDC 2003; DTS, 2004.

toxin with a molecular weight of 64,000 Daltons (DA). It consists of two peptide chains (each ca. 32,000 DA) linked by a disulfide bond. A similar toxin, abrin (CAS 1393-62-0), is recovered from the *Abrus precatorius* seed (common names: jequirity bean, Indian licorice seed, rosary pea, Mienie-Mienie) (NIOSH, 2003). Even though ricin is recovered from a pharmaceutical plant, unintentional intoxication is unlikely. There are no antidotes for either ricin or abrin. These both work by inhibiting protein synthesis and have potential medical applications in the treatment of bone marrow transplant and cancer patients. Routes of exposure include oral, dermal, and parenteral, through the skin, skin-to-eye contact, and injection. A lethal dose can be as little as 500 μg (by injection). Toxicity from inhalation or ingestion is high. Ricin can remain in an aerosol dispersion for several hours.

These features support the fear that ricin could be a vehicle for a terrorist attack on food. Initially, ricin poisoning would resemble gastroenteritis (1 to 4 hours onset) or a respiratory illness, making it difficult to discern from other illnesses (MMWR, 2003). Misdiagnosis is likely, and if intake is sufficient, death would occur within hours.

Blood Agents

A number of cyanide (AN)-containing compounds (CAS 74-90-8, 143-33-9, 151-50-8, 592-01-8, 544-92-3, 506-61-6, 460-19-5, and 506-77-4) (ATSDR, 1997) have been developed for military applications (CDC, 2003). Hydrogen cyanide (Zyklon B) was employed by the Germans as an agent of mass extermination of civilians. It was also a likely warfare agent in the Iran–Iraq War and in the Kurdish (Halabja) uprisings in Iraq in the 1990s. Cyanides are highly toxic and act by depriving cells of oxygen, causing death. Other possible cyanide-containing agents include cyanogen chloride (CK), sodium cyanide (NaCN), bromobenzylcyanide (CA), and potassium cyanide.

Cyanides occur naturally in cassava root, lima beans, and seeds (almond and apricot pits — the source of laetrile, a cancer treatment from the 1970s and 1980s). Cyanides are also produced as a metabolic by-product by a number of terrestrial and aquatic microbes. These compounds are also products of combustion from cigarettes and burning plastics. The simplest chemical form is hydrogen cyanide (HCN), a colorless gas with a faint bitter almond aroma. Potassium cyanide (KCN) is a white solid with a similar aroma.

Cyanide poses a risk as a food contamination agent because it is widely available, with numerous important uses. It is widely used in the photography, metal plating, textile, paper, and plastic industries. Cyanide-containing compounds are used as fumigants in buildings and ships. It is a potentially persistent environmental contaminant with a half-life in air of 1 to 3 years.

Most cyanides entering water will form hydrogen cyanide and evaporate; however, cyanides can leach into groundwater, where they would be retained for long periods. High levels of cyanide can be found around waters near gold mining operations.

Cyanide causes neurological difficulties and cardiovascular damage, coma, and death. Exposure to low levels causes respiratory difficulties, heart pain, vomiting, changes in blood chemistry, headaches, and enlargement of the thyroid gland. Survivors can suffer long-term heart and brain damage. Initial symptoms of intoxication would occur within minutes and could include rapid breathing, restlessness, dizziness, weakness, headache, nausea and vomiting, and a rapid heart rate. Higher levels of exposure would lead to convulsions, low blood pressure, slow heart rate, loss of consciousness, lung injury, and respiratory failure.

Nerve Agents

Sarin (or GB) (CAS 107-44-8) is a nerve agent and among the most toxic and rapidly acting of the chemical warfare agents. This agent was developed specifically in the late 1930s by the Germans for mass extermination of undesirable civilians in gas chambers. It is an organophosphate and causes symptoms similar to those of insecticide poisoning. Like other nerve agents of similar chemical composition, exposure is cumulative. Sarin is a clear, odorless, colorless, and tasteless water-soluble liquid. Sarin is a potential intentional food or water contaminant because it is highly toxic and its presence would be difficult to detect.

Sarin vapor is heavier than air and will sink, making it an ideal agent to use in enclosed spaces. It evaporates quickly and poses little long-term environmental threat. Sarin can be absorbed through the skin, and a single small drop on the skin can cause uncontrollable muscle twitching and sweating. Larger doses can cause loss of consciousness, convulsions, paralysis, or respiratory failure and death. People suffering mild or moderate exposure usually recover, and unlike with exposure to organophosphate pesticides, neurological symptoms tend to last no more than 1 to 2 weeks after exposure. Although there are antidotes to sarin, the best means of avoiding inhalation exposure is to leave the affected area and quickly escape to fresh air seeking the highest ground possible. If skin or clothing contact is suspected, clothing should be removed and the body washed with soap and water. To avoid contact with eyes or mouth, clothing should be cut off and not removed over the head (CDC, 2003). For an outline of the symptoms, toxicity, and stability of these agents, refer to Table 2.3.

VX (CAS 50782-69-9) is an odorless, tasteless amber-colored nerve agent with a low volatility, most commonly distributed as an oily liquid. It is one of the most potent of the nerve agents (CDC, 2003). It can be inhaled or ingested as a mist, absorbed through the eyes, or transferred into the body by skin contact. Unlike sarin, it is more toxic by skin entry and somewhat more toxic by inhalation. Any skin contact with VX could be lethal. People could be exposed to VX by consuming contaminated food. Toxicity via ingestion of contaminated water is less likely because VX is not water soluble.

Tabun (CAS 77-81-6) is a somewhat volatile (upon heating), water-soluble, clear, colorless, tasteless liquid with a faint fruity odor. It was used as a warfare agent in Iraq and could be used to poison food or water (CDC, 2003) because its presence would be relatively difficult to detect and it is soluble in water. Exposure through skin contact is also dangerous. Like sarin, tabun- or VX-contaminated clothing can release the agent for 30 minutes or more.

Symptoms from inhalation exposure appear within a few seconds or minutes. Symptoms from exposure by ingestion or skin contact take a few minutes to 18 hours for both VX and tabun. Long-term neurological complications for either of these agents are similar to those of sarin. Likewise, precautionary measures are very similar. Also, like sarin, any volatilized tabun or VX would tend to sink to low-lying areas. Because these are not very volatile, environmental persistence is high and can last for days or months. Any contaminated surfaces should be treated as long-term hazards (CDC, 2003).

Soman (GD) (CAS 96-64-0) is a water-soluble, clear, colorless, tasteless liquid with a slight camphor or rotting-fruit aroma. It can be vaporized by heating (CDC, 2003). It has similar health effects from exposure as the other nerve agents, but is less likely to be used in food unless its sensory characteristics could be masked.

Blister or Vesicant Agents

Mustard gas or mustard agent (CAS 505-60-2) (ATSDR, 2003; CDC, 2003) refers to a number of different chemicals widely used during World War I and more recently in the Iran–Iraq War. Mustard agent is a colorless, odorless, oily-textured liquid at room temperature, but when mixed with other chemicals, it turns brown and has a garlic-like sulfur smell. Mustard agent can also be a solid. Its vapors can be carried long distances through the wind, settling in low-lying areas since it is denser than air. Fortunately, mustard gas evaporates from soil and water and breaks down in the environment in a matter of minutes. However, less volatile mustard agent components can persist in the environment for days under normal weather conditions, or for weeks or months in very cold environments.

Mustard gas or agents cause second- or third-degree skin burns, blisters, and long-term scarring. They cause greater harm in hot, humid environments than in drier or more temperate ones. Mustard agent is a powerful irritant to the eyes and lungs, causing burning and irritation, coughing, bronchitis, and long-term respiratory problems. Permanent blindness can result. Long-term exposure may cause reduced sperm counts because the material can damage DNA. Individuals exposed to these agents may not be immediately aware of it because there can be no immediate symptoms.

Nitrogen mustards (CAS 538-07-8 (HN-1), 51-75-2 (HN-2), and 555-77-1 (HN-3)) (ATSDR, 2002; CDC, 2003) were first produced in the 1920s and 1930s as potential warfare agents. These materials are colorless, yellow or pale amber, oily liquids with limited volatility. They can also be produced in a solid form. HN-1 has a faint fishy or musty odor, HN-2 a soapy aroma at low concentrations and a fruity aroma at higher concentrations, and HN-3 a butter almond aroma. To date, these have not been used as chemical warfare agents. Nitrogen mustards will persist in the environment for a number of days in the air and break down more quickly in water or moist soil. Like other mustard agents, these are strong irritants and can cause serious damage to the skin, eyes, and respiratory tract. Ingestion would cause burns in the esophagus and digestive tract. Symptoms of illness are similar to those of other mustard agents (Table 2.2).

Lewisite (L) (CAS 541-25-3) or lewisite–mustard mixtures (HL or HD) (ATSDR, 2002) were developed as a chemical weapon in 1918 and would have been used during World War I if greater quantities had been available (CDC, 2003). Lewisite was used by the Japanese in the 1930s against Chinese forces during the occupation preceding World War II. Lewisite contains arsenic and in a pure form is an oily, colorless liquid with a geranium-like odor. Less pure forms are amber to black. Lewisite persists longer in the environment than other blister agents because it has a low volatility. It remains liquid from below freezing to very hot ambient temperatures. Lewisite causes less skin blistering than other blister agents, but it damages bone marrow and blood vessels, causes a drop in blood pressure, and leads to greater digestive system damage, resulting in nausea and vomiting and bloody stools. These added symptoms mean that it is probably more lethal than the other blister agents.

Mustard-based agents are not likely candidates for food contamination because of the distasteful aroma (and presumed bad flavor) these materials would impart to foods. It would also be difficult to mask the color of mustard agents in some foods.

Urticant or Nettle Agents

Phosgene oxime (CAS 1794-86-1) (ATSDR, 2002) has been manufactured as a chemical agent, but has not to our knowledge been used in war. It is a colorless, crystalline solid, or a yellowish brown liquid, somewhat volatile at ambient temperatures. Phosgene oxime is water soluble. It has a disagreeable penetrating odor, making its use as a food contaminant unlikely. Like nerve agents, it is heavier than air and will settle in low-lying areas. It is stable enough to cause severe localized health damage by contact with skin or eye tissue. It can penetrate clothing and rubber faster than other chemical agents. Phosgene oxime can persist in water or air and can be transferred through the soil into groundwater. It can be slowly degraded by water or soil bacteria.

Toxic effects are through skin contact, ingestion, or inhalation. Skin contact causes swelling and itching hives and immediate and painful skin injury. Eye contact causes severe pain and conjunctivitis. Inhalation causes severe bronchitis and pulmonary edema. Ingestion causes swelling and bleeding of the GI tract lining. Exposure to high levels can be fatal. There is no information on the long-term health effects of this compound. It is an unlikely food-contaminating agent, but its use should not be ruled out.

Choking Agents

Phosgene (CG) (CAS 75-44-5) was widely used as a chemical warfare agent in World War I and in later conflicts. It degrades slowly in the environment. The toxic effects of phosgene are immediate and there is no antidote. It can cause flammable materials to burn, creating additional risks. Phosgene is not likely to be used as a food contaminant because of its sensory characteristics, but it would be effective if its features could be masked.

White phosphorus (CAS 7723-14-0) is a possible agent and causes lung and throat irritation, and prolonged inhalation can lead to poor wound healing and loss of integrity of bones, particularly the jaw (ATSDR, 1997). Ingestion can cause liver, heart, or kidney damage. The affect of long-term consumption of small amounts of material is not known, and because of its relatively low toxicity (compared to other possible agents), its use is not likely.

Toxic Metals

Arsenic (CAS 7440-38-2) and arsine (AsH_2) (CAS 7440-38-2) are highly toxic teratogenic compounds and can bioaccumulate in body tissues, causing long-term health risks, including increased cancer risk (ATSDR, 2001). Inorganic forms of arsenic tend to pose a greater health risk than organic forms. The major effect is to destroy red blood cells. Inhaling arsenic can cause throat and lung irritation. Ingestion causes nausea and vomiting, damage to blood vessels and heart tissue, an abnormal heart rhythm, and reduced production of red and white blood cells. Long-term neurological damage occurs and a "pins and needles" sensation in the feet and hands is common. Long-term inhalation or ingestion can cause a darkening of the skin and the formation of small warts, or corns, on the palms of the hands, soles of the feet, and torso. Skin contact causes redness or swelling. Environmental exposure to arsenic is common. Some areas have naturally high levels of arsenic in rock or in groundwater. Arsenic is used as a pesticide, and transfer of it onto food or into water supplies is a legitimate concern. Because it is widely available and low doses are effective, arsenic has been used historically to poison food.

Arsine (also arsenic or arsenious hydride, hydrogen arsenide, arsenic trihydride, and arseniuretted hydrogen (CAS 7784-42-1; UN2188)) (ATSDR, 2003) is a colorless, flammable, highly toxic gas with a fishy or garlic-like odor at high levels (>0.5 ppm). It is water soluble and can outgas in the respiratory or digestive system. It is heavier than air and if released will tend to settle in enclosed or low-lying areas. A number of arsine derivatives are possible: diphenylcyanoarsine (DC), diphenylchloroarsine (DA), ethyldichloroarsine (ED), methyldichloroarsine (MD), and phenodichloroarsine (PD). Unlike arsenic, there are no specific antidotes for arsine poisoning. Arsine is commercially available and is used as a dopant in the semiconductor industry and metal plating applications.

There may be no immediate symptoms from exposure to arsine compounds. Inhalation is the most common form of exposure, with an onset of symptoms such as malaise, dizziness, shivering, headache, nausea, thirst, abdominal pain, dyspnea, difficulty breathing, and impacted hemoglobin function occurring within 1 to several hours at exposure levels around 3 ppm or higher. Nausea, vomiting, and cramping abdominal pain tend to be the first symptoms, appearing as early as a few minutes to 24 hours following exposure. Jaundice occurs in 1 to 2 days, along with a characteristic bronze tint to the skin, an abnormal urine color (brown, red, or green), and possibly a red staining or green discoloration to the conjuctiva. Skin contact with liquified arsine causes symptoms similar to those of frostbite. Death can result from kidney failure. Long-term neurological affects include peripheral nerve damage, heart, liver, spleen and bone marrow damage, skeletal muscle injury or necrosis, and hypotension. Its use is less likely than that of arsenic as a food poison because of aroma problems and difficulty with dispersement of a gas into most food products.

Mercury (CAS 7439-97-6) (ATSDR, 1999) is a neurotoxic element and is present naturally as a metal and as a component of inorganic and organic compounds. The metal is a shiny, silvery, odorless liquid at room temperature. It can volatilize to a colorless, odorless gas. Mercury salts of chlorine, sulfur, and oxygen are common. Industrial and medical uses of mercury are common, and it is present in thermometers, batteries, dental fillings, and medicinal creams. Bacteria convert mercury to methylmercury, a compound that can pass up the food chain. Methylmercury and mercury gas are more harmful than other forms of the element, as these pass more easily to the brain, possibly inhibiting protein synthesis. Methylmercury crosses the blood–brain and placental barriers as a complex with free L-cysteine, mimicking the amino acid methionine. Many individuals suffering from mercury poisoning go through a symptomless latent period that can last for weeks or months. When symptoms of intoxication do emerge, these include lung damage, nausea, vomiting, diarrhea, hypertension, increased heart rate, skin rashes, and eye irritation. Long-term exposure can cause neurological damage through loss of nerve cells, especially in the cerebellum and cerebrum, as well as kidney damage, birth defects, and possible increase in cancer risk. Long-term neurological effects of mercury poisoning involve irritability, shyness, tremors, numbness, loss of the ability to speak, reduced coordination, changes in vision and hearing, and loss of memory. Mercury has been used in a number of recent intentional contamination incidents of fruit in the Middle East. Illnesses have resulted, but the primary motivation behind these incidents has been to damage trade from Israel to Europe, including the famous 1978 incident in which the Arab Revolutionary Council contaminated Israeli oranges injuring at least 12 people and reducing orange exports to Europe by 40% (Monke, 2004).

Thallium (CAS 7440-28-0) is a by-product of metal smelting and is used in the manufacture of electronic components. It is widely available and can be ingested, inhaled, or contacted through the skin. Inhalation may cause central nervous system damage. Thallium enters plants through water and soil and is routinely consumed in small levels. However, consumption of large amounts causes vomiting, diarrhea, temporary hair loss, loss of nerve function, and damage to the heart, lungs, liver, and kidneys. Fetal development may be affected (ATSDR, 1995). A level of 15 mg/m^3 is considered an immediate health danger. Thallium has been used as a food contaminant by terrorists, and because of a lack of familiarity with it in the medical community, misdiagnosis of intoxication is likely.

Pesticides

The intentional adding of pesticides to foods at unsafe levels probably has a higher likelihood of use by terrorists to contaminate food compared to other materials that are more difficult to obtain. The use of pesticides against agricultural targets includes the 1997 attack by Israeli settlers against Palestines in which pesticides were sprayed on grapes in two Palestinian villages, destroying up to 17,000 metric tons of grapes (Monke, 2004). Pesticides can be highly toxic at low levels, and many would contribute little if any sensory changes to the food. Only a few pesticides are mentioned here, to give the reader a short overview of the type of products available and their properties.

Phosphine (CAS 7803-51-2) (ATSDR, 2002) is a fumigant for stored grain, as a flame retardant, and in the semiconductor and plastics industries. It is a colorless, flammable, and explosive gas that has the odor of garlic or rotten fish. The compound breaks down rapidly in the environment. Exposure can be through inhalation, ingestion, or skin contact. Early symptoms include pain in the diaphragm, nausea, vomiting, excitement, and a phosphorus smell on the breath. High levels of exposure cause weakness, bronchitis, pulmonary edema, shortness of breath, convulsions, and death. Long-term effects of exposure include pulmonary edema, convulsions, liver damage, anemia, bronchitis, gastrointestinal problems, and visual, speech, and motor problems. Skin contact with liquified phosphine causes frostbite. Ingestion of metal phosphides will release phosphine in the stomach, causing nausea, vomiting, abdominal pain, and diarrhea.

Paraquat is one of the most widely available herbicides. It is marketed in the U.S. containing a blue dye, a pungent odorant, and an agent to induce vomiting. However, this is not the case in other markets. Upon ingestion, pain and swelling of the mouth occur, followed by nausea, vomiting, abdominal pain, and bloody diarrhea. Dehydration, an electrolyte imbalance, and low blood pressure may also result. Within a few hours to a few days, a victim may experience pulmonary edema, confusion, coma, seizures, muscle weakness, heart injury, or organ scarring. Exposure can be by ingestion or absorption through the skin. Long-term exposure causes lung scarring, esophageal scarring, and liver and kidney damage. Death results from heart, kidney, or respiratory failure.

Sanitizers

Chlorine (CAS 7782-50-5) is a common sanitizing agent used to kill microbes in food processing and clinical settings and is the active component in household bleach (ATSDR, 2002). Chlorine is a yellowish green, pungent, irritating gas at ambient temperature. It is heavier than air. Chlorine was used as a chemical warfare agent in World War I and, because it is widely available, could be used in a terrorist attack (CDC, 2003). Chlorine causes damage primarily through skin or eye contact and inhalation. When the gas comes into contact with moist tissue, it forms corrosive hypochlorous and hydrochloric acids, which burn and scar lung tissue and digestive tract lining. Immediate exposure at levels from 1 to 10 ppm causes coughing, sore throat, and eye and skin irritation. At higher levels, rapid breathing, bronchial constriction, and difficulty breathing can occur. In sensitive individuals, reactive airways dysfunction syndrome (RADS), a type of asthma, is common. Other symptoms of intoxication are pulmonary edema (2 to 4 hours), a burning sensation in the nose, throat, and eyes, watery eyes, blurred vision, nausea, and vomiting. Skin can develop a bluish tinge. Skin contact causes redness, burning pain, and blistering. Long-term health complications from chlorine exposure include eye and skin burns and bronchitis from individuals who may have developed pneumonia. Death results from respiratory failure. Chlorine is so common in food processing environments that risks associated with exposure deserve mentioning. Foods contaminated with chlorine and ammonia (a common food refrigerant) have been subject to recall, and unintentional food contamination incidents from oven cleaners and unapproved detergents have resulted in personal injury lawsuits. Although these materials, if present in food at dangerous levels, would be easy to detect, intentional addition to cause economic harm is always a possibility.

Managing Exposure to an Unknown Chemical

When managing the health risk from exposure to an unknown chemical, the first objective is to determine the risk of secondary contamination (ATSDR, 2003). Victims exposed only to gas or vapors or who do not have a large deposition of the chemical on their body or clothing pose little risk of secondary contamination to others, including rescue or hospital personnel. High levels of contamination on clothing pose a risk from direct contact or from offgassing vapor. Vomit from the victim could pose a similar contamination risk.

Contaminated clothing and personal belongings should be removed from the victim and double bagged. Skin and hair should be rinsed in water for 3 to 5 minutes. If the chemical is oily, or strongly adherent, then a mild detergent containing a surfactant should be used when rinsing the hair. If the eyes are affected, flush with water or saline. If the chemical has been ingested, vomiting should not be induced, but instead, 4 to 8 ounces of water should be administered to dilute stomach contents, if the victim is conscious and able to swallow. If a slurry of activated charcoal is available, this should be used instead of water unless ingestion of a corrosive material is suspected.

The intentional contamination of food with dangerous chemicals is probably more likely than the use of biological agents if recent history is to be our guide. Motivation would be important to a terrorist in designing a food contamination incident, and although the publicity from a small-scale incident involving sarin or ricin might be greater, the overall damage, from both a public health and economic perspective, caused by using a less exotic material is just as great. Expediency, availability, ease of handling, and ability to escape detection would be other important factors to consider when developing a preparedness program and planning a response.

References

ACI, 2000. Product Tampering and Accidental Contamination Conference and Workshop. American Conference Institute. June 12–14, 2000. San Francisco, CA.

Adams, MR and Moss, MO. 2000. *Food Microbiology,* 2nd ed. Royal Society of Chemistry, London.

Anderson, C. 2003. Terrorist may try to improvise weapons. www.fbi.gov.

Anon. 2003. *Biological Agents/Diseases.* Centers for Disease Control, http://www.bt.cdc.gov.

Arnon, SS, Schechter, R, Inglesby, TV, Henderson, DA, Bartlett, JG, Ascher, MS, Eitzen, E, Fine, AD, Hauer, J, Layton, M, Lillibridge, SR, Osterholm, MT, O'Toole, T, Parker, G, Perl, TM, Russell, PK, Swerdlow, DL, and Tonat, K. 2001. Botulinum toxin as a biological weapon. Medical and public health management. *JAMA* 285:1059–1070.

ATSDR. 1995–2003. Agency for Toxic Substances and Disease Registry. 1995–1997. http://www.ushhs.gov. Accessed December 2003.

Barton, R. 1998. The application of the UNSCOM experience to international biological arms control. *Critical Reviews in Microbiology.* 24(3):219–233S.

Berg, R. 2004. Terrorism response and the environmental health role: The million dollar question. *J. Environmental Health.* 67(2):29–39.

Berns, KL, Atlas, RM, Cassel, G and Shoemaker, J. 1998. Preventing the misuse of microorganisms: the role of the American Society of Microbiology in protecting against biological weapons. *Critical Reviews in Microbiology.* 24(3):273–280.

Borio, L, Inglesby, TV, Schmaljohn, AI, Hughes, JM, Jahrling, PB, Ksiazek, T, Johnson, KM, Meyerhoff, A, O'Toole, T, Ascher, MS, Bartlett, J, Breman, JG., Eitzen, E, Hamburg, M, Hauer, J, Henderson, DA, Johnson, RJ, Kwik, G, Layton, M, Lillibridge, S, Nabel, GJ, Osterholm, MT, Perl, TM, Russell, PK, and Tonat, K. 2002. Hemorrhagic fever viruses as biological weapons, medical and public health management. *JAMA* 287:2391–2405.

Carus, WS. 1998. Biological warfare threats in perspective. *Critical Reviews in Microbiology.* 24(3):149–155.

CDC. 2002. Facts about Chemical Agents. www.cdc.gov. Accessed December 2003.

Dennis, DT, Inglesby, TV, Henderson, DA, Bartlett, JG, Ascher, MS, Eitzen, E, Fine, AD, Friedlander, AM, Hauer, J, Layton, M, Lillibridge, SR, McDade, JE, Osterholm, MT, O'Toole, T, Parker, G, Perl, TM, Russell, PK, and Tonat, K. 2001. Tularemia as a biological weapon. *JAMA* 285:2763–2773.

DHS. 2003. Homeland Security Advanced Research Projects Agency. http://www.dhs.gov/dhspublic/interapp/editorial_0344.xml

DTS. 2004, 1998. US Army Center for Health Promotion and Preventive Medicine. Deputy for Technical Services. http://chppm-www.apgea.army.mil/dts/dtchemfs.html. USACHPPM. Aberdeen Proving Ground, MD.

FBI, 2000. FBI Sponsors genetic engineering ecoterrorism conference in Berkeley, CA. January 26, 2000 in conjunction with the National Institute of Justice, Berkeley and Davis police departments.

Fergenson, DP, Tobias, HJ, Steele, PT, Czerwieniec, GA, Russell, SC, Lebrilla, CB, Horn JM, Coffee, KR, Srivastava, A, Pillai, SP, Shih, MTP, Hall, HL, Ramponi, AJ, Chang, JT, Langlis, RG, Estacio, PL, Hadley, RT, Frank, M, and Gard, EE 2004. Reagentless detection and classification of individual bioaerosol particles in seconds. *Anal Chem.* 76(2):373–378.

Fitch, JP, Raber, E and Imbro, D. 2003. Technology challenges in responding to biological or chemical attacks in the civilian sector. *Science.* 302:1350–1354.

Henderson, DA, Inglesby, TV, Bartlett, JG, Ascher, MS, Eitzen, E, Jahrling, PB, Hauer, J, Layton, M, McDade, JE, Osterholm, MT, O'Toole, T, Parker, G, Perl, TM, Russell, PK, and Tonat, K. 2001. Smallpox as a biological weapon, medical and public health management. *JAMA* 281:2127–2137.

Hosenball, M. 2004. IEDs secret sarin supply. *Newsweek* CXLII (24):8. June 14, 2004.

Inglesby, TV, Dennis, DT, Henderson, DA, Bartlett, JG, Ascher, MS, Eitzen, E, Fine, AD, Friedlander, AM, Hauer, J, Koerner, JF, Layton, M, Lillibridge, SR, McDade, JE, Osterholm, MT, O'Toole, T, Parker, G, Perl, TM, Russell, PK, Shoch-Spana, M, and Tonat, K. 2001. Plague as a biological weapon, medical and public health management. *JAMA* 283:2281–2290.

Inglesby, TV, O'Toole, T, Henderson, DA, Bartlett, JG, Ascher, MS, Eitzen, E, Friedlander, AM, Gerberding, J, Hauer, J. Hughes, J, McDade, JE, Osterholm, MT, Parker, G, Perl, TM, Russell, PK, and Tonat, K. 2002. Anthrax as a biological weapon, 2002. Updated recommendations for management. *JAMA* 287:2236–2252.

Jernigan, JA, Stephens, DS, Ashford, DA, Omenaca, C, Topiel, MS, Galbraith, M, Tapper, M, Fisk, TL, Zaki, S, Popovic, T, Meyer, RF, Quinn, CP, Harper, SA, Fridkin, SK, Sejvar, JJ, Shepard, CW, McConnell, M, Guarner, J, Shieh, WJ, Malecki, JM, Gerberding, JL, Jughes, JM, and Perkins, BA. 2001. Bioterrorism related inhalation anthrax: the first 10 cases reported in the United States. *CDC* 7:1–22.

Martinez, I. 2003. Rhodesian anthrax: the use of bacteriological and chemical agents during the Liberation War of 1965–1980. *Indiana International and Comparative Law Review.*13:447–475.

Merrill, RA and Francer, JK. 2000. Organizing federal food safety regulation. 31 *Seton Hall Law Review,* 61–170.

MMWR. 2001a. Morbidity and mortality weekly report. Investigation of anthrax associate with intentional exposure and interim public health guidelines, October 2001. *JAMA* 286:2086–2090.

MMWR. 2001b. Morbidity and mortality weekly report. Update: investigation of bioterrorism-related anthrax and interim guidelines for exposure management and antimicrobial therapy, October 2001. *JAMA* 26:2226–2232.

MMWR. 2003. Investigation of a ricin-containing envelope at a postal facility, South Carolina, 2003. *Morbidity and mortality weekly report.* 52:1129–1131.

Monke, J. 2004. *Agroterrorism: Threats and preparedness.* CRS Report for Congress. Order code RL 32521. Aug. 13. 2004. 45 pp.

NIOSH. 2003. Emergency Response Card Biotoxin Abrin. cdc.gov/agent/abrin. Accessed December 2003.

Okumura, T. 1998. Tokyo subway sarin attack: disaster management. Part 2: hospital response. *Academic Emergency Medicine.* 5:681–624.

Primmerman, CA. 2000. Detection of biological agents. *Lincoln Laboratory J.* 12:3–31.

Shilbey, GR and McKay, R. 2003. Having a Conversation. Statewide Meeting, Food Security Agenda, September 9–11, 2003, Eugene, OR.

Stern, J. 1999. *The Ultimate Terrorist.* Harvard University Press, London.

Tubbs, D. 2000. 4/13/00: face of ecoterrorism. Elf defines use of violence.

Vastag, B. 2001. Experts urge bioterrorism readiness. *JAMA* 285:30–33.

Weise, E. 2004. Toxic attack on cows points to reporting problem. Time-lags called disturbing in light of September 11 and food supply concerns. *USA Today.* June 24, 2004.

Woodall, J. 1998. The role of computer networking in investigating unusual disease outbreaks and allegations of biological and toxin weapon use. *Critical Reviews in Microbiology.* 24(3):255–272.

Zilinskas, RA. 1998. Verifying compliance to the biological and toxin weapons convention. *Critical Reviews in Microbiology.* 24(3):195–218.

Notes

1. In *Making the Nation Safer: The Role of Science and Technology in Countering Terrorism.* Committee on Science and Technology for Countering Terrorism, National Research Council of the National Academies. The National Academies Press, Washington, DC, 2002, p. 108.
2. There are at least five antigentically distinct enterotoxin proteins (SEA, SEB, SEC, SEE, TSS, and T-1) all of which are heat stable. Staphylococcal enterotoxin B (SEB) has a molecular weight of 28,000 Da.

3. Among these are: China, Egypt, Iran, Iraq, Libya, Russia, Syria, Taiwan, North Korea, South Africa and Cuba. The US stopped offensive biological weapons research in the 1969. Other possible countries with weapons capabilities are: Bulgaria, Laos, Zimbabwe, Romania, South Africa, Vietnam and possibly Laos, (Carus, 1998; Martinez, 2003).
4. There are seven related protein toxins, designated A through G, produced from various strains of *Clostridium botulinum*. These have a molecular weight of approx. 150,000 Daltons (Da). Botulinum type A is most toxic to humans and is composed of two proteins, a 97,000 Da chain and a 53,000 Da chain linked by at least one disulfide bond.

3 Bioterrorism Regulations and Their Impact on the Safety of the Food Supply and Trade

Bad law is the worst form of tyranny.

— **Blaise Pascal**

No man is above the law and no man is below it; nor do we ask any man's permission when we ask him to obey it.

— **Theodore Roosevelt**

The Public Health Security and Bioterrorism Preparedness and Response Act of 2002 (Bioterrorism Act: PL107-188) signed into law on June 12, 2002, has a direct and substantial impact on the entire U.S. food[1] industry, particularly upon the imported segment,[2] and will have far-reaching international trade ramifications. Food is broadly defined under the act and includes food and animal feed, nutritional supplements, food contact materials, live animals, bottled water, and alcoholic beverages (69 Federal Register (FR) 31670). congressional intent and the key provisions of this bill were to provide a coordinated national preparedness program for bioterrorism with an emphasis on public health and health services, including pediatric care, controls on dangerous biological agents, and improved protection for the food and water supplies. Unfortunately, in the resultant regulations, this focus appears to have been lost,[3] at least in the segments of the bill intended to increase protection of the food supply. Instead, it may well be that the U.S. Food and Drug Administration (FDA) has used this act as an opportunity to exploit public and congressional concern over the threat of bioterrorist acts to expand the agency's jurisdiction and control over the U.S. food supply and its distribution system. Unfortunately, these regulations come at a substantial cost and may do little to improve the safety and security of the food supply.

Key regulations promulgated from the Bioterrorism Act became effective on December 12, 2003. According to the FDA, the paper trail (68 FR 58896) resulting from these rules is likely to discourage potential criminals intent upon contaminating food because additional evidence generated from the newly required records could then be used against them. In the age of suicide bombers, is this a credible argument? These regulations focus the agency's attention and limited resources upon the detection of corporate fraud and upon employees who handle and process paperwork, not upon the activities of individuals determined to contaminate our food supply. The provisions do nothing to affect the activities of an individual, acting alone or as part of a small conspiracy, who will intentionally contaminate food with poisonous materials, as there is little chance they would intentionally generate a written or electronic record related to their terrorist activity. The goal of a terrorist in this situation, unless clearly perpetrated to increase the impact of a terrorist act, is to leave no tracks at all.

Unfortunately, the administrative burden of these regulations may actually result in a dilution of effective food safety implementation measures on the part of the FDA and state and other governmental entities. The new regulations clearly have a negative impact on industry, as compliance requires companies to establish new and relatively complicated procedures, taking efforts away from other food safety-monitoring activities and thus diluting preparedness activities.

In addition, the regulations fail to meet the standards outlined by the Office of Management and Budget (OMB) for regulatory impact. OMB reviews regulations and mandates that regulatory alternatives have the lowest net cost to society.[4] These new rules do not meet the criteria outlined under the five elements of the OMB Regulatory Impact Analysis, as follows.

First, the agency has failed to adequately consider the needs and consequences of these regulations to society in general.

Second, the agency has failed to show that the potential benefit of these new regulations outweighs the costs. The relatively detailed economic analysis provided by the agency as part of the preamble to the proposed and final rules failed to accurately estimate certain costs and did not provide a full-cost analysis of the regulatory impact. Economic analyses of certain segments, specifically small business impact and imported foods, are clearly flawed.

Third, the regulatory objectives chosen by the agency were not selected to maximize net benefits to society, but instead to further an expansion of agency jurisdiction.

Fourth, regulatory alternatives that have the lowest net cost to society were not selected as per the OMB mandate. This includes the alternative of *not* to regulate, when this option is clearly permissible by statute. Obviously, having no new regulations would come at no new cost either to the government or to the regulated industry. Any argument presented on behalf of the agency that the net cost to society incorporating public health and other social factors into this analysis is not valid, as the regulatory measures do little to increase overall food security, and thereby public health.

And finally, the agency, through these regulations, has failed to consider the condition of the affected food industries, potential future regulatory actions, and the weak state of the national economy as required.

While Congress clearly stated in the Bioterrorism Act that the purpose of it was to address "public health security and bioterrorism preparedness and response," the FDA was quick to state that the provisions would be implemented and enforced across the board and applied to all elements within the food system without consideration of whether any alleged incident of food adulteration was tied to a bioterrorism act or a possible security threat[5] (see, for example, 68 FR 31662, where the agency clearly states that the Bioterroism Act does not limit the agency's authority to only those situations involving intentional adulteration (68 FR 31663)). This is accomplished by tying enforcement of the act to an "event of food borne disease" (see, for example, 68 FR 58893) or another "food related emergency" (68 FR 58895), neither of which is clearly defined.

The regulations and stated expanded areas of control imposed by the FDA are overreaching and challenge long-established federal and state jurisdictional boundaries.[6] Clearly, under these new regulations, FDA is moving into areas delegated to state control under the enabling statute and the 10th Amendment to the U.S. Constitution.[7] However, by imposing this regulatory scheme, the agency can avoid and circumvent the very safeguards established to limit expansion of federal jurisdiction. Most certainly, if Congress and the public had foreseen the scope of FDA's proposed jurisdictional expansion, the act would have received considerably more analysis and scrutiny.

Does this mean that food security measures are not needed? To the contrary, the threat of purposeful contamination is real. Development of reasonable preventive measures and appropriate responses, including rational governmental activities that are effective within every facet of the food system, is critical for protecting public safety (Bledsoe and Rasco, 2003). However, to be effective, these measures must be driven by the public and the food industry itself. As with all food safety programs, the most workable and effective measures are market and not regulatory driven.

Regulatory actions under the Bioterrorism Act should be focused upon incidents tied to threat events, as Congress intended. However, the new regulations encompass *any possible* perceived incident of food adulteration that may "present a threat of serious adverse health consequences to humans or animals to be defined in later rulemaking"[8] under a plethora of new and loosely defined legal standards,[9] including "credible evidence or information," a term that the agency refuses to define (69 FR 31673). Thus, the primary objective of the act (preventing and reacting to a bioterrorist act) becomes buried and subordinated to an almost inconsequential role when compared to the other agency prerogatives incorporated within these regulations.

Furthermore, the new regulations are burdensome and overlap with current requirements under 21 Code of Federal Regulations (CFR) Parts 7 and 110, and others such as Hazard Analysis Critical Control Point (HACCP) under 21 CFR Parts 123 and 1240. Proper implementation and enforcement of provisions under regulations in existence prior to passage of this act would have been more than adequate to address concerns the agency may have with rapid location of affected product and ingredient traceability, both of which are legitimate and important concerns.[10]

Particularly under some of the provisions of the Bioterrorism Act, the agency had the discretion *not* to regulate, but chose to do so without strong justification; and when proposing regulations, the agency refused to select the least burdensome alternative.

The huge cost of these regulations in terms of human capital and the cost to taxpayers in implementation and enforcement is simply not justified. Furthermore, the cost to the regulated industry will be staggering.[11] The agency is now in a position to take various actions against a food, but has no responsibility to determine who is responsible for the threat (69 FR 31667). Neither will the agency take any responsibility in determining whether the threat is real or only a hoax. The predicament of "involuntary noncompliance" by small business, in particular from a lack of understanding as to what the rules require, will lead to a loss of many small firms and the accompanying employment. Tied directly to this is a loss of an interesting and eclectic selection of regional and imported specialty foods (Anon., 2003).

More importantly, the political implications of these new regulations on international trade are significant. The agency notes that it has complied with its international trade obligations under applicable World Trade Organization (WTO) agreements and the North American Free Trade Agreement (NAFTA) (see, for example, 69 FR 31662, 31667, 31668). However, many foreign governments consider provisions such as those for records and registration (68 FR 58900) as violations of national sovereignty and have stated that they will not comply. The Swiss government was quick to notify the U.S. that the proper venue for international activities such as these is the State Department, not the

FDA. Recent botched investigations and product retentions at the border have reduced the confidence of our trading partners. The impact of these regulations on both the cost and availability of food in the U.S. is huge. Using increased costs of airport security as a possible scenario, it is feasible that these new rules could increase the retail costs of food in the U.S. by 10%. Nearly 20% of all U.S. imports are food products, and these regulations will have serious negative impacts on these food imports. In addition, there are negative ramifications for U.S. food exports. Trading partners in Asia, the European Council, and Canada express the concern that these new requirements are unfair, unworkable, and punitive. Increased tariffs and nontariff trade barriers are currently under construction as a direct result of these regulations, seen overseas as a sign of U.S. isolationism and an overly zealous nationalistic reaction to weak international support of the Afghanistan campaign and Iraq invasion in 2002 and 2003. Retaliatory trade restrictions placed upon U.S. exports may be a direct result of the regulatory requirements generated from this act.

In short, these new rules may cause more harm than good and serious consideration should be focused upon eliminating major components of this regulatory scheme, as well as streamlining the rules to address only what is clearly mandated by statute. These modifications should be accomplished in a manner posing the least administrative burden upon both the government and the regulated industry.

Specific provisions of the regulations in effect and proposed by the FDA to implement the new Bioterrorism Act are outlined below, accompanied by comments regarding potential economic and trade implications. The FDA states that it has made significant changes in the interim final rules based on comments received on the proposed rule for registration of food facilities and for prior notice. These are noted here. The final rules from this act had not been issued by the date of the final revision of this manuscript (September 2004); however, the information presented here is current through this date.

Registration (Section 305 of the Act;[12] Interim Final Rule, 21 CFR Part 1, Section 1.225 *et seq.*; 68 FR 58893-58974)

Mandatory registration of food facilities is supposed to allow the FDA to more rapidly track food products implicated in a food-borne outbreak. This sounds rational. However, in the regulation FDA also claims that registration will deter intentional contamination of food because "persons who might intentionally contaminate the food supply would be deterred from entering the food production chain" (Anon., 2003). It is truly doubtful that this regulation would provide any deterrent to an individual or group of individuals determined to perpetrate an intentional act to contaminate food.

Registration actually creates a false impression of security. The requirements could be easily circumvented or fraudulently complied with.

Regardless of the necessity of this provision, effective December 12, 2003, all foreign[13] (21 CFR § 1.227(ii) (68 FR 58961)) and domestic[14] (21 CFR Section 1.227(i)) food manufacturing[15] or processing operations, including distributors, (re)packagers, labelers (21 CFR § 1.227(6)), holders (e.g., warehouses, silos, bulk grain storage for domestic and export shipments (21 CFR § 1.277(5) (68 FR 58900)), cold storage facilities, and liquid storage tanks), and foreign government facilities holding food (68 FR 58900), were required to be registered with the FDA. Fishing vessels are exempt if they do not engage in processing activities as outlined under the seafood HACCP regulation[16] (21 CFR § 1.226 (f)); however, registration is likely in any effect, as agency interpretation of the seafood HACCP regulations and the resulting guidance have been used to expand agency activities into these venues.

In addition, facilities regulated jointly by the FDA and U.S. Department of Agriculture (USDA) must register (21 CFR § 1.226(g)).

Registration of food transport turns on whether or not the FDA defines the business as a "mobile facility"[17] (68 FR 58904). Transport vehicles are *not* facilities if they hold food only in the usual course of business as carriers (68 FR 58895). However, trucker/dealers who purchase grain, take title to grain, and hold grain in a transportation conveyance until it can be sold to another processor, storage facility, or end user must register (68 FR 58904). This requirement for trucker/dealers appears to be predicated upon the *ownership* of the grain, and not upon the food safety risk associated with holding and transporting grain, creating an inconsistency in the registration requirement that should be corrected.

Transshippers of food across the U.S. for consumption outside the U.S. do not have to register, although they are covered by prior notice requirements (68 FR 58898). This provision may well divert "European land bridge" traffic for Asian products from West Coast U.S. ports, such as Seattle/Tacoma, WA, to nearby Vancouver, BC, at a cost of several million dollars in annual lost revenues to American ports, as well as the revenues tied to the associated rail transit operations. Large job losses for workers in these sectors are likely, as mixed cargo loads transporting food along with other materials will be diverted in their entirety.

A number of domestic firms not normally considered to be "food establishments" also have to register. These include farm and retail and food service engaged in certain specific activities. Whether a food establishment falls within a category that will trigger registration depends not upon a potential food safety risk, but instead upon ownership of the product (in the case of farms[18]), the nature of the sales (in the case of retail establishments[19]), or how the food is prepared and distributed (for restaurant operations[20]).

Nonprofit food establishments such as food banks, church groups and charities,[21] and pet shelters (68 FR 58913) are exempt from registration; however, it is unclear how the food safety or security risks posed of these nonprofit and animal feeding establishments would differ significantly from those of for-profit restaurants or retail establishments engaged in the same activities that are required to register (21 CFR § 1.226). In fact, nonprofit or animal facilities may pose a higher potential food security risk. Generally, these facilities are operated by individuals with less experience in food safety than individuals in the for-profit sector. Second, the political stance of a church or charity may make it a target for extremists, and food-related activities may be among the most expedient means of attacking an organization. Similarly, the nature of an organization's activities (e.g., holding animals "captive") could make it a target for terrorists.

Simply put, the registration requirement will impose substantial cost and will negatively impact trade, particularly trade in boutique and specialty food items. These are foods produced in small volume, generally from small producers. These firms, whether domestic or foreign, will have neither the sophistication nor the resources to comply with the registration provision and will no longer be able to sell their products in U.S. markets (see, for example, 68 FR 58946).

Registration also has a disproportionate effect upon U.S. businesses compared to foreign competitors. For example, all U.S. businesses engaged in the activities listed above must register. However, a foreign facility must register only if food from the facility is exported to the U.S. for consumption without further processing or packaging. This provision means that only *the last holder of goods* must comply (68 FR 58901). Despite this discrepancy, compliance by foreign manufacturers also has a major financial cost to them and will lead to the loss of variety and availability of foods on our markets.

The FDA estimates that roughly 270,000 U.S. firms (revised downward to 216,271 in the interim final rule (68 FR 58936)) and 280,000 foreign food facilities (revised downward to 205,405) will have to register (68 FR 5387 *et seq.*) The numbers presented for small domestic and foreign facilities are probably low (68 FR 58933). The economic analysis also undercounts direct marketers (68 FR 58934) and mixed-use operations, including affected farming operations, meaning that the number of affected businesses is more likely to exceed the 540,000 originally estimated.

A U.S. agent may register for a foreign facility and is the contact of choice by the agency if an emergency arises (68 FR 59986). Who this individual must be (68 FR 58916), what the requirements, qualifications, and liability are of a U.S. agent (68 FR 58915), and whether foreign governmental officials can fulfill this function are not clear. Regardless, the requirement for having a U.S. agent creates an additional expense and unexpected hurdle for a potential importer. A requirement of English language records and dubious Internet access to government sites create further hurdles.

FDA cost estimates to *simply register* U.S. businesses exceed $19 million. The estimated cost to register foreign businesses exceeds $423 million, with ongoing annual expenses in excess of $227 million (68 FR 5387 *et seq.*). However, the actual costs of compliance, particularly when one considers multiplier effects, will certainly be greater. A current rule of thumb in the U.S. business community suggests that each page of documentation costs a minimum of $5.00 to generate and that each signature costs an additional $2.50. Even though there is no cost or fee to access the FDA electronic site for registration, requirements for initial compliance, plus the costs to maintain and update registration information as required by the regulation, guarantee that this provision, which is simple on its face, will be a significant ongoing cost for business. Updates to the required information on a registration must be made within 60 days (21 CFR § 1.234), meaning that facility registration is not a one time event. Interesting little quirks (e.g., format of passwords) in the current registration program can make registration a somewhat frustrating experience, particularly for registrants with limited English literacy whether they represent domestic or foreign businesses.

In the rule, the FDA downplays the significance of the costs of the registration to the regulated community[22] (68 FR 58932). First, the agency presumes that at least 90% of facilities have computers with appropriate software and Internet access and, therefore, will not incur any capital costs (68 FR 58939). The FDA estimated that it would take a domestic company with highly literate employees fluent in English 3 hours to comply with registration requirements initially, and 1 hour to collect information and register each additional facility. Other costs, such as port delays tied to checking whether a valid registration exists, confirmation of the registration of suppliers (68 FR 58940), and loss of product value from delays and possible market withdrawals, were not considered in agency cost estimates.

Furthermore, the costs tied to hiring and retaining a U.S. agent (estimated at $1000.00 in the proposed rule (68 FR 58945)) are likely underestimated. The costs for an agent were further reduced to $700.00 in the interim final rule (68 FR 58945), with no justification presented for reducing this cost figure. We have seen charges posted by agents to simply process the registration for a business exceeding $500 for each registration. Further costs for the services of an agent, for example, retaining a law firm, start at $150 per hour and increase from there. Compliance with the agent provision for foreign business is projected by the agency at $220.5 million to $364.6 million in the first year and $144.6 million to $267.4 million in subsequent years. Because of the uncertainty with cost estimates, the agency is seeking comments on the following: cost of compliance by U.S. small business; the cost (and underlying assumptions) of hiring and retaining a U.S. agent; the number of foreign firms hiring U.S. agents; the number of exporters ceasing to export to the U.S. due to the registration requirement and the impact that this has on U.S. businesses; and the distribution of costs between submitting registrations and other services provided by a U.S. agent (68 FR 58959). Unfortunately, the agency is not seeking information on the economic impact on domestic businesses and the resultant loss of U.S. jobs and U.S. small businesses tied to the added costs of this component of the registration requirement.

Any food facility must have a registration number before it can sell products in the U.S. Food will be "misbranded" if it is not from a registered facility, making its sale illegal in the U.S. Registration means that the FDA will have the power to suspend or revoke a facility's registration if the business fails to permit an inspection or if, in the opinion of the FDA, suspending a registration is likely to prevent a significant risk of adverse health consequences. This provision effectively authorizes the FDA to directly embargo food and shut an operation down, circumventing the current requirements of a court order or involvement of state authorities.

Imported product not in compliance with this regulation will be held at the port of entry in a secure location at the owner's, consignee's, importer's, or purchaser's expense. Both civil and criminal charges can be brought against companies failing to register and could lead to "permissive debarment," another new provision under this act, prohibiting a company or certain individuals from importing food into the U.S. altogether.

The ability to "pull" a registration provides numerous opportunities for overreaching agency action. Personal discretion on the part of agency employees may lead to determinations that certain cultural foods, methods of food packaging, labeling provisions, etc., are "unsafe," with little credible information to support those claims, thereby preventing the sale of these products. Recently, Asian, eastern European, and Native American cultural foods, products such as cold smoked fish, cured or noncured ready-to-eat aquatic foods in sealed cans or plastic containers, foods treated with "tasteless smoke," and fishery products held refrigerated under vacuum, have been targets of heightened regulatory scrutiny even though the foods in question have not been tied to incidents of food-borne illness and were later shown by approved analytical methods to meet regulatory requirements for safe foods.

With a broad application of these new rules, we may see food activists suing the government under the act to prohibit foods that some activists consider to be unhealthy — for example, products containing meat, trans-fatty acids or another food constituent of perceived high risk that poses a "significant risk of adverse health consequences" in the opinion of the activist. Similarly, registration could be used to prevent the importation or sale of targeted foods containing the "scary chemical" of the week, be it chloramphenicol, Alar, benzo(a)pyrene, or acrylamide. Animal rights or environmental activists could attempt to extend the scope of "significant health risk" to the already highly politicized harvest of large pelagic fish, such as swordfish (e.g., mercury), salmon (e.g., parasites, colorants added to feed, or environmental contaminants (PCBs)), or tuna (e.g., histamine), in an attempt to remove these fishery products from commerce. Other foods could be similarly impacted, depending upon which direction the political winds are blowing at the time.

Registration information must be provided in English and include name, address, and contact information for the facility and parent company; emergency contact information; all trade names used at the facility; and product categories. A facility has only 60 days (21 CFR § 1.234(a)) to report any changes in registration information, including changes in product categories. Failing to make these changes places a firm out of compliance. The rigidity and detail of these requirements increase cost and lead to involuntary noncompliance.

By FDA's own estimates, the greatest potential economic loss from the registration requirement is the elimination of small importers, about 16% of the current total, who import 10 or fewer line entries per year. It is quite likely that the registration requirement alone will also lead to the loss of small niche markets and the small businesses integral to those markets in the U.S. Most disconcerting will be the loss of these often unique and regional products to U.S. consumers. Again, what will the retaliatory actions of the governments of the affected exporters be toward U.S. producers — food products or otherwise?

There is no question that the increased regulatory burden will force small U.S. producers out of the industry as well. Feeling victimized by the law and diverting energies from the operation of the business, small food business owners will simply adopt defensive measures designed solely to avoid tripping over rules that seem to exist only because someone put them there (Howard, 1994). Is this the climate we want for one of the most successful segments of small business in this nation, particularly in rural communities where small food processing businesses are crucial to the viability of the local economy?

Records and Record Retention[23] (Section 306 of the Act; Proposed Regulations, 21 CFR Part 1, Section 1.326 *et seq.*)

New records[24] and records retention requirements are designed to improve food traceability and provide the agency with the "ability [to] effectively and efficiently … respond to bioterrorist threats and other food related emergencies." Currently the normal response time for requested records is 2 to 3 days. The agency's goal is to speed up this process. Objectives are to make these particular regulations performance-based and provide flexibility to use existing recordkeeping systems. However, most existing recordkeeping systems do not conform to regulatory requirements and must be modified to incorporate new materials or to permit tracking of products in certain ways.

Numerous unsuspecting souls not normally swept up in the food processor dragnet have to comply with these new record requirements. Importers and companies relabeling or repacking foods,[25] retailers,[26] restaurants with retail activities, mixed-use operations (such as the 30,000+ farms that package food in the field for retail sale), and food producers under concurrent jurisdiction of the FDA and USDA must comply. Any foreign facility that must register will have to comply with these records requirements as well. The FDA estimates that roughly 841,000 facilities operated by 646,000 firms are covered by this rule. An estimated 225,000 foreign operations are also affected.[27] Unfortunately, compliance is a serious economic burden to the industry, particularly on small firms,[28] and it remains to be proven what specific and concrete increases to the overall safety of the food supply will result from this new requirement.

Under the records regulations, the FDA requires a business to be able to trace the immediate previous source and immediate subsequent recipient in the distribution chain,[29] trace distribution, and be able to provide the actual physical location of the food. These records require providing the name of the company and responsible individual; address, phone, and other contact information on immediate prior source; type and quantity of food; date of receipt; name, address, and phone of transporter; contact information for the transporter's immediate subsequent recipient; identifier; code or lot number for the food; quantity; and identification of each and every mode of transportation of the food from the time it is received until it is delivered. Transporters must establish and maintain records for each shipment to include mode(s) of transportation, responsible individuals throughout the entire transportation process, the consignor from whom the food shipment was received, and the consignee to whom the food is delivered. New traceability devices such as radio frequency identification tags (RFIDs) will help to provide tracking information; however, these systems are not yet widely available and are still relatively expensive. Further, initial implementation would pose a significant capital expense.

It should not be a surprise that the new records requirements are redundant. Companies are currently required to have traceability provisions as part of the federal requirement for recall plans under 21 CFR Part 7. More importantly, though, food producers keep similar records in the normal course of business, to track inventory, source, and disposition of product. In addition, food producers and sellers keep such records for liability reasons. Food manufacturers, distributors, and sellers are strictly liable under state products liability statutes or under breach of implied warranty provisions of the Uniform Commercial Code for injuries that can be traced to the food they sell and recognize the necessity to ensure product safety by having suitable tracking programs in place (Buzby et al., 2002).

Under the new rule, records for perishable products must be kept for 1 year, and for other food products, 2 years. There is also an implicit requirement for traceability back to individual growers, both here and under prior notice provisions for importers. Records must be made available within 4 hours of an FDA request if the request is made between 8 A.M. and 6 P.M. local time Monday through Friday except holidays and within 8 hours if made at any other time. Details on how it will be possible to obtain compliance with this requirement from foreign processors are sketchy at best. The regulations on recordkeeping are also weak in addressing how products produced in a foreign country from components generated by yet other countries are to be handled and whether there is any U.S. jurisdiction over such activities. Do these regulations now attempt to impinge upon the sovereignty of other nations by imposing U.S. law on companies in any country to keep and then be obligated to produce records on demand to the agency, regarding foods or ingredients, if that country wishes to export any portion of that production to the U.S.? In short, where is the origin of this recordkeeping requirement, both of these regulations and under international treaty obligations, positioned in the complex spider web?

Under these regulations, any entity that manufactures, processes, packs, distributes, receives, holds, transports, or imports a food must provide access to records related to that food when the agency has a "reasonable belief" that a food is adulterated or presents a threat of serious adverse health consequences or death to humans or animals or both. Failure to maintain or produce records is a violation and can lead to civil and criminal penalties.

These regulations increase cost[30] to the private sector through a *de facto* requirement to limit or prohibit commingling of ingredients from different lots or suppliers in bulk storage (e.g., corn sweetener, flour, oil) as a means of reducing records requirements.

Although current records (e.g., bills of lading, purchase orders, or invoices) can be used in part to meet requirements, modifications are required; thus, new costs will be incurred. The FDA claims that the learning curve for the registration and recordkeeping components of these regulations overlap, reducing implementation costs; however, the initial implementation costs and ongoing costs for compliance remain high. The FDA estimates that learning costs (interpreting how to adopt the regulations by the company) alone will exceed 5.5 million hours or $72 million for domestic firms and

$201 million for foreign operations. This is in addition to the FDA-estimated costs to redesign records for compliance, which are in the range of $4000 to $12,000 per firm, with an estimated time commitment industrywide of over 11 million hours. While the FDA-estimated full costs per hour of roughly $25 for administrative employees and $57 for management are reasonable, the time commitments do not appear to be realistic. Nor is there provided a reasonably identifiable split in effort between clerk, administrative, and managerial time requirements in every phase from interpretation through implementation of this new rule.

The large number of firms involved may well cause havoc with initial implementation. For example, a six-step supply chain would generate at least three separate sets of records. A more complicated supply chain, e.g., (1) farmer, (2) transporter, (3) bulk storage (e.g., grain silo), (4) transporter, (5) processor, (6) transporter, (7) warehouse, (8) transporter, and (9) retailer, would generate at least four sets of bills of lading and four sets of invoices. The agency predicts that compliance with this recordkeeping requirement will involve a commitment of 15 to 30 minutes per week for a facility. This is based upon agency predictions for HACCP compliance (that proved to be quite inaccurate) and upon surveys of handling practices of purchase orders for a single large retailer. The agency also predicts that records handling and retention costs would be minimal. "Minimal" to a federal agency may be quite a bit larger than what a smaller producer might define as such. Unfortunately, these estimates do not appear to be realistic for food processors tracking foods with multiple ingredients or for companies shipping several dozen orders per day, particularly when considering the multitude of small businesses involved in the U.S. food industry.

To its credit, the agency did recognize some of these issues through its initial outreach efforts and, as a result, will not require companies to be able to tie specific raw material inputs to specific finished products. However, the intent of the rule is to take us there anyway[31] at an estimated cost exceeding $87 billion annually. The FDA adds that additional costs to maintain these segregated records would run in excess of $730 million annually at a time commitment exceeding 9 million hours per year.

Furthermore, the costs of *accessing* the records upon an FDA request and, more importantly, the costs tied to changes in business practices to allow for rapid response in case a request is made are unrealistic, except perhaps for a firm with a round-the-clock workforce and a dedicated data processing system. Compliance with the record retrieval provisions of these regulations leads to incurred costs regardless of whether the agency ever requests records. The FDA predicts that it will take a single person at a business 6 hours to develop an access plan and that this is a one-time fixed cost based upon the person's experience with juice HACCP regulations (assuming that it will take 1/10 the time to develop a records access plan as it did to create a HACCP plan) or $300 to $2600 per firm. The agency's prediction that it will cost roughly $100 to $900 per year to maintain these records is also unrealistic. Experience with the FDA-projected costs in terms of both time and money for seafood HACCP compliance (21 CFR Parts 123, 1240) shows that they were low by as much as a factor of 10, and we would predict that the FDA cost estimates for compliance with these regulations are similarly underestimated.

The agency also estimates that the costs of a 4- to 8-hour response and 24-hour response are solely involved in the costs associated with preplanning a response, ignoring the additional personnel requirements necessary to respond to a crisis on short notice. Even without this factor taken into consideration, the cost to the industry for a more rapid response time is estimated by the FDA at $715 million, with future annual costs associated with recordkeeping (year 3 and later) projected in excess of $210 million.

While the FDA estimates that roughly 207,000 U.S. and 280,000 foreign food facilities will have to meet the registration requirements, the FDA further states that it assumes that 1,300,000 facilities and 960,000 firms will be affected by recordkeeping requirements. This number includes domestic facilities that manufacture, process, transport, distribute, pack, receive, hold, or import food or food packaging, and foreign facilities performing any of these activities on food or food packaging destined for consumption or use in the U.S.

As with the new registration requirements, the agency also predicts that at least 16% of the foreign firms will be driven out of business by this regulatory burden. (The FDA did not, however, estimate how many domestic firms would be driven out of business by this same regulatory burden.) The loss of these businesses will not only lead to a loss of products on our markets, but also increase the cost of goods sold abroad.

The agency furthermore claims that the regulations, though costly, should not substantially affect national productivity, jobs, growth, or full employment. Furthermore, "the total costs will be small relative to the economy and will be offset by benefits." Benefits such as "the improved ability to respond to, and contain serious adverse health consequences thereby resulting in less illness and fewer sick days taken by employees, and lower adjustment costs by firms that would otherwise need to hire replacement workers." This may be true for the businesses that are still left standing after regulations are implemented. It would certainly be interesting to evaluate the compiled data underlying the economic assumptions used by the agency for conducting the cost–benefit analysis and to develop a variance analysis on what is actually experienced when these regulations finally go into effect.

Prior Notice for Imported Food (Section 307 of the Act;[33] Pending 21 CFR Part 1, Section 1276 *et seq.*; Interim Final Rule, 68 FR 58975-59007)

The purpose of the prior notice is to enable inspection[34] of food to "protect consumers in the United States from food imports that may be at risk of intentional adulteration or that may pose other risks."[35] This rule is designed to provide the FDA with the "ability to detect accidental and deliberate contamination of food and deter deliberate contamination" and permit the agency to ensure that products "appear to have been safely produced, contain no contaminants or illegal additives or residues, and are properly labeled" (68 FR 58976). The agency can take action under the statute when it receives "credible evidence" that an entry represents a "serious threat to human or animal health." Prior notice information will undergo both a validation process and screening in the FDA Operational and Administrative System for Import Support (OASIS) for food safety and security criteria (68 FR 58976).

The FDA is required to provide notice to states regarding imported food within its territory if the food poses a serious health risk under Section 310 of the act.[36]

Food for which inadequate (21 CFR § 1.283(a)(1)(i)), inaccurate (21 CFR § 1.283(a)(1)(ii)), or untimely (21 CFR § 1.283(iii)) prior notice has been provided will be refused entry and considered to be general order merchandise (under the Tariff Act of 1930, 19 U.S.C. 1490; 21 CFR § 1.283(a)(2)) and be segregated from other products (21 CFR § 1.283(a)(3)) and then stored, transported under bond most likely at the port of entry[37] (21 CFR § 1.283(a)(2)(i)), and ultimately reexported (21 CFR § 1.283(a)(5, 6)) or destroyed at the owner's, shipper's, or importer consignee's expense (21 CFR § 1.283). Most likely, the common carrier will be notified of the refusal (68 FR 59017). Prior notice can be corrected and resubmitted under certain circumstances. Even some foods carried by individuals on their person or in their baggage are subject to prior notice (21 CFR § 1.283(b)); this includes quality control, marketing, and test samples.

Bizarre as it may seem, noncommercial shipments of food sent by international mail are also covered (21 CFR § 1.283(e), § 1.285(k)) unless these are gifts that the mailer produced in his or her own home. Notice through the the FDA's Prior Notice System Interface (FDA PN System Interface) must be sent before the food is mailed (68 FR 59024), and if registration information (21 CFR § 1.285) is not provided, the gift may be returned to the sender, but will most likely be destroyed even if the food sent was produced in an FDA-registered facility (68 FR 59024). Food imported from unregistered facilities is refused entry and is treated in the same manner as food imported without adequate prior notice (21 CFR § 1.285). Such facility may register and resubmit a product for import (21 CFR § 1.285(i)(2)).

Food imports are valued at close to $50 billion annually and are increasing both in quantity and in value. Prior notice has already had a major impact on the cost and shipment of foods into the U.S. These new importer prior notice requirements have tight deadlines and require submission of detailed information on product quantity, type, and source prior to arrival of product into a U.S. port, even with a relaxation of the notice requirements between the interim and the earlier proposed rules. Fortunately, either the Bureau of Customs and Border Protection[38] (CBP) Automated Broker Interface (ABI) of the Automated Commercial System (ACS) or the FDA PN System Interface (68 FR 58975) can be used for filings to reduce redundancy and delays. Despite these provisions, there are questions within the regulated community as to whether enough flexibility is being provided to deal with the vagaries of transportation and difficulties with perishable food shipping. Clearly, the FDA does not have the staff to police the 5.2 million to 6.5 million entry lines of food (FY 2002) entering at 250 ports (68 FR 59024, 59027), and it is unlikely that Customs and Border Patrol agents will be given adequate training to inspect food shipments or agricultural products (Monke, 2004). It is also unlikely that the agency will be able to monitor the 25,000+ daily prior notices that will result from these regulations and ensure that necessary inspections can be scheduled and conducted (68 FR 59027), particularly since the scope of commercial food shipments is broad and international mail, including mailing of personal gifts of food, is covered under this prior notice requirement (68 FR 78977, 21 CFR § 1.277).[39]

With prior notice, the purchaser or importer or agent residing in the U.S. must provide notice to the FDA. Prior notice must be received and confirmed by the FDA for review (21 CFR § 1.279). Notification is as follows: no less than 2 hours for food arriving by land by road at the port of arrival[40] (21 CFR § 1.279(a)(1)), no less than 4 hours for food by land by rail (21 CFR § 1.279(a)(2)), no less than 4 hours for food arriving by air (21 CFR § 1.279(a)(3)), and no less than 8 hours for food arriving by water (21 CFR § 1.279(a)(4)), but no more than 5 calendar days (21 CFR § 1.279(b)) prior to arrival to provide the FDA with an opportunity to inspect a shipment.[41] Food, food additive, and drug components[42] are covered.

The identity of *each article* of food (21 CFR § 1.281(b)(5)), not just a general description of the content of the shipment, and its location must be provided (21 CFR § 1.281(a)), including the name and contact information and registration number of the individual submitting the prior notice (21 CFR § 1.281(a)(1)), the name and contact information and registration number for the transmitter if different from those of the submitter (21 CFR § 1.281(a)(2)), shipper[43] (21 CFR § 1.281(a)(9)), entry type (21 CFR §

1.281(a)(3)), CBP entry identifier (21 CFR § 1.281(a)(4)), the complete FDA product code (21 CFR § 1.281(a)(5)(i)), common or usual name or market name (21 CFR § 1.281(a)(5)(ii)), estimated quantity, from smallest package size to largest container (21 CFR § 1.281(a)(5)(iii)), and lot codes or numbers (21 CFR § 1.281(a)(5)(iv)). In addition, for food no longer in its natural state,[44] the manufacturer[45] and its registration number must be provided (21 CFR § 1.281(a)(6)). For food in its natural state, the grower, if known,[46] must be provided (21 CFR § 1.281(a)(7)). For all foods, the following must be provided: the country of production[47,48] (21 CFR § 1.276(b)(4)); country from which the food is shipped (21 CFR § 1.281(a)(10)); anticipated arrival information, including port, date, and time of arrival; name and address of importer; name of owner and ultimate consignee; mode of transportation; and planned shipment information (21 CFR § 1.281(a)(11)). If the same food item is shipped, but each is from a different manufacturer, then two notices are required. Similarly, if product forms are different, e.g., frozen vs. canned, then a separate notice is required for each. Although amendments were permitted under the proposed rule, under the interim prior rule no amendments to the prior notice are permitted. This change is due to the shortened notice requirements between the proposed and interim final rules (21 CFR § 1.279).

There is civil liability (21 CFR § 1.284(b)(1) under 21 U.S.C. 331, 333) and criminal liability (21 CFR § 1.284(b)(2) under 21 U.S.C. § 333), including permissive disbarment for failure to comply with these provisions (21 CFR § 1.284(c) under 21 U.S.C. § 335a).

Although the changes in prior notice provisions between the interim and proposed rules have somewhat lessened barriers to trade, problems remain. Perishable foods such as fresh seafood (catch of the day) and produce where the variety and exact quantity to be shipped will vary experience the greatest impact. The quantity of affected food is astronomical. In 2001 alone, firms imported 22.6 billion pounds of 48 common types of produce. Similarly, shipments by truck, rail, or air are affected more so than those by sea.

Cost of compliance with these new prior notice requirements is high and will provide little additional food security over current inspection practices.[49] The FDA estimates that each prior notice would cost $75 to $110 per notice (68 FR 59027), for an annual cost to the industry of $187.5 million (presuming 6.5 million entries and 2.6 entries per notice at $75 per notice). Additional costs for high volume importers under CBP BRASS (Customs & Border Protection, Border Release Advance Screening and Selectivity), a system specifically established for high-volume repetitive shipments, are estimated at $48 million annually (68 FR 59029).

The FDA estimates that for industry (77,427 importing firms) to learn the requirements of the new rule would require one manager (at $56.74 per hour) and two subordinates (at $25.10 per hour) from each importing firm to attend an 8-hour training session (68 FR 59026) at a projected first year cost of $66 million. Ongoing compliance training costs for new entrants into the industry are predicted at $6.6 million per year. However, this is not a full-cost estimate. The agency estimate presumes that training is conducted locally, with no travel required. Furthermore, costs to cover expenses and labor for the employees while they are at the training are not factored in. The agency also presumes that training will "stick" and that this 8 hours is a full accounting of the effort required for understanding compliance requirements. The additional costs tied to the development of new infrastructure, computer systems, forms, and paperwork are not considered. Also, only the training costs for *importers* are factored in. There is no cost estimate provided by the agency for suppliers, transporters, and carriers who have compliance requirements under this rule as well. There is a perfunctory assessment that these costs will be restricted to 16 hours of an administrative assistant ($25.10 per hour) to coordinate prior notice activities and establish new business practices at an initial industry cost of $31 million (68 FR 59028).

Agency predictions for the cost of lost product are also low, particularly for perishable foods. Most importantly, their costs do not include the cost to cover orders and other transaction costs for impacted customers due to the loss of sales caused from delays of foods held at the border because of inevitable paperwork snafus. Under the proposed rule, estimated losses in value resulting from a 4-hour delay at the border were predicted for Mexico at $16 million for produce and $2 million for seafood, and for Canada at $31 million for seafood and $2 million for produce. Earlier agency estimates for a delay of 12 hours would reduce the value of Mexican products by $110 million ($98 million produce, $12 million fish) and Canadian foods by $200 million (approximately $190 million fish, $10 million produce) annually. In the interim final rule, these costs (for truck shipments) were reduced to $1.8 billion for imported Canadian and Mexican produce and $860 million for aquatic foods (68 FR 59039), presuming that a 2-hour notice would reduce most delays and reduce product losses.

Although CBP personnel will most likely be granted inspection authority under these regulations, still a minimum of two federal agencies will be involved with imported food shipments, and the reporting systems, although coordinated, are far from operating seamlessly (68 FR 59028). It will take years for the kinks to be worked out in this system. FDA prior notice start-up costs were estimated at $13 million. There will be additional government infrastructure costs as the current ABI/ACS is replaced by a new Automated Commercial Entry (ACE) system over the next few years. Besides failing to estimate the full development costs, the agency does not include personnel costs within costs of compliance and fails to factor in the cost of existing border inspection personnel (FDA, CBP, USDA, state agriculture or public health, law enforcement, and port staff), plus the efforts of the newly hired 300 FDA border inspection staff (68 FR 59028), who will be primarily responsible for implementing provisions of this act.

Costs to trade overall were not estimated by the agency, but the impact is clearly, even based upon agency estimates, in the billions of dollars per year.

Administrative Detention[50] (21 CFR Part 1, Section 1.377 *et seq.*, Parts 10, 16; 69 FR 31702–31705)

Under these new regulations, the FDA would have expanded authority to embargo food directly. Prior to this act, if the FDA wanted to keep a food out of the stream of commerce, the agency would request that a company conduct a voluntary recall of the food, have the food seized by the Justice Department upon a federal court order, or request that a state embargo the food under its police power (21 United States Code (U.S.C.) § 334, as amended). The agency claims these new powers are necessary to speed actions and maximize security (69 FR 31660). Time frames for action are concrete, and the necessity for collaboration with other agencies is lessened. This new regulation chips away at concurrent jurisdiction and reduces the historically important role states have had in controlling the safety of products within their borders.

The FDA now has expanded authority to detain food regardless of whether intentional contamination is suspected.[51] Conventional foods, including those jointly regulated by the USDA and FDA, and animal feed products could be detained as well as food packaging material, dietary supplements, bottled water, infant formulas, and alcoholic beverages.

Under these regulations, the FDA can detain a food for up to 20 days after a detention order has been issued (21 CFR § 1.379(a)) if the agency has "credible evidence or information" that the food "presents" a threat of serious adverse health consequences or death to humans or animals (21 CFR § 1.378).[52] Foods can be detained for an additional 10 calendar days if a greater period is required to institute a seizure or injunction action (21 CFR §1.379). A detention order terminates when the FDA terminates it or when the detention period expires (21 CFR § 1.384). Detention orders can be appealed (21 CFR §§ 1.401–1.405).[53]

These provisions are similar to what is already in place for the administrative detention of medical devices that the agency has had some success implementing. However, the analogy between medical devices and foods does not hold. Detaining fresh salmon fillets or lettuce for up to 30 days has a much greater impact on the marketability of the product than, say, detaining hip joints or heart valves for the same period. Foods, for the most part, are perishable, shipped in large volumes from numerous suppliers and places, and have a relatively low per-unit cost. Medical devices, on the other hand, are nonperishable, high in value, and shipped in small volumes from a limited number of very sophisticated suppliers. The optimism expressed by the agency regarding the successful implementation of these regulations is misplaced, as the provisions in new rules will be less workable for foods than medical devices.

The agency can require that the food be marked as detained, removed to a secure facility, and held at the private party's expense (21 CFR § 1.381). Removing the food or detention markings before a food is released (if it is released) is a violation of the act, punishable by fines or imprisonment. The Bioterrorism Act (Section 303[54]) also provides that the agency can temporarily hold a food at a point of entry.

Historically, food seizure actions are challenged 65% of the time, and challenges to detention orders will probably remain the same or possibly increase due to the ambiguity of the legal standard proposed for detention. Fortunately, the hearing process for a product detention is expedited in order to preserve the quality and integrity of the food. However, it is not short enough. For a perishable food,[55] a claimant must file appeal of a detention order within 2 days (21 CFR § 1.383). If a hearing is granted by the FDA, it will be held within 2 days of the appealing being filed. A decision will then be issued within 5 days (total time for the hearing process, 4 to 10 days after detention notice has been received). Although this is a quick administrative process, it still poses serious delays for fresh produce, fluid milk, live fish, and seafood, and will effectively leave the food with no remaining market value.[56,57] The owner, operator, or agent in charge of the place where the food is located has the responsibility of ensuring that the food is segregated and held in a secure facility, and if necessary, under proper temperature and humidity control and other conditions to maintain food safety, wholesomeness, quality, and value (21 CFR § 1.380). The FDA will also be able to issue detention orders to common carriers, complicating the issues of product segregation and further increasing costs (21 CFR § 1.392(b)).

Costs of these regulations are similar for domestic and imported foods and include the costs of segregating, transporting, and storing the affected food, canceling previously scheduled transportation and later rescheduling shipments, covering orders and mitigating damages including loss of product value during the detention period, as well as costs associated with appeals, hearings, and enforcement actions.

The agency predicts that it may initiate up to 223 administrative detention actions per year, with the potential that up to 50% of the detained food would be released at annual costs of up to $50 million (69 FR 31700). The agency estimates that transportation costs associated with these actions will reach $4 million annually; storage costs, $2 million; loss of value costs, $15 million[58]; marking or relabeling costs, $1 million; and appeals costs, $16 million. The potential direct cost to a firm for a single incident could range from $20,000 to $330,000 and is probably no greater than what would otherwise be involved in a product seizure or a voluntary recall.

However, there is actually a greater likelihood that more detention actions will be initiated as the cost to the agency of an administrative detention will be less than that with current enforcement strategies. The costs to the industry will be a different matter. This would certainly be an area for the General Accounting Office to monitor in detail and on a full-cost basis at least initially. The requisite involvement of a state or the Department of Justice in a seizure action has the effect of dampening overly zealous agency action. The fact that 65% of seizure actions are challenged, and that half of these are successful, indicates that the processes in place for detecting an adulterated food are far from foolproof.

One of the first cases in this area occurring immediately after requirements went into effect in December 2003 involved detention of refrigerated, canned ikura (salmon caviar) from Russia, a traditional popular item consumed primarily during Christmas and New Year's celebrations. Many cases of the product were detained at the port of entry under suspicion that they were "decomposed" (note — there was no indication that the product had been intentionally contaminated or that it posed a risk to human health). Subsequent investigations proved that the product was not adulterated and did not pose a risk to human health. It was released, but only after the Christmas holidays.

After this release, a casual discussion of this incident between federal employees resulted in a demand by the DC office that the product once again be detained and inspected, even though there are no provisions in the regulations for a reseizure. Once again, the product was retested and found to be in compliance with federal law and re-released in mid-January, but not until the importer had missed the entire holiday season for which the product was intended and which constitutes over 3/4 of the annual sales volume for this food product. Ikura, even canned, is perishable, and the seizure caused a loss of roughly 1/4 of its total shelf-life.

Marking Articles (Section 308 of the Act[59])

Under the act, the FDA can mark articles refused admission into the U.S. with all expenses born by the owner or consignees of the involved food[60] and drugs.[61] Any food marked "United States: Refused Entry" is considered to be misbranded.[62] There are specific provisions in the act against "port shopping."[63] The law requires the FDA to provide notices to the states where the food is held, or in which a manufacturer, packer, or distributor is located when the agency has credible information or evidence that a food shipment or a portion of it may pose adverse health consequences to humans or animals. The FDA may make grants to states for conducting food inspections or investigations and may also assist states with the costs of taking appropriate action to protect the public health in response to the provisions of this new law, including planning and other types of preparedness.

Permissive Debarment (Section 304 of the Act[64])

This is a provision in the act intended to prevent individuals who have engaged in criminal activity regarding food imports from participating in the industry (permissive debarment). Failure to register a facility, provide proper notice of an import shipment, or maintain and permit access to certain records could lead to criminal sanctions and potentially debarment.

Under the act, the FDA can prohibit certain individuals from importing food (permissive debarment) when the person has been convicted of a felony for conduct relating to food imports; or the person has repeatedly imported or offered for importation adulterated food, and the person knew or should have known that the food was adulterated (proposed amendments to Section 306(b) of the Federal Food, Drug and Cosmetic Act (21 U.S.C. 335a(b))). A felony could consist of intentionally misleading the FDA regarding the contents of a prior notice (e.g., Haggen and Harris, 2003). Similar provisions have not been proposed for purveyors of domestic manufactured product, although the potential safety concerns are the same for both situations.

In addition to the Bioterrorism Act, several other recent federal laws deal with response preparedness by governmental agencies with little emphasis on prevention (Bledsoe and Rasco, 2003). There is virtually no governmental funding for private entities, particularly for instituting preventative measures to improve food security. Yet industry bears by far the larger economic burden of defense against and from the impact of intentional or unintentional contamination incidents. These regulations substantially increase the burden for food security upon the food industry, particularly upon smaller companies, who may not have the resources to comply with these new provisions. Although there are grants and other programs within the bill for providing funds for improving security, analytical capability, employee training, and enforcement and implementation for state and local governmental agencies, there are no similar provisions for the regulated businesses to cover the increased costs these new regulations entail.

Although the risk of intentional contamination of the food supply by international and domestic terrorists is real, the provisions of this legislation and the associated regulations do very little to improve the safety of the food supply against such acts. The FDA states in each preamble to these series of new regulations that the probability of a deliberate contamination incident to the food supply is *low*, but the potential cost is high. Further, the accidental food-borne outbreak incidents cited as justification for the provisions in the proposed rules either are outside FDA jurisdiction or describe situations that could have been adequately addressed by more vigilant compliance with provisions of 21 CFR Part 110 or appropriate HACCP regulations in place at the time of the incident.

As illustrations of the potential risk and impact of intentional contamination, the agency discusses the *Salmonella typhimurium* contamination incident of restaurant salad bars in The Dalles, OR, in 1984 by members of a religious cult (Miller et al., 2001). Criminal investigation of this incident ground to a halt when federal (Centers for Disease Control and Prevention (CDC)) and state experts ignored the experience of local public health officials regarding food-borne outbreak patterns and the overall food safety practices in the community. The experience of local law enforcement and their evidence of the cult's intentional contamination of water a year earlier using the same pathogen were also discounted (Miller et al., 2001). Interestingly, the activity of the affected businesses (restaurants) clearly falls within the *restaurant exception* of these new rules and would not be covered by these new regulations. Similarly, an intentional incident involving contamination of muffins with *Shigella dysentriae* type 2 in a laboratory break room was also outside FDA jurisdiction. The 1988 *Shigella sonnei* contamination of a tofu salad prepared by volunteer food handlers at an outside music festival would also fall within the catering or restaurant exemption of these new rules.

The 1994 *Salmonella enteritidis* outbreak tied to ice cream premix in the midwestern U.S. cited by the FDA as justification for necessity of the new regulations is an incident that could have been prevented by closer monitoring of critical processing and sanitation controls, which were customary practices at that time in the dairy products industry.

Only in the case of *Cyclospora cyatanenis* contamination of imported raspberries in 1996, cited by the agency, could the provisions have possibly prevented an illness. However, the imported product would have had to have "appeared" to be adulterated when arriving at the port of entry, and then been selected for inspection. Registration would have done little to prevent this outbreak, as contamination occurred during cultivation of the crop and foreign farms do not have to register. If the records and traceability provisions had been in place, more cases may have been detected early on, but this is unlikely. At the time of this incident, cyclospora was an emerging organism of public health concern and few labs had the ability to test for it. Therefore, more rapid access to records would not have been helpful.

Interestingly, the 1989 cyanide grape hoax, which is clearly within the scope of agency jurisdiction, was not cited as a justification for the new rules, even though the projected costs to the industry from this incident was in the range of $100 million to $300 million.

In summary, these regulations are not practical and fail to consider the usual and customary business practices already in place in the food industry to protect the integrity of food shipments and the safety of food products in commerce. The proposed regulations are ineffective and inefficient, while posing a substantial burden to the regulated industries and to the governmental agencies tasked with enforcement (Willard, 2003). Regulations must be workable. Ideally, they should also make a difference in our ability to prevent, prepare for, and respond to security threats to the food supply (Mikesell, 2003). These are not. Further, the incremental increase in the level of food security to be achieved under the most optimistic agency projections does not justify the increased costs or the potential damage to U.S. domestic and international trade.

It should also be noted that while the act was intended to apply to both domestic and imported products, enforcement will not be equivalent. Regardless of how intensive the FDA attempts to make its efforts toward imported products, the agency does not have the resources (nor could we as a nation afford to provide it with the requisite resources) to apply the same intensive level of regulatory scrutiny experienced by domestic producers through the network of federal, state, and local authorities.

Further, the bill will have a disproportionate impact upon smaller producers, distributors, transporters, etc., who do not possess the staff, data processing capabilities, and other supporting infrastructure that larger companies have. Indeed, larger companies may even benefit from these new regulations, as these regulations will eliminate smaller competitors who are unable to economically comply with the new regulations.

The greatest negative impact to trade is that provisions of the act are exercised *in general* and would not have to be related to a real or threatened food terrorism incident (Fraser, 2003). In short, FDA's interpretation of the act expands its authority to the greatest extent possible. It is clear that congressional intent within the act was to provide for preventing, detecting, and reacting to *intentional* adulteration of food (Section 302(a,b), Section 303(a)(4)(e)), not just any old run-of-the-mill, garden-variety food contamination incident.

The vague legal standards for removing a food from commerce under the act, such as "credible ... information" and "threat of serious adverse health consequences," are not clearly defined. In a similar vein, imported food can be refused entry if it "appears" to be adulterated. Under current recall regulations, a Class I recall involves "a *reasonable probability* that the use of, or exposure to, a violative product will cause serious adverse health consequences or death" (21 CFR § 7.3(m)(1)), but what will constitute a *threat*? How does the law differentiate between a *threat* and a *hoax*, or will it matter? In trial practice, we understand what meets the evidentiary standards under federal and state rules of evidence for *credible evidence*, but what is *credible information*? Will the first test case be a terrorist contaminating the water and food, reminiscent of the *Salmonella* sp. incident in The Dalles, OR, in 1984, or the cyanide in grapes hoax in 1989? Or will the incident be a political action by an eco- or political terrorist, contaminating fish with an exotic zoonotic agent? Perhaps it will come in the form of an anonymous, fraudulent tip. Hoaxes from an activist group objecting to one of many food industry activities is a likely scenario. Just as likely though, the full force of the federal government will come crashing down on some poor soul just scratching by and unfortunate enough to produce a ready-to-eat product unintentionally contaminated with *Listeria monocytogenes.*

> Law cannot save us from ourselves. Waking up every morning, we have to go out and try to accomplish our goals and resolve disagreements by doing what we think is right. That energy and resourcefulness, not millions of legal cubicles, is what was great about America. Let judgment and personal conviction be important again. There is nothing unusual or frightening about it. Relying on ourselves is not, after all, a new ideology. It's just common sense. (Howard, 1994)

References

Anon. 2003. Registration of food facilities under the Public Health Security and Bioterrorism Preparedness and Response Act of 2002. Registration: Section 305; 68 FR 108, January 29, 2003; 68 FR 5378, February 3, 2003. See also for Prior Notice of Imported Food (Section 307; 68 FR 5428, February 3, 2003); for Administrative Detention of Food (Section 303, 68 FR 25242, May 9, 2003; Final Rule, 69 FR 31660, June 4, 2004); and for Records and Record Retention (Section 306, 68 FR 25188, May 9, 2003).

Bledsoe, GE and Rasco, BA. 2003. Effective food security plans for production agriculture and food processing. *Food Protection Trends* Feb.:130–141.

Buzby, JC, Frenzen, P, and Rasco, B. 2002. Jury decisions and awards in personal injury lawsuits involving foodborne pathogens. *Journal of Consumer Affairs* 36:220–238.

Fraser, L. 2003. Center for Food Safety and Applied Nutrition. Food and Drug Administration satellite broadcast on proposed rules, May 8, 2003. http://www.fda.gov.

Haggen, J and Harris, M. 2003. Teleconference on provisions of the Bioterrorism Bill, June 18, Pullman, WA. http://ext.wsu.edu/nwfpa.

Howard, PK. 1994. *The Death of Common Sense. How Law Is Suffocating America.* Random House, New York, chap. 1, pp. 48, 187.

Miller, J, Engelberg, S, and Broad, W. 2001. "The Attack." In *Germs: Biological Weapons and America's Secret War.* Simon & Schuster, New York, chap. 1, pp. 15–24.

Mikesell, L. 2003. NFPA Views New FDA Bioterrorism Proposals with Cautious Optimism. May 6. www.nfpa-food.org.

Monke, J. 2004. Agroterrorism: Threats and Preparedness. CRS Report for Congress. Order code RL 32521. Aug. 13, 2004. 45 pp.

Willard, T. 2003. Proposed Bioterrorism Regulations "Must Not Unnecessarily Burden the Food Industry or the FDA." April 3. www.nfpa-food.org.

Notes

1. *Food*, under the interim final rules, does not include food contact substances under Section 409(h)(6) of the Food, Drug and Cosmetic Act or pesticides as defined under the Federal Insecticide, Fungicide and Rodenticide Act (98 FR 58898).

   Food is defined as under Section 201(f)(3) of the Food, Drug and Cosmetic Act and includes, but is not limited to, fruits; vegetables; fish; live food animals; dairy products; eggs; raw agricultural commodities for use as food or components of food; animal feed, including pet food and food and feed ingredients; additives, including substances that migrate into food from food packaging and other articles that contact food; dietary supplements and dietary ingredients; infant formula; beverages, including alcoholic beverages and bottled water; bakery goods; snack food; candy; and canned food (68 FR 58908).
2. The bulk of the regulations associated with the Bioterrorism Act were in place by December 12, 2003. Provisions of the act are self-executing (69 FR 31661 (21 CFR Parts 1, 10, and 16; Administrative Detention of Food for Human or Animal Consumption under the Public Health Security and Bioterrorism Preparedness and Response Act of 2002; Final Rule) June 4, 2004). The regulations on registration and prior notice will take effect whether or not final rules have been issued. (See Interim Final Rule: Registration of Food Facilities under the Public Health Security and Bioterrorism Preparedness and Response Act of 2002, 68 FR (No. 197), October 10, 2003, pp. 58893–58974. See also Interim Final Rule: Prior Notice of Imported Food under the Public Health Security and Bioterrorism Preparedness and Response Act of 2002, 68 FR (No. 197), October 10, 2003, pp. 58975–59077.
3. In the interim final rule for registration, the FDA has included instructions for filing forms in English, Spanish, and French, the official languages of the WTO (68 FR 58920). However, forms must still be completed and submitted in English.

Interestingly, the FDA estimates the cost of reading and interpreting the pending regulations in the proposed rule. For example, the records proposed rule (21 CFR Part 1, § 1.32 *et seq.*) is at 3 hours and 18 minutes for a native English speaker (at a rate of $25 to $57 per hour, depending upon whether an administrative worker or manager reads the rule). The agency also estimates that 16% of foreign manufacturers have staff competent enough in English to comply with the provisions of the records regulation, leaving the remaining 84% of the foreign firms in quite a pickle. Projections are that time to simply read the rule would be 8 hours. Understanding what it means is another matter.

4. See, for example, the proposal to establish procedures for the safe processing and importing of fish and fishery product codified at 21 CFR Part 123 *et seq.*, commonly known as the Seafood HACCP Regulation. See generally W. F. Fox. 1992. *Understanding Administrative Law*, 2nd ed. Matthew Bender & Co. Inc., New York, chaps. 1–3. The Office of Management and Budget conducts an economic assessment of agency rules using the following for a Regulatory Impact Analysis:
   1. Insure that agency decisions are based on adequate consideration of the need and the consequences.
   2. Refuse to let the agency take action unless the potential benefit to society outweighs the potential cost to society.
   3. Force selection of regulatory objectives that maximize net benefits to society.
   4. Choose the alternative that involves the lowest net cost to society if there is more than one alternative.
   5. Set regulatory priorities that take into consideration the condition of a particular industry, potential future regulatory action, and the state of the national economy.
5. If congressional intent is clear, then the agency must "effect the unambiguously expressed intent" of Congress. Here the intent of Congress was to control acts involving *intentional contamination* of food. An agency's interpretation of a statute is permissible if it "substantially complies" with the statute through the regulations it promulgates. Courts will uphold an agency's determination unless it is "arbitrary, capricious or manifestly contrary to the statute" (*Chevron v. National Resources Defense Council, Inc.*, 467 U.S. 837, 104 S.Ct. 2778, 81 L. Ed.2d 694 (1984)). As part of the court's determination, it will determine whether the agency examined relevant data, and whether the agency can articulate a satisfactory explanation for its action, including a rational connection between the facts found and the choice the agency made (467 U.S. at 843-844; 104 S.Ct. at 2782).
6. In the June 4, 2004, preamble to the final rule on administrative detention, the FDA "does not change its conclusion that it has the authority to detain food administratively that does not enter interstate commerce" (68 FR 31663). In earlier presentation, for example, the May 7, 2003, Food and Drug Administration, 21 CFR Part 1 [Dockets 02N-0275 and 02N-0277], Proposed

Regulations Implementing Title III of the Public Health Security and Bioterrorism Preparedness and Response Act, satellite downlink public meeting, are statements by FDA attorneys Robert Lake and Leslye Fraser that the law shall be applied to products solely in intrastate commerce. The agency argues that if Congress had specifically intended the provision to apply only to items in *interstate* commerce, it would have incorporated a specific interstate commerce nexus to these new provisions of the Food, Drug and Cosmetic Act (68 FR 58989). Citing *United States v. Lopez* (514 U.S. 549, 567 (1995)) and by reference *Wickard v. Filbert* (317 U.S.C. 111 (1942)), the agency argues that although an individual's contribution to interstate commerce may be trivial (in this case the sale of an individual farmer's wheat), his activities are not enough to remove him from the scope of federal regulation, where his contribution along with many others similarly situated would be far from trivial. In the preamble to Interim Final Rule on Facilities Registration (68 FR 58893 *et seq.*), the agency acknowledges the restrictions to expansion of governmental authority as cited in *United States v. Morrison* (529 U.S. 598,618 (2000)) but argues that limits to federal power must be construed in light of relevant and enduring precedents. This interpretation of the commerce clause is very broad, making any exclusion for businesses engaged in intrastate commerce evaporate. This is unfortunate.

7. The agency has failed to conduct an adequate analysis of the federalism issues tied to the bioterrorism regulations under Executive Order 13132, and wrongfully concludes that the provisions in the registration (68 FR 58959) and prior notice (68 FR 59070) interim final rules do not contain policies that have substantial direct effects on the states, on the relationship between the national government and the states, or on the distribution of power and responsibilities among the various levels of government.
8. FDA intends to define "serious adverse health consequences" in a separate rulemaking (69 FR 31673).
9. See 211 U.S.C. § 334(h)(1)(A). For example, the agency seeks total discretion and disagrees that "clear evidence," such as laboratory analyses to confirm the presence of an adulterant, corroborated before a detention order is made, or affidavits sworn under penalty of perjury, should be required before a food is detained (69 FR 31673), and that the agency should have the flexibility to conduct fact-specific inquiries, which provides them with the maximum interpretive discretion.
10. The agency blew off these and similar comments (see Comment 14, 69 FR 31666) as not being based upon the substantive standard for administrative detention under Section 303 of the act. In this and in other justification for the rules under this act, the agency has failed to consider the overlap between new regulatory provisions and those existing under prior law.
11. Despite legitimate concerns about the cost of these regulations, this is not the agency's concern. There are no provisions in the Food, Drug and Cosmetic Act or the Bioterrorism Act for damages (69 FR 31666).

12. Section 320. Registration of Food Facilities.
    (a) In General — Chapter IV of the Federal Food, Drug and Cosmetic Act (21 U.S.C. 341 *et seq.*) is amended by adding at the end the following:

Section 415. Registration of Food Facilities.

(a) Registration —
  (1) In General — The Secretary shall by regulation require that any facility engaged in manufacturing, processing, packing, or holding food for consumption in the United States be registered with the Secretary. To be registered —
    (A) For a domestic facility, the owner, operator, or agent in charge of the facility shall submit a registration to the Secretary; and
    (B) For a foreign facility, the owner, operator, or agent in charge of the facility shall submit a registration to the Secretary and shall include with the registration the name of the United States agent for the facility.
  (2) Registration — An entity (referred to in this section as the "registrant") shall submit a registration under paragraph (1) to the Secretary containing information necessary to notify the Secretary of the name and address of each facility at which, and all trade names under which, the registrant conducts business and, when determined necessary by the Secretary by guidance, the general food category (as identified under Section 170.3 of Title 21, Code of Federal Regulations) of any food manufactured, processed, packed, or held at such facility. The registrant shall notify the Secretary in a timely manner of changes to such information.
  (3) Procedures — Upon receipt of a completed registration described in paragraph (1), the Secretary shall notify the registrant of the receipt of such registration and assign a registration number to each registered facility.
  (4) List — The Secretary shall compile and maintain an up-to-date list of facilities that are registered under this section. Such list and any registration documents submitted pursuant to this subsection shall not be subject to disclosure under section 552 of title 5 United States Code. Information derived from such list or registration documents shall not be subject to disclosure under section 552 of title 5 United States Code, to the extent that it discloses the identity or location of a specific registered person.

(b) Facility — For purposes of this section:
  (1) The term "facility" includes any factory, warehouse, or establishment (including a factory, warehouse, or establishment of an importer) that manufactures, processes, packs, or holds food. Such term does not include farms; restaurants; other retail food establishments; nonprofit food establishments in which food is prepared for or served directly to the consumer; or fishing vessels (except such vessels en-

gaged in processing as defined in section 123.3(k) of title 21 Code of Federal Regulations).

(2) The term "domestic facility" means a facility located in any of the States or Territories.

(3)(A) The term "foreign facility" means a facility that manufactures, processes, packs, or holds food, but only if food from such facility is exported to the United States without further processing or packaging outside the United States.

(B) A food may not be considered to have undergone further processing or packaging for purposes of subparagraph (A) solely on the basis that labeling was added or that any similar activity of a de minimis nature was carried out with respect to the food.

Section 307. Prior Notice of Imported Food Shipments.

(a) In General — Section 801 of the Federal Food, Drug and Cosmetic Act (21 U.S.C. 381), as amended by section 305(e) of this Act, is amended by adding at the end the following subsection: (m)(1) In the case of an article of food that is being imported or offered for import into the United States, the Secretary, after consultation with the Secretary of the Treasury, shall by regulation require, for the purpose of enabling such article to be inspected at ports of entry into the United States, the submission to the Secretary of a notice providing the identity of each of the following: the article; the manufacturer and shipper of the article; if known within the specified period of time that notice is required to be provided, the grower of the article; country from which the article originates; the country from which the article is shipped; and the anticipated port of entry for the article. An article of food imported or offered for import without submission of such notice in accordance with the requirements under this paragraph shall be refused admission into the United States. Nothing in this section may be construed as a limitation on the port of entry for an article of food.

(2)(A) Regulations under paragraph (1) shall require that a notice under such paragraph be provided by a specified period of time in advance of the time of the importation of the article of food involved or the offering of the food for import, which period shall be no less than the minimum amount of time necessary for the Secretary to receive, review, and appropriately respond to such notification, but may not exceed five days. In determining the specified period of time required under this subparagraph, the Secretary may consider, but is not limited to consideration of, the effect on commerce of such period of time, the locations of various ports of entry into the United States, the various modes of transportation, the types of food imported into the United States, and any other such consideration. Nothing in the preceding sentence may be construed as a limitation on the obligation of the Secretary to receive, review, and appropriately respond to any notice under paragraph (1).

(B) (i) If an article of food being imported or offered for import into the United States and a notice under paragraph (1) is not provided in advance in accordance with the requirements under paragraph (1), such article shall be held at the port of entry for the article, and may not be delivered to the importer, owner, or consignee of the article, until such notice is submitted to the Secretary, and the Secretary examines the notice and determines that the notice is in accordance with the requirements under paragraph (1). Subsection (b) does not authorize the delivery of the article pursuant to the execution of a bond while the article is so held. The article shall be removed to a secure facility as appropriate. During the period of time that such article is so held, the article shall not be transferred by any person from the port of entry into the United States for the article, or from the secure facility to which the article has been removed, as the case may be.
(ii) In carrying out clause (i) with respect to an article of food, the Secretary shall determine whether there is in the possession of the Secretary any credible evidence or information indicating that such article presents a threat of serious adverse health consequences or death to humans or animals.

(3)(A) This subsection may not be construed as limiting the authority of the Secretary to obtain information under any other provision of this Act.

(B) This subsection may not be construed as authorizing the Secretary to impose any requirements with respect to a food to the extent that it is within the exclusive jurisdiction of the Secretary of Agriculture pursuant to the Federal Meat Inspection Act (21 U.S.C. 601 *et seq.*), the Poultry Products Inspection Act (21 U.S.C. 451 *et seq.*), or the Egg Products Inspection Act (21 U.S.C. 1031 *et seq.*).

(b) Prohibited Act — Section 301 of the Federal Food, Drug and Cosmetic Act (21 U.S.C. 331), as amended by section 305(b) of this Act, is amended by adding at the end the following:

(ee) The importing or offering for import into the United States of an article of food in violation of the requirements under section 801(m).

(c) Rulemaking; Effective Date —

(1) In General — Not later than 18 months after the date of the enactment of this Act, the Secretary of Health and Human Services shall promulgate proposed and final regulations for the requirement of providing notice in accordance with section 801(m) of the Federal Food, Drug and Cosmetic Act (as added by subsection (a) of this section). Such requirement of notification takes effect —

(A) upon the effective date of such final regulations; or

(B) upon the expiration of such 18-month period if the final regulations have not been made effective as of the expiration of such period, subject to compliance with the final regulations when the final regulations are made effective.

(2) Default; Minimum Period of Advance Notice — If under paragraph (1) the requirement for providing notice in accordance with section 801(m) of the Federal Food, Drug and Cosmetic Act takes effect without final regulations having been made effective, then for purposes of such requirement, the specified period of time that the notice is required to be made in advance of the time of importation of the article of food involved or the offering of the food for import shall be not fewer than eight hours and not more than five days, which shall remain in effect until the final regulations are made effective.

Section 801 of the Federal Food, Drug and Cosmetic Act (21 U.S.C. 381(a)) as amended by section 306(a) of this Act is amended ...:

(l)(1) If a food has been refused admission under subsection (a) other than such a food that is required to be destroyed, and the Secretary determines that the food presents a threat of serious adverse health consequences or death to humans or animals, the Secretary may require the owner or consignee of the food to affix to the container of the food a label that clearly and conspicuously bears the statement: "UNITED STATES: REFUSED ENTRY."

(2) All expenses in connection with affixing a label under paragraph (1) shall be paid by the owner or consignee of the food involved, and in default of such payment, shall constitute a lien against future importations made by such owner or consignee.

13. (c) Rule of Construction — Nothing in this section shall be construed to authorize the Secretary to require an application, review, or licensing process.

(d) Prohibited Acts — Section 301 of the Federal Food, Drug and Cosmetic Act (21 U.S.C. 331), as amended by section 304(d) of this Act, is amended by adding at the end the following:

(e) The failure to register in accordance with section 415.

(f) Importation; Failure to Register — Section 801 of the Federal Food, Drug and Cosmetic Act, as amended by section 304(e) of this Act, is amended by adding at the end the following subsection:

(l)(1) If an article of food is being imported or offered for import into the United States, and such article is from a foreign facility for which a registration has not been submitted to the Secretary under section 415, such article shall be held at the port of entry for the article, and may not be delivered to the importer, owner, or consignee of the article, until the foreign facility is so registered. Subsection (b) does not authorize the delivery of the article pursuant to the execution of a bond while the article is so held. The article shall be removed to a secure facility, as appropriate.

During the period of time that such article is so held, the article shall not be transferred by any person from the port of entry into the United States for the article, or from the secure facility to which the article has been removed, as the case may be.

(g) Electronic Filing — For the purposes of reducing paperwork and reporting burdens, the Secretary of Health and Human Services may provide for, and encourage the use of, electronic methods of submitting to the Secretary registrations required pursuant to this section. In providing for the electronic submission of such registration, the Secretary shall ensure adequate authentication protocols are used to enable identification of the registrant and validation of the data as appropriate.

(h) Rulemaking; Effective Date — Not later than 18 months after the date of the enactment of this Act, the Secretary of Health and Human Services shall promulgate proposed and final regulations for the requirement of registration under section 415 of the Federal Food, Drug and Cosmetic Act (as added by subsection (a) of this section). Such requirement of registration takes effect —

a. Upon the effective date of such final regulations; or
b. Upon the expiration of such 18-month period if the final regulations have not been made effective as of the expiration of such period, subject to compliance with the final regulations when the final regulations are made effective.

Foreign facility means a facility other than a domestic facility that manufactures/processes, packs, or holds food for consumption in the U.S. (Interim Proposed Rule: 21 CFR Section 1.227(ii)).

14. *Domestic facility* means any facility located in any state or territory of the United States, the District of Columbia, or the Commonwealth of Puerto Rico that manufactures/processes, packs, or holds food for consumption in the United States (Interim Final Rule at 68 FR 58961).
15. *Manufacturing/processing* means making food from one or more ingredients, or synthesizing, preparing, treating, modifying, or manipulating food, including food crops or ingredients. Examples of manufacturing/processing activities are cutting, peeling, rimming, washing, waxing, eviscerating, rendering, cooking, baking, freezing, cooling, pasteurizing, homogenizing, mixing, formulating, bottling, milling, grinding, extracting juice, distilling, labeling, or packaging (21 CFR § 1.277(6)).
16. This provision is identical to the definition of exempt fishing vessels under the Seafood HACCP regulation under 21 CFR Part 123,1240. However, a number of other verbs are included here describing what constitutes processing of seafood products. Fishing operations, including those that not only harvest and transport fish, but also engage in practices such as heading, eviscerating, or freezing intended solely to prepare fish for holding on board a harvest vessel, are exempt. However, any fishing vessels otherwise engaged in processing fish are subject to this subpart. For the purposes of this section, *processing* means handling, storing, preparing, shucking, changing into dif-

ferent market forms, manufacturing, preserving, packing, labeling, dockside unloading, holding or heading, eviscerating, or freezing other than solely to prepare fish for holding on board a harvest vessel (21 CFR § 1.226(f)).

17. *Facility* means any establishment, structure, or structures under one ownership at one physical location or, in the case of a mobile facility, traveling to multiple locations that manufactures/processes, packs, or holds food for consumption in the U.S. Transport vehicles are not facilities if they hold food only in the usual course of business as carriers (21 CFR § 1.227(2)).
18. The definition of *farm* is as follows: a facility in one general physical location devoted to the growing and harvesting of crops, the raising of animals (including seafood), or both. Washing, trimming of outer leaves of, and cooling produce are considered part of harvesting (see 21 CFR § 1.227(3); at 68 FR 58895). However, some odious provisions remain. For example, a farm includes facilities that pack or hold food provided that all the food used in such activities is grown, raised, or consumed on that farm or another farm under the same ownership. Id. at (i). What this means then is if holding facilities (such as silos) on a farm are under different ownership (e.g., a coop or food processing company), then the farm or the holding facility would have to register. Furthermore, if a farmer leases storage to another, or stores the crop of another on his land, then that farming operation would have to be registered.
19. A *retail establishment* is defined under the interim final rule as an establishment that sells food products directly to consumers as its primary function (21 CFR § 1.227 (11)). A retail establishment may manufacture, process, pack, or hold food if the establishment's primary function is to sell from that establishment food that it manufactures, processes, packs, or holds directly to consumers. A *retail food establishment's* primary function is to sell food directly to consumers if the annual monetary value of sales of food products directly to the consumers exceeds the annual monetary value of sales of food products to all other buyers. The term *consumer* does not include businesses. A retail food establishment includes grocery stores, convenience stores, and vending machine locations (68 FR 58896). Therefore, club stores or food warehouses that sell to other businesses may or may not be exempt. It is not clear from the regulation whether an operation can "cancel" a registration for an upcoming year based upon a shift in the nature of its sales from wholesale to consumer purchases.
20. *Commissaries or central kitchens* that provide food to restaurants are not exempt. *Restaurant* is limited to establishments that prepare and sell food directly to consumers for immediate consumption (21 CFR § 1.227(10); 68 FR 58913).
21. A *nonprofit food establishment* is a charitable entity that prepares or serves food directly to the consumer or otherwise provides food or meals for consumption by humans or animals in the U.S. (21 CFR Section 1.227(7); 68 FR 58902). It includes central food banks, soup kitchens, and nonprofit food

delivery programs. Any such program must meet the requirements for a charity under Section 501(c)(3) of the U.S. Internal Revenue Code (26 U.S.C. 501(c)(3)).

22. Executive Order 12866 requires all agencies to assess *all costs* and benefits of available regulatory alternatives and, when regulation is necessary, to select regulatory approaches that maximize net benefits (including economic, environmental, public health and safety, and other advantages; distributive impacts and equity) (68 FR 58931). Economic analysis (PRIA) for the proposed rule is cited at 68 FR 5387–5413.
23. Section 306. Maintenance and Inspection of Records for Foods.
    (a) In General — Chapter IV of the Federal Food, Drug and Cosmetic Act, as amended by section 305 of this Act, is amended by inserting before section 415 the following section:

    Section 414. Maintenance and Inspection of Records.

    (a) Records Inspection — If the Secretary has a reasonable belief that an article of food is adulterated and presents a threat of serious adverse health consequences or death to humans or animals, each person (excluding farms and restaurants) who manufactures, processes, packs, distributes, receives, holds, or imports such article shall, at the request of an officer of employee duly designated by the Secretary, permit such officer or employee, upon presentation of appropriate credentials and a written notice to such person, at reasonable times and within reasonable limits and in a reasonable manner, to have access to and copy all records relating to such article that are needed to assist the Secretary in determining whether the food is adulterated and presents a threat of serious adverse health consequences or death to humans or animals. The requirement under the preceding sentence applies to all records relating to the manufacture, processing, packing, distribution, receipt, holding, or importation of such article maintained by or on behalf of such person in any format (including paper and electronic format) and at any location.

    (b) Regulations Concerning Recordkeeping — The Secretary, in consultation and coordination, as appropriate with other Federal departments and agencies, with responsibilities for regulating food safety, may by regulation establish requirements regarding the establishment and maintenance, for not longer than two years, of records by persons (excluding farms and restaurants) who manufacture, process, pack, transport, distribute, receive, hold, or import food, which records are needed by the Secretary for inspection to allow the Secretary to identify the immediate previous sources and the immediate subsequent recipients of food, including its packaging, in order to address credible threats of serious adverse health consequences or death to humans or animals. The Secretary shall take into account the size of a business in promulgating regulations under this section.

(c) Protection of Sensitive Information — The Secretary shall take appropriate measures to ensure that there are in effect effective procedures to prevent the unauthorized disclosure of any trade secret or confidential information that is obtained by the Secretary pursuant this section.

(d) Limitation — This section shall not be construed
  (1) to limit the authority of the Secretary to inspect records or to require establishment and maintenance of records under any other provision of this Act;
  (2) to authorize the Secretary to impose any requirements with respect to a food to the extent that it is within the exclusive jurisdiction of the Secretary of Agriculture pursuant to the Federal Meat Inspection Act (21 U.S.C. 601 *et seq.*), The Poultry Products Inspection Act (21 U.S.C. § 452 *et seq.*), or the Egg Products Inspection Act (21 U.S.C. 1031 *et seq.*)
  (3) to have any legal effect on Section 552 of Title 5 United States Code, or Section 1905 of Title 18 United States Code; or
  (4) to extend to recipes for food, financial data, pricing data, personnel data, research data, or sales data (other than shipment data regarding sales).

(e) Factory Inspection — Section 704(a) of the Federal Food, Drug and Cosmetic Act (21 U.S.C. 374(a)) is amended
  (1) in paragraph (a), by inserting after the first sentence the following new sentence: "In the case of any person (excluding farms and restaurants) who manufactures, processes, packs, transports, distributes, holds, or imports foods, the inspection shall extend to all records and other information described in Section 414 when the Secretary has a reasonable belief that an article of food is adulterated and presents a threat or serious adverse health consequences or death to humans or animals, subject to the limitation established in section 414(d)"; and
  (2) in paragraph (2), in the matter preceding subparagraph (A), by striking "second sentence" and inserting "third sentence."
    (i) Prohibited Act — Section 301 of the Federal Food, Drug and Cosmetic Act (21 U.S.C. § 331) is amended —
  (1) in paragraph (e) —
    (A) by striking "by section 412, 504, or 703" and inserting "by section 412, 414, 504, 703, or 704 (a)"; and
    (B) by striking "under section 412" and inserting "under section 412, 414(b)"; and
  (2) in paragraph (j), by inserting "414" after "412."

(f) Expedited Rulemaking — Not later than 18 months after the date of the enactment of this Act, the Secretary shall promulgate proposed and final regulations establishing recordkeeping requirements under subsection 414(b) of the Federal Food, Drug and Cosmetic Act (as added by subsection (a)).

24. Protected records would include recipes, financial data, pricing data, personnel data, research data, or sales data other than shipping data.
25. Foreign companies "further manufacturing/processing," including *de minimis* activities such as labeling an outer package or packaging individual units for sale (e.g., placing plastic rings around a set of beverage bottles), would have to maintain records. In addition, the facility immediately prior to the packager/labeler would be required to both register and maintain records.
26. Retailers have a limited exclusion when food is sold directly to a consumer.
27. Firms exporting from the EU are already subject to similar recordkeeping requirements under EU regulation (178/2002. Article 18: Traceability).
28. Essentially all (85 to 95%) food manufacturers, wholesalers, warehouses, packagers, transporters, packers, grocery and other retail, and convenience stores have less than 20 employees. Most of these businesses (73 to 90%) have 10 or fewer employees. No data were available for mixed-type farm operations or importers.
29. A transporter is not an immediate previous source or immediate subsequent recipient.
30. Although the FDA is unable to quantify the benefits of this regulation, it considers them to be substantial.
31. Identity preservation is already in place for organic/non-GMO foods, kosher foods, and some specialty bulk products. The cost of identity preservation is high and consists of the cost to segregate crops to prevent commingling and the cost of tracking individual ingredients. Segregating and handling corn is estimated at $0.17 per BU, and for soy, $0.48 per BU, or roughly a 10% premium. The total cost to eliminate commingling of grains, milk, fruits, and vegetables, among other products, is $87.9 billion per year, with premiums for different products ranging from 5.0 to 10.5%, for a total cost of $6.4 billion annually.
32. Sec 307. Prior Notice of Imported Food Shipments.
    (a) In General – Section 801 of the Federal Food, Drug and Cosmetic Act (21 U.S.C. §381), as amended by section 305(e) of this act, is amended by adding at the end the following subsection:
    (m)(1) In the case of an article of food that is being imported or offered for import into the United States, the Secretary, after consultation with the Secretary of the Treasury, shall by regulation require, for the purpose of enabling such article to be inspected at ports of entry into the United States, the submission to the Secretary of a notice providing the identity of each of the following: the article; the manufacturer and shipper of the article; if known within the specified period of time that notice is required to be provided, the grower of the article; country from which the article originates; the country from which the article is shipped; and the anticipated port of entry for the article. An article of food imported or offered for import without

submission of such notice in accordance with the requirements under this paragraph shall be refused admission into the United States. Nothing in this section may be construed as a limitation on the port of entry for an article of food.

(2) (A) Regulations under paragraph (1) shall require that a notice under such paragraph be provided by a specified period of time in advance of the time of the importation of the article of food involved or the offering of the food for import, which period shall be no less than the minimum amount of time necessary for the Secretary to receive, review and appropriately respond to such notification, but may not exceed five days in determining the specified period of time required under this subparagraph, the Secretary may consider, but is not limited to consideration of, the effect on commerce of such period of time, the locations of various ports of entry into the United States, the various modes of transportation the types of food imported into the United States, and any other such consideration. Nothing in the preceding sentence may be construed a limitation on the obligation of the Secretary to receive, review and appropriately respond to any notice under paragraph (1).

(B)(i) If an article of food being imported or offered for import into the United States and a notice under paragraph (1) is not provided in advance in accordance with the requirements under paragraph (1), such article shall be held at the port of entry for the article, and may not be delivered to the importer, owner or consignee of the article, until such notice is submitted to the Secretary, and the Secretary examines the notice and determines that that notice is in accordance with the requirements under paragraph (1). Subsection (b) does not authorize the delivery of the article pursuant to the execution of a bond while the article is so held. The article shall be removed to a secure facility as appropriate. During the period of time that such article is so held, the article shall not be transferred to any person from the port of entry into the United States for the article, or from the secure facility to which the article has been removed, as the case may be.

(ii) In carrying out clause (i) with respect to an article of food, the Secretary shall determine whether there is in the possession of the Secretary any credible evidence or information indicating that such article presents a threat of serious adverse health consequences or death to humans or animals.

(3) (A) This subsection may not be construed as authorizing the Secretary to impose any requirements with respect to a food to the extent that it is within the exclusive jurisdiction of the Secretary of Agriculture pursuant to the Federal Meat Inspection Act (21 U.S.C. §601 *et seq*), the Poultry Products Inspection Act (21 U.S.C. §451 *et seq*) or the Egg Products Inspection Act (21 U.S.C. §1031 *et seq*).

(b) Prohibited Act – Section 301 of the Federal Food, Drug and Cosmetic Act (21 U.S.C.§ 331), as amended by section 305(b) of this Act, is amended by adding at the end the following:
(ee)The importing or offering for import into the United States of an article of food in violation of the requirements under section 801(m).
(c) Rulemaking – Effective Date
(1) In General – Not later than 18 months after the date of the enactment of this Act, the Secretary of Health and Human Services shall promulgate proposed and final regulations for the requirement of providing prior notice in accordance with section 801(m) of the Federal Food, Drug and Cosmetic Act (as added by subsection (a) of this section). Such requirement of notification takes effect –
(A) upon the effective date of such final regulations; or
(B) upon the expiration of such 18 month period if the final regulations have not been made effective as of the expiration of such period, subject to compliance with the final regulations when the final regulations are made effective.
(2) Default; Minimum Period of Advance Notice – If under paragraph (1) the requirement for providing notice in accordance with section 801(m) of the Federal, Food, Drug and Cosmetic Act takes effect without final regulations having been made effective, then for the purposes of such requirement, the specified period of time that the notice is required to be made in advance of the time of importation of the article of food involved or the offering of the food for import shall not be fewer than eight hours and not more than five days, which shall remain in effect until the final regulations are made effective.

33. PROHIBITED ACT — Section 301 of the Federal Food, Drug and Cosmetic Act, as amended by section 305(b)(1) of this Act is amended by adding at the end the following: "(i.e.) The importing or offering for import into the United States of an article of food in violation of regulations under section 801(k)."
34. This section may not be construed as authorizing the Secretary of Health and Human Services to impose any requirements with respect to a food to the extent that it is within the exclusive jurisdiction of the Secretary of Agriculture pursuant to the Federal Meat Inspection Act (21 U.S.C § 601 *et seq.*), the Poultry Products Inspection Act (21 U.S.C. § 451 *et seq.*), or the Egg Products Inspection Act (21 U.S.C. § 1031 *et seq.*).
35. Section 310. Notice to States Regarding Imported Food.

    Chapter IX of the Federal Food, Drug and Cosmetic Act (21 U.S.C. 391 *et seq.*) is amended by adding at the end the following section:

    Section 908. Notices to States Regarding Imported Food.

    (a) In General — If the Secretary has credible evidence or information indicating that a shipment of imported food or portion thereof presents a threat of serious adverse consequences or death to humans or animals, the Secretary shall provide notice regarding such threat to the States in which the food is held or will be held, and to the States in which the manufacturer, packer, or distributor of the food is located, to the extent that the Secretary has knowledge of which States are so involved. In providing notice to a State, the Secretary shall request the State to take such action as the State considers appropriate, if any, to protect the public health regarding the food involved.

(b) Rules of Construction — Subsection (a) may not be construed as limiting the authority of the Secretary with respect to food under any other provision of this Act.

36. As U.S. Customs bonded warehouses may not be equipped to handle perishable goods, new complications may arise.
37. The Bureau of Customs and Border Protection was created on March 1, 2003, as part of the Department of Homeland Security and includes all 18,000 customs, immigration, and agriculture inspectors at more than 300 ports of entry into the U.S. (Anon. 2003. P03-44, FDA and Bureau of Customs and Border Protection announce steps to streamline collection of information on food imports. *FDA News*, May 27. www.fda.gov).
38. Section 1.277. What is the scope of this subpart?

    (a) This subpart applies to all food for humans and other animals that is imported or offered for import into the United States for use, storage or distribution in the United States, including food for gifts and trade and quality assurance/quality control samples, food for transshipment through the United States to another country, food for future export, and food for use in a U.S. Foreign Trade Zone.

    Excluded products are: (1) food for individual personal use carried or accompanying an individual entering the United States; (2) food made by an individual in a personal residence and sent as a personal gift; (3) food imported and then exported without leaving the port of arrival; (4) meat; (5) poultry; and (6) egg products under exclusive jurisdiction of the USDA (21 CFR § 1.277(b)(1–6)).
39. *Port of arrival* means the water, air, or land port at which the article of food is imported or offered for import into the U.S., i.e., the port where the article of food first arrives in the U.S. This port may be different from the port of consumption or warehouse entry or foreign trade zone. Admission documentation is presented to the U.S. Bureau of Customs and Border Protection (CBP) (21 CFR § 1.276(9)).
40. Notice of a predicted arrival time may not be sent more than 5 days in advance for international mail (21 CFR § 1.279(b)). The addition of prior notice for international mail is a new addition in the interim prior rule compared to the prior rule. This provision covers shipments of gifts, if the food has not been made in a home kitchen (68 FR 58977).
41. An importer of a food additive, color additive, or dietary supplement that will be used as a component in a processed food must provide a statement

that: (1) the article is to be further processed by the initial owner or consignee, (2) identifies the manufacturer, and each processor, packer, distributor, carrier, or other entity that had possession of the article in the chain of possession of the article from the manufacturer to the importer, and (3) the imported material will be used as per its original intent (e.g., as a food component).

Certificates of analysis may be required: (4) the importer must maintain records of the use of the material or its destruction and be able to submit these records to the FDA upon request, and (5) the importer must be able to account for any exports of the material. A bond must be posted that is sufficient to cover liquidated damages of the Department of Treasury in event of a default by the importer or consignee. The FDA will have authority to demand any records associated with food imports and to review and copy these records (proposed amendments to 21 U.S.C. § 341 *et seq.*). The agency claims that food recipes, financial data, pricing data, personnel data, research data, or sales data (other than shipping data regarding sales) will be protected as sensitive information.

42. A *shipper* is the owner or exporter of an article of food who consigns and ships the article from a foreign country or the person who sends an article of food by international mail to the U.S. (21 CFR § 1.276(b)(12)).
43. Food *no longer in its natural state* means an article of food that has been made from one or more ingredients or synthesized, prepared, treated, modified, or manipulated. Examples of activities that render food no longer in its natural state are cutting, peeling, trimming, washing, waxing, eviscerating, rendering, cooking, baking, freezing, cooling, pasteurizing, homogenizing, mixing, formulating, bottling, milling, grinding, extracting juice, distilling, labeling, or packaging. Crops that have been cleaned (e.g., dusted, washed), trimmed, or cooled attendant to harvest or collection or treated against pests, waxed, or polished are still in their natural state for purposes of this subpart. Whole fish, eviscerated, or frozen attendant to harvest are still in their natural state for purposes of this subpart (21 CFR § 1.276(8)).
44. A registration number is not required if the food is to be transshipped, stored, and then exported, or further manipulated and then exported. For personal gifts, the name and address of the firm may be provided (21 CFR § 1.281(a)(7)).
45. Under the proposed rule in Section 1.288(k)(1)(b) regarding CHANGES to a prior notice, submission specifically requires that the identity of the grower, including name, address, phone number, fax number, and e-mail of all growers and growing locations (if different from business address), be provided. There is some confusion in the regulation as to how this provision relates to earlier language in the regulation that requires that grower information be provided "if known." Clearly, a greater degree of traceability of the grower will be required as time goes on.

    Under the interim final rule (68 FR 59071) a *grower* is a person who engages in growing and harvesting or collecting crops (including botanical), raising animals (including fish and other aquatic animals or plants), or both (21 CFR § 1.276(b)(6)).

46. The *FDA country of production* means:
    (i) For an article of food that is in its natural state, the country where the article of food was grown, including harvested or collected and readied for shipment to the United States. If an article of food is wild fish, including seafood that was caught or harvested outside the waters of the United States by a vessel that is not registered in the United States, the FDA country of production is the country where the vessel is registered. If an article of food that is in its natural state was grown, including harvested or collected and readied for shipment, in a territory, the FDA country of production is the United States.
    (ii) For an article of food that is no longer in its natural state, the country where the article was made; except that, if an article of food is made from wild fish, including seafood, aboard a vessel, the FDA country of production is the country in which the vessel is registered. If an article of food that is no longer in its natural state was made in a territory, the FDA country of production is the United States (21 CFR § 1.276(4)).
47. Thankfully, between the proposed rule and the interim final rule, the agency clarified issues surrounding the originating country, dropping the term entirely and substituting it with *country of production*. This has reduced confusion with the term *country of origin*, which is covered by a completely different set of regulations. In the proposed rule, the originating country may not have been the same as the country of origin traditionally used by U.S. Customs. For example, for cultured shrimp from Thailand, processed and packed in China, the *originating country* is Thailand, where the product was grown, but the *country of origin* is China, because this is where the article of food was produced.
48. Ironically, the costs of options tied to the timing of prior notice are in different computer formats. The tabular data outlining FDA's selected option (option 6) (68 FR 59044 *et seq.*) are a TIFF file, and not downloadable from their website without appropriate software. Data for other options transfer through as simple text files. Similar problems are experienced with the retrieval of cost information for the registration interim final rule (68 FR 58947 *et seq.*).
49. Section 303. Administrative Detention.
    (a) Expanded authority — Section 304 of the Federal Food, Drug and Cosmetic Act (21 U.S.C. 334) is amended by adding at the end of the following subsection:
    (h) Administrative Detention of Food —
        (1) Detention Authority —
            (A) In General — An officer or qualified employee of the Food and Drug Administration may order the detention, in accordance with this subsection, of any article of food that is found during an inspection, examination, or investigation under this Act con-

ducted by such officer or qualified employee, if an officer or article or qualified employee has credible evidence or information indicating that such article presents a threat of serious adverse health consequences or death to humans or animals.

(B) Secretary Approval — An article of food may be ordered detained under subparagraph (A) only if the Secretary or an official designated by the Secretary approves the order. An official may not be designated unless the official is the director of the district under this Act in which the article involved is located, or is an official senior to such director.

(2) Period of Detention — An article of food may be detained, not to exceed 20 days, unless a greater period, not to exceed 30 days, is necessary, to enable the Secretary to institute an action under subsection (a) or section 302. The Secretary shall by regulations provide for procedures for instituting such action on an expedited basis with respect to perishable foods.

(3) Security of Detained Articles — An order under paragraph (1) with respect to an article of food may require that such article be labeled or marked as detained, and shall require that the article be removed to a secure facility, as appropriate. An article subject to such an order shall not be transferred by any person from the place at which the article is ordered detained, or from the place to which the article is so removed, as the case may be, until released by the Secretary or until the expiration of the detention order period applicable under such order, whichever occurs first. This subsection may not be construed as authorizing the delivery of the article pursuant to the execution of a bond while the article is subject to the order and section 8021(b) does not authorize the delivery of the article pursuant to the execution of a bond while the article is subject to the order.

(4) Appeal of Detention Order —

(A) In General — With respect to an article of food ordered detained under paragraph (1), any person who would be entitled to be a claimant for such article if the article were seized under subsection (a) may appeal the order to the Secretary. Within 5 days after such an appeal is filed, the Secretary, after providing opportunity for an informal hearing, shall confirm or terminate the order involved, and such confirmation by the Secretary shall be considered a final agency action for purposes of section 702 of title 5 United States Code. If during such 5-day period the Secretary fails to provide such an opportunity, or to confirm or terminate such order, the order is deemed to be terminated.

(B) Effect of Instituting Court Action – The process under subparagraph (A) for the appeal of an order under paragraph (1) terminates if the Secretary institutes an action under subsection (a) or section 320 regarding the article of food involved.

(b) Prohibited Act — Section 301 of the Federal Food, Drug and Cosmetic Act (21 U.S.C. 331) is amended by adding at the end the following:
  (bb) The transfer of an article of food in violation of an order under section 304(h) or the removal or alteration of any mark or label required by the order to identify the article as detained.

(c) Temporary Holds at Ports of Entry — Section 801 of the Federal Food, Drug and Cosmetic Act as amended by section 302(d) of this Act, is amended by adding at the end the following:
  (j) (1) If an officer or qualified employee of the Food and Drug Administration has credible evidence or information indicating that an article of food presents a threat of serious adverse health consequences or death to humans or animals, and such officer or qualified employee is unable to inspect, examine, or investigate such article upon the article being offered for import at a port of entry into the United States, the officer or qualified employee shall request the Secretary of the Treasury to hold the food at the port of entry for a reasonable period of time, not to exceed 24 hours, for the purpose of enabling the Secretary to inspect, examine, or investigate the article as appropriate.
  (2) The Secretary shall request the Secretary of the Treasury to remove an article held pursuant to paragraph (1) to a secure facility as appropriate. During the period of time that such article is so held, the article shall not be transferred by any person from the port of entry into the United States for the article has been removed, as the case may be. Subsection (b) does not authorize the delivery of the article pursuant to the execution of a bond while the article is so held.
  (3) An officer or qualified employee of the Food and Drug Administration may make a request under paragraph (1) only if the Secretary or an official designated by the Secretary approves the request. An official may not be so designated unless the official is the director of the district under this Act in which the article involved is located, or is an official senior to such director.
  (4) With respect to an article of food for which a request under paragraph (1) is made, the Secretary, promptly after the request is made, shall notify the State in which the port of entry involved is located that the request has been made, and as applicable, that such article is being held under this subsection.

50. There is also no requirement that a food producer be negligent, only that the food product "appears" to the FDA to be adulterated or misbranded (see 21 U.S.C. § 801).
51. PROHIBITED ACT — Section 301 of the Federal Food, Drug and Cosmetic Act (21 U.S.C. § 331) is amended by adding at the end the following: "(bb) The transfer of an article of food in violation of an order under section 304(h) or the removal or alteration of any mark or label required by the order to identify the article as detained."

52. Confirmation of a detention order by an FDA Regional Food and Drug Director or another FDA official senior to an FDA District Director (21 CFR § 1.404) is final agency action (21 CFR § 1.405(f)) and becomes subject to judicial review at this point (21 CFR § 10.45).
53. The FDA plans to issue a guidance on temporary hold provisions at a later date. Therefore, these are not part of this regulation.
54. Perishable food is defined as a food that is not heat treated, not frozen, and not otherwise preserved in a manner so as to prevent the quality of the food from being adversely affected if held longer than 7 calendar days under normal shipping and storage conditions (21 CFR § 1.377).
55. As part of the detention procedure for perishable food, the FDA sends a seizure recommendation to the Department of Justice within 4 calendar days of detention order (21 CFR § 1383).
56. Time for appeal of a detention order for nonperishable food is longer, with notice of intent to appeal and request a hearing filed within 4 calendar days; appeals must be filed within 10 days, with a hearing scheduled within 3 days of that. The FDA's decision is then issued 5 days later.
57. These figures are based upon an estimated 1 million pounds of food at $0.73 per pound. For highly perishable foods, particularly items such as seafood and live fish, this value is unrealistically low. Live fish have a wholesale value of $5 per pound on average. Any administrative detention of live fish would be a total loss.
58. Section 308. Authority to Mark Articles Refused Admission into the United States.

    (a) In General — Section 801 of the Federal Food, Drug and Cosmetic Act (21 U.S.C. § 381(a)), as amended by section 307(a) of this Act, is amended by adding at the end the following:

    (n)(1) If a food has been refused admission under subsection (a) other than such a food that is required to be destroyed, the Secretary may require the owner or consignee of the food to affix to the container of the food a label that clearly and conspicuously bears the statement: "UNITED STATES: REFUSED ENTRY."

    (2) All expenses in connection with affixing a label under paragraph (1) shall be paid by the owner or consignee of the food involved, and in default of such payment, shall constitute a lien against future importations made by such owner or consignee.

    (3) A requirement under paragraph (1) remains in effect until the Secretary determines that the food involved has been brought into compliance with this Act.

    (b) Misbranded Foods — Section 403 of the Federal Food, Drug and Cosmetic Act (2 U.S.C. § 343) is amended by adding at the end the following:

    (v) If

    (1) it fails to bear a label required by the Secretary under section 801 (n)(1) (relating to food refused admission into the United States);

(2) the Secretary finds that the food presents a threat of serious adverse health consequences to humans or animals; and

(3) upon or after notifying the owner or consignee involved that the label is required under section 810, the Secretary informs the owner or consignee that the food presents such a threat.

(c) Rules of Construction — With respect to articles of food that are imported or offered for import into the United States, nothing in this section shall be construed to limit the authority of the Secretary of Health and Human Services or the Secretary of the Treasury to require the marking of refused articles of food under any other provisions of law.

59. (3) A requirement under paragraph (1) remains in effect until the Secretary determines that the food involved has been brought into compliance with the Act.

Provisions for drugs include annual registration of foreign manufacturers, shipping, and drug and device listing.

60. (b) MISBRANDED FOODS — Section 403 of the Federal Food, Drug and Cosmetic Act (21 U.S.C. § 343) as amended by section 305(b)(2) of this Act, is amended by adding at the end the following: "(u) If it fails to bear a label required by the Secretary under section 801(l)(1) (relating to food refused admission into the United States)."

61. Section 402 of the Federal Food, Drug and Cosmetic Act (21 U.S.C. § 342) is amended by adding at the end the following: "(h) If it is an article of food imported or offered for import into the United States and such article has previously been refused admission under section 801(a), unless the person reoffering the article affirmatively establishes, at the expense of the owner or consignee of the article, that the article is not adulterated, as determined by the Secretary." Temporary holds at a port of entry of 24 hours are permissible to allow for the agency to inspect the food. The FDA has the authority to detain the food, have it held in a secure facility, and contact the state regarding the detention.

62. Section 304. Debarment for Repeated or Serious Food Import Violations.

(a) Debarment Authority —

(1) Permissive Debarment — Section 306(b)(1) of the Federal Food, Drug and Cosmetic Act (21 U.S.C. § 335a(b)(1)) is amended —

(A) in subparagraph (A) by striking "or" after the comma at the end;

(B) in subparagraph (B) by striking the period at the end and inserting ", or"; and

(C) by adding at the end the following subparagraph: "(C) a person from importing an article of food or importing such an article for import into the United States."

(2) Amendment Regarding Debarment Grounds — Section 306(b) of the Federal Food, Drug and Cosmetic Act (21 U.S.C. § 335a(b)) is amended —

(A) in paragraph (2), in the matter preceding subparagraph (A), by inserting "subparagraph (A) or (B) of" before paragraph (1);

(B) by redesignating paragraph (3) as paragraph (4); and

(C) by inserting after paragraph (2) the following paragraph:
(3) Persons Subject to Permissive Debarment; Food Importation — A person is subject to debarment under paragraph (1)(C) if —
(A) the person has been convicted of felony for conduct relating to the importation in the United States of any food; or
(B) the person has engaged in a pattern of importing, or offering for import, adulterated food that presents a threat of serious adverse health consequences or death to humans or animals.
(b) Conforming Amendments — Section 306 of the Federal Food, Drug and Cosmetic Act (21 U.S.C. § 335a) is amended —
(1) in subsection (a) in the heading for the subsection, by striking "Mandatory Debarment —" and inserting "Mandatory Debarment; Certain Drug Applications —";
(2) in subsection (b) —
(A) in the heading for the subsection, by striking "Permissive Debarment —" and inserting "Permissive Debarment; Certain Drug Applications; Food Imports —"; and
(B) in paragraph (2), in the heading for the paragraph, by striking "Permissive Debarment —" and inserting "Permissive Debarment; Certain Drug Applications —";
(3) in subsection (c)(2)(A)(iii), by striking "subsection (b)(2)" and inserting "paragraph (2)(A) or (3) of subsection (b)";
(4) in subsection (d)(3) —
(A) in subparagraph (A)(i), by striking "or (b)(2)(A)" and inserting "or paragraph (2)(A) or (3) of subsection (b)";
(B) in subparagraph (A)(ii)(II), by inserting "in applicable cases" before "sufficient audits";
(C) in subparagraph (b) in each of clauses (i) and (ii), by inserting "or subsection (b)(3)" after "subsection (b)(2)(B)"; and
(D) in subparagraph (B)(ii), by inserting before the period the following: "or the food importation process, as the case may be."
(c) Effective Dates — Section 306(l)(2) of the Federal Food, Drug and Cosmetic Act (21 U.S.C. § 335a(l)(2)) is amended —
(1) in the first sentence —
(A) by striking "and" after "subsection (b)(2)"; and
(B) by inserting ", and subsection (b)(3)(A)" after "subsection (B)(2)(B)"; and (2) in the second sentence by inserting ", subsection (b)(3)(B)" after "subsection (b)(2)(B)."
(d) Prohibited Acts — Section 301 of the Federal Food, Drug and Cosmetic Act, as amended by adding at the end the following:
(cc) The importing or offering for import into the United States of an article of food by, with the assistance of, or at the direction of a person debarred under section 303(b)(3).
(e) Importation by Debarred Persons — Section 801 of the Federal Food, Drug and Cosmetic Act, as amended by section 303(c) of this Act, is amended by adding at the end the following subsection:

(k)(1) If an article of food is being imported or offered for import into the United States, and the importer, owner, or consignee of the article is a person who has been debarred under section 306(b)(3), such article shall be held at the port of entry for the article, and may not be delivered to such person. Subsection (b) does not authorize the delivery of the article pursuant to the execution of a bond while the article is so held. The article shall be removed to a secure facility, as appropriate. During the period of time that such article is so held, the article shall not be transferred by any person from the port of entry into the United States for the article, or from the secure facility to which the article has been removed, as the case may be.

(2) An article of food held under paragraph (1) may be delivered to a person who is not a debarred person under section 306(b)(3) if such person affirmatively establishes, at the expense of the person, that the article complies with the requirements of this Act, as determined by the Secretary.

63. With the greatest priority given to inspections to detect the intentional adulteration of food (Section 302(2a)(h)(1)); regarding making improvements to information systems to: "detect the intentional adulteration of food" (Section 302(b)(2)); assessment of threat: "is being conducted on the threat of the intentional adulteration of food is completed;" (Section 303(4)(3)(1)).
64. Even the agency acknowledges problems with this legal standard and recognizes that no precise definition of the standard exists. It invokes a standard of "worthy of belief or confidence, trustworthy." Determinations will be made on a case-by-case basis, considering factors such as reliability, reasonableness, and the totality of the facts and circumstances (preamble to Proposed Rule Section 1.378). See also Anon., 2003.

4

Effective Food Security Strategies and Plans for Production Agriculture and Food Processing

The way we produce things makes it somewhat easy for a terrorist to infiltrate our food supply.

— Lawrence Dyckman,
Head of the Natural Resources and Environment Section,
U.S. General Accounting Office, February 2004

Introduction

A February 3, 2004, study suggests that the U.S. food sector is vulnerable to attack. With U.S. officials concerned that terrorist groups, specifically Al-Qaeda, may be plotting attacks against food and agricultural sectors (Drees, 2004), as evidenced through the findings of hundreds of documents implicating U.S. agriculture in their safehouses and caves, food has become more vulnerable. As Lawrence Dyckman, Head of the Natural Resources and Environment Section of the U.S. General Accounting Office (GAO) stated: "The way we produce things makes it somewhat easy for a terrorist to infiltrate our food supply, whether it is live animals or the manufacturing process. So this is a real issue. This is not a hypothetical situation."

The USDA and FDA have stated in various forums that vulnerability assessments of the food industry have been conducted. Specific threats cannot be released, as these have been classified. This is an unfortunate situation for individuals involved in the production and distribution of safe food, as those with the greatest need to know will not have the information they need to conduct an appropriate risk assessment. Unfortunately, as security tightens up in other areas and from "spectacular high casualty, high profile attacks, the focus of terrorist strikes will be softer, primarily economic targets" (Drees, 2004). Fears of the intentional introduction of a devastating plant or animal disease top the list, and the gaps in security and inherent weakness of the farm sector make it fairly easy to introduce disease. A list of possible diseases and disease agents is provided in footnote 1 at the end of this chapter. These risks were recognized in 2001 when the president added agriculture and food industries to the list of critical infrastructure sectors needing protection from a terrorist attack (GAO, 2003).

The GAO noted a set of desirable characteristics for an effective national security strategy incorporating cost–benefit analyses and other economic criteria. These strategies developed for a range of different security issues could serve as a model for the development of a food security plan at a lower level of government or in the private sector. These recommendations for developing a security program are based upon the recently issued GAO directive and are as follows:

1. *Statement of purpose, scope, and methodology*: Addresses why the strategy was produced, the scope of its coverage, and the process by which it was developed.
   Elements: Statements of broad purpose
   Statements of narrow purpose
   How this strategy compares with others
   The major functions, mission areas, or activities it covers
   Principles or theories guiding development
   Impetus for strategy: event, statutory or regulatory requirement
   Process for developing strategy: task force, government, local or private input
   Definition of key terms
2. *Problem definition and risk assessment*: Addresses the particular (national, in the GAO case) problems and threats the strategy is directed toward.
   Elements: Discuss or define problems, their causes
   Define operating environment
   Conduct risk assessment, including analysis of threats and vulnerabilities
   Determine quality of available data, constraints, deficiencies, and unknowns
3. *Goals, subordinate objectives, activities, and performance measures*: Addresses what the strategy is trying to achieve, steps to achieve those

results, and the priorities, milestones, and performance measures to gauge results.

Elements: Determine overall desired result
Develop hierarchy of strategic goals and subordinate objectives
Develop specific activities to achieve results
Set priorities, milestones, and outcome-related performance measures
Determine process for monitoring and reporting on progress
Determine limitations on indicators of progress

4. *Resources, investments, and risk management*: Addresses what the strategy will cost, the sources and types of resources and investments needed, and where resources and investments should be targeted based on balancing risk reductions with costs.

   Elements: Determine type of resources available (e.g., budgetary, human, capital, information technology, research and development, contacts)
   Determine amount of resources available
   Determine which resource allocation mechanism to use (government) or which are available (private sector) (e.g., grants, in-kind services or support, loans, user fees)
   Evaluate government incentives or disincentives and government mandates
   Determine types and sources of investments
   Evaluate economic principles involved
   Conduct cost–benefit analysis

5. *Organizational roles, responsibilities, and coordination*: Addresses who will be implementing the strategy, what their roles will be compared to others, and mechanisms for them to coordinate their efforts.

   Elements: Determine roles and responsibilities of specific offices/divisions and individuals/job functions
   Evaluate role of government officials, other private-sector parties, domestic and international (if applicable)
   Determine who or what has lead, support, or partner roles and what these roles are
   Develop accountability and oversight framework
   Attempt to predict how critical changes to organizational will affect coordination and organizational effectiveness; conduct similar analysis for lead, support, and partner entities
   Develop specific processes for coordination and collaboration
   Determine how conflicts will be resolved

6. *Integration and implementation*: Addresses how a national strategy relates to other strategies' goals, objectives, and activities, and to subordinate levels of government and their plans to implement the strategy. A national strategy, in particular, would benefit from ad-

dressing how intergovernmental and private-sector initiatives can be operationally coordinated and integrated. Specifically, development of an "overarching, integrated" framework would be important for helping to deal with issues of potential duplication, overlap, and conflict (GAO, 2004a). Along these lines, recommendations of the Gilmore Commission, defining a "new normalcy" in governmental operations of vertical and horizontal sharing of information and intelligence, with integration of these operations wherever and whenever possible, should be followed.

Elements: Develop horizontal integration (with other entities similarly situated)
Develop means for vertical integration with implementing organizations (higher up, lower down, or in government)
Correlate relevant documents from implementing organizations
Details of specific federal, state, local, or private strategies or plans
Implementation guidance
Details on subordinate strategies and plans for implementation (e.g., human capital and enterprise architecture)

Besides the GAO report, there has been a great deal of activity at federal agencies on security programming in general and, more recently, on protecting infrastructure, which now includes food security. Under presidential directive, HSPD-9 (January 30, 2004), U.S. policies have been developed to protect agriculture and food from terrorist attack, and during major disasters and other emergencies, while ensuring that the federal departments and agencies involved do not diminish the overall economic security of the U.S. This directive includes the following objectives and goals:

1. Identifying and prioritizing sector-critical infrastructure and key resources for establishing protection requirements
2. Developing awareness and early-warning capability to recognize threats
3. Mitigating vulnerabilities at critical production and processing nodes
4. Enhancing screening procedures for domestic and imported products
5. Enhancing response and recovery procedures.

The Department of Homeland Security (DHS) is to lead, integrate, and coordinate the efforts of the federal governments, state and local governments, and private sector to protect critical infrastructure and key resources.

Awareness and Warning

Awareness and warning objectives under directive HSPD-9 will be tasked primarily to the Departments of Interior, Agriculture (USDA), Health and

Human Services (DHHS), and the Environmental Protection Agency (EPA) and will build upon current food safety monitoring and public health surveillance programs. Specific objectives are to:

1. Develop robust, comprehensive, and fully coordinated surveillance and monitoring systems that also include international information for animal disease, plant disease, wildlife disease, food, public health, and water quality and provide early detection and awareness of disease, pests, or poisonous agents.
2. Develop systems that track specific animals and plants, as well as specific commodities and food.
3. Develop nationwide laboratory networks for food, veterinary and plant health, and water quality that integrate existing federal and state laboratory resources and are interconnected and utilize standardized diagnostic protocols and procedures.

Intelligence Operations and Analysis Capabilities

In addition, the Justice Department, Central Intelligence Agency (CIA), and Department of Homeland Security, in conjunction with the USDA, FDA, DHHS, and EPA, will develop and enhance intelligence operations and analysis capabilities, focusing on agriculture, food, and water security. These intelligence capabilities will include collection and analysis of information concerning threats, delivery systems, and methods that could be directed against these sectors.

Also, the DHS will be creating a new biological threat awareness capability that will enhance detection and the ability to characterize an attack in coordination with the USDA, DHHS, EPA, and other agencies. This new capacity will build upon improved and upgraded surveillance systems and will integrate and analyze domestic and international surveillance and monitoring data collected from human health, animal health, plant health, food, and water quality systems. Vulnerability assessments would be part of the awareness programs and would identify requirements for agriculture and food that should be part of the DHS National Infrastructure Protection Plan.

Mitigation Strategies

Mitigation strategies would coordinate efforts of the DHS with the Justice Department, USDA, DHHS, EPA, CIA, and other agencies to prioritize, develop, and implement strategies to protect vulnerable critical nodes in production or processing from the introduction of disease, pests, or poisonous agents. In addition, the USDA, DHHS, and DSH will build upon existing efforts to expand the development of common screening and inspection procedures for agriculture and food imports and maximize domestic inspection activities for food items.

Response Planning and Product Recovery

Response planning and product recovery will again involve a coordinated effort of the DHS, USDA, DHHS, EPA, and state and local responders to ensure that response capabilities are adequate to respond quickly and effectively to a terrorist attack, a major disease outbreak, or another type of disaster affecting the national agriculture or food infrastructure. These efforts will be integrated with other DHS response efforts under HSPD-8 on national preparedness.

An outcome of these efforts will be the development of a coordinated agriculture and food-specific standardized response plan that will be integrated into the National Response Plan program under HSPD-8. The plan will ensure a coordinated response to an agriculture or food incident and will delineate the appropriate role of federal, state, local, and private-sector partners and will address risk communication to the general public. As part of this, the USDA and DHHS, in conjunction with the DHS and EPA, shall enhance recovery systems that are able to stabilize agricultural production, the food supply, and the economy; rapidly remove and effectively dispose of contaminated agriculture and food products or infected plants and animals; and decontaminate affected premises.

As an assistance to the private sector, the USDA will make recommendations to the DHS regarding the use of existing and creation of new financial risk management tools encouraging self-protection for agriculture and food enterprises vulnerable to losses due to terrorism. As with all the programs to date, there are no funds allocated to the private sector through tax incentives or otherwise to fund security improvements. Furthermore, these risks are increasingly self-insured, as liability policies add exclusions for terrorist activity and damage caused by environmental agents (e.g., environmental mold) and etiologic agents (e.g., food-borne or waterborne bacteria) (Rasco, 2001).

The USDA and DHS are to work with state and local authorities and the private sector to develop:

1. A National Veterinary Stockpile (NVS) containing sufficient amounts of animal vaccine, antiviral, or therapeutic products to appropriately respond to the most damaging animal disease affecting human health and the economy and that will be capable of deployment within 24 hours of an outbreak. The NVS shall leverage resources of the strategic national stockpile.
2. A National Plant Disease Recovery System (NPDRS) capable of responding to a high-consequence plant disease with pest control mea-

> sures and the use of resistant seed varieties within a single growing season to sustain a reasonable level of production for economically important crops. The NDRS will utilize the genetic resources contained in the U.S. National Plant Germplasm System, as well as the scientific capabilities of the federal–state–industry agricultural research and extension system. The NPDRS shall include emergency planning for the use of resistant seed varieties and pesticide control measures to prevent, slow, or stop the spread of a high-consequence plant disease, such as wheat smut or soybean rust.

Outreach and professional development are also key components of this directive and will provide an effective information-sharing and analysis mechanism between federal and private-sector entities. The directive provides for higher education programs (undergraduate and graduate) to address protection of the food supply. This will include capacity-building grants to universities for interdisciplinary programs that combine training in food science, agriculture sciences, medicine, veterinary medicine, epidemiology, microbiology, chemistry engineering, and mathematics (statistical modeling) to prepare food defense professionals. The DHS, USDA, and DHHS have or will establish university-based centers of excellence in agriculture and food security.

Government-sponsored research and development efforts will accelerate and expand development of current and new countermeasures against intentional introduction or natural occurrence of catastrophic animal, plant, or zoonotic diseases. The DHS will coordinate these activities. Efforts will also include countermeasures research and the development of new methods for detection, prevention technologies, agent characterization, and dose–response relationships for high-consequence agents in the food and water supply. The USDA and DHS will develop and plan to provide safe, secure, and state-of-the-art agriculture biocontainment laboratories that improve the national capacity for research into and development of diagnostic capabilities for foreign animal and zoonotic diseases.

The focus of this particular federal program is upon expanded and improved *government* programs, with the overall intent of protecting human health and limiting the impact of a plant or animal disease outbreak. A program goal is to better manage disasters affecting agriculture or food infrastructure in the U.S. The Department of Health and Human Services (of which the FDA is a part) has a budget of $3.776 billion for these activities for FY 2004, and the USDA has a budget of $368 million for combating terrorism. This is out of the total federal "terrorism" budget of $52.732 billion (GAO, 2004a). Although the physical infrastructures strategy (which encompasses food security) provides specific guidelines for resource allocation, no cost estimates are provided in the aggregate or for specific goals or objectives. Neither is there a detailed discussion of risk management. It is the opinion of GAO that more guidance on resource, investment, and risk management would help parties that must implement these strategies to determine how to allocate resources and investment according to priorities and constraints, track costs and performance, and shift the available investments and resources to where they would be most effective.

Although private-sector input is woven into these objectives, the primary focus in these strategies and initiatives is on the role of government in characterizing an incident and then developing an appropriate response to it. Private-sector preparedness programs are hinted at, but as with earlier government-proposed programs, this has not been an important focus — an unfortunate continuing trend. Private-sector preparedness was virtually ignored in the Public Health Security and Bioterrorism Preparedness and Response Act of 2002 (Bioterrorism Act: PL 107-188), with a myriad of regulations and unfunded mandates that substantially increase the cost of food safety and security programs, some provisions of which do little to improve either the security or safety of the food supply (see Chapter 3).

Identifying and prioritizing critical infrastructures and the key resources needed to protect them are the focus of the December 17, 2003, Homeland Security Presidential Directive HSPD-7 (which supercedes Presidential Decision Directive NSC-63, May 22, 1998, Critical Infrastructure Protection). This directive was designed to improve the internal management of the executive branch and is a national policy recognizing that terrorists may seek to destroy, incapacitate, or exploit critical infrastructure and key resources as a means to undermine national security. Food is part of the critical infrastructure of the U.S., although it was not initially on the list for consideration prior to the 2003 directive. Other critical sectors include: water, energy, health care, banking and finance, transportation, communications and cyber systems, and the defense industrial base. Protection of national monuments and icons is also included within this critical infrastructure directive. Under HSPD-7, the objectives of a terrorist attack would be to:

1. Cause mass casualties or a catastrophic health impact comparable to that of weapons of mass destruction
2. Impair the ability of the federal government to perform its essential mission or to ensure public health and safety
3. Undermine state and local government capacity to maintain order and deliver a minimum of public services
4. Damage the private sector's capability to ensure the orderly functioning of the economy and provide essential services

5. Have a negative effect on the economy through the cascading disruption of other critical infrastructure and key resources
6. Undermine the public's morale and confidence in our national economic and political institutions

Because American society is open and technologically complex, there is a wide array of possible targets in both the physical and virtual worlds, making it impossible to protect or eliminate all vulnerabilities. Therefore, this directive was designed to bolster security through strategic improvements that make it more difficult for a terrorist attack to succeed and also to lessen the impact of a terrorist attack should one occur. Tactical security improvements that can be rapidly implemented to deter, mitigate, or neutralize potential attacks would be developed as part of this program.

In this infrastructure strategy there is a discussion on organizing response and planning and a determination of the roles and responsibilities of federal, state, and local governments, the private sector, and international sectors, along with identifying the need to create collaborative mechanisms for government–industry planning. These entities would work together to identify, prioritize, and coordinate infrastructure protection efforts, but so that efforts do not dilute the overall economic security of the nation. Under the infrastructure strategy, DHS is the primary liaison and facilitator for cooperation between all relevant parties (GAO, 2004a). Unfortunately, the roles, responsibilities, and relationships between emergency response, law enforcement, and the private sector are not clear, particularly where there is an overlap in responsibilities. How these strategies would be implemented in a crisis is also not clear, because although generalities are presented, roles and responsibilities again are not clearly defined (GAO, 2004a).

Sharing of information about physical and cyber threats, vulnerabilities, incidents, and possible protective measures and best practices is to be developed with industry input, with greater focus put upon the infrastructure and resources in densely populated areas. Critical infrastructure would be geospatially mapped. Further, an understanding of baseline infrastructure operations, the identification of indicators and precursors to an attack, and a surge capacity for detecting and analyzing patterns of potential attack would be developed. For the agriculture and food sector, the development of such information will take substantial effort, particularly since so much of our commerce in food involves imported products and a complex distribution system. With food, as with other sectors, there is the potential that intellectual property will be jeopardized for the sake of national security. The major concern with this directive for agriculture and food businesses is that, as in earlier directives, food and agriculture will not receive the attention it deserves, because either the special needs of this sector are not well understood, a catastrophic impact is not envisioned from targeting the food supply, or the rural and diffuse nature of the industries involved will yield a cost–benefit analysis that would sacrifice security of the food sector for a sector that is more urban, "sexier," or perceived as having a greater long-term impact on the economy. Again, as after the Bioterrorism Response Act, food producers are more or less on their own with regard to development of food security measures, as there is a strong likelihood that there has been little support or useful guidance for the food industry resulting from this directive.

To address some of the concerns raised in Chapter 1 and Chapter 2, the objective of this chapter is to provide strategies a company can use to develop a food security preparedness plan using concepts from the most recent promulgated strategies and guidance publicly available, as well as concepts already familiar to food safety professionals, including Hazard Analysis Critical Control Point (HACCP) principles and a related strategy, Organizational Risk Management (ORM). Implementation strategies and developmental approaches for a food security program will be outlined. These models are applicable to production agriculture, food processing, food distribution, and food service and interface with the current HACCP (e.g., for fishery products, 21 Code of Federal Regulations (CFR) Part 123; meat and poultry products, 9 CFR Part 301 *et seq.*, Part 381 *et seq.*), good manufacturing practices (GMPs) (21 CFR Part 110), and recall programs (21 CFR Part 7).

The Impact of Small Strategic Attacks

The malicious contamination of food for terrorist purposes is a real and current threat, with the deliberate contamination of food at one location having the potential for a global public health impact (WHO, 2002; Appendix E, this volume). We are presuming here that terrorism will track the GAO definition of politically motivated violence to coerce a government or civilian population (GAO, 2004a). Even though weapons of mass destruction remain a potential threat and have been the focus of numerous preparedness and response strategies, they are not the most likely risk to food systems or to the public at large because they are relatively difficult to stabilize, transport, and effectively disseminate on a large scale. A simpler and more likely form of attack involves the limited or individual use of pathogens or toxicants (such as ricin) developed specifically for biological warfare or terrorist purposes, as well as common bacterial food-borne or zoonotic agents.

The most likely possibility for a food terrorist attack involves the use of food-borne pathogens or chemicals that are relatively easy to grow or obtain. Readily available toxic chemicals such as pesticides, heavy metals, and industrial chemicals and naturally occurring microbiological pathogens are likely agents (WHO, 2002). The effectiveness of an attack would depend upon the potential public health impact of the agent, the food used for dissemination, and the point of introduction into the food chain. Those agents causing rapid acute effects, death, paralysis, incapacitating symptoms, or long-term consequences would be most effective.

Attacks could be targeted at a specific commercial entity or industry segment involving the real or threatened introduction of an animal or plant pathogen (or its genetic material) at a production, distribution, or retail facility. These attacks would most likely involve agents disseminated in a somewhat crude form and possibly in a manner in which a small number of what initially appear to be isolated incidents would surface within a short period. Very crude forms of microbial contaminants have been used to contaminate food as evidenced by incidents in New York City (2001) and Canada (2004) where human feces or urine were introduced into salad bar and produce items at retail. Recent incidents involving commercial pesticides include one in Michigan (2004), where a supermarket employee deliberately contaminated 200 pounds of beef with the insecticide Black Leaf 40, poisoning 92 people (Gisborne, 2004). Deliberate attacks on food, compared to air or water, would be easier to control, making food a relatively reliable vehicle (WHO, 2002). Decreasing the impact of an attack on the food supply, however, includes private-sector and government control on both the production and distribution of food (reducing opportunities for attack) and assumes a certain degree of dietary diversity and multiple food choices. These factors would mean that every individual in an area would not be affected, diluting the overall public health impact of a food contamination incident. However, food is highly vulnerable because of the diversity of sources of food, the distribution of food in global markets, and the complexity of the supply chain, which make prevention of an intentional contamination incident difficult, if not impossible, to prevent. Furthermore, many developing countries still lack basic food safety infrastructure, putting the food supply of large numbers of people at risk (WHO, 2002).

The impact of an intentional well-orchestrated contamination incident can be extrapolated from recent unintentional food-borne illness incidents. Major recent incidents include the illness of 170,000 people in 1985 from *Salmonella typhimurium*-contaminated milk in the U.S. Midwest and a second outbreak in 1994 with *Salmonella enteritidis* in ice cream mix, affecting 224,000 people in 41 states. Possibly the largest food-borne outbreak in recorded history was from hepatitis A in Shanghai, China, in 1991, affecting 300,000 people. Some of the deadliest incidents involved the chemical contamination of food, e.g., the 1981 incident in Spain that resulted in the death of 800 and the illness and permanent injury of 20,000 individuals from consumption of contaminated cooking oil. These incidents illustrate how a carefully planned, simultaneous attack at multiple locations could have an impact similar to or greater than the distribution of a single contaminated product over a wide geographical area.

A single isolated incident, even one with no apparent human health impact, can have devastating consequences. The single mad cow in Washington State in December 2003 led to widespread product market withdrawal and major disruptions in trade, including a loss of over 90% of the U.S. export market for beef as of March 2004. This supports the contentions of many experts that a deliberate contamination incident with a zoonotic agent would have enormous economic implications, even if the disease episode is relatively minor (Monke, 2004). Since economic disruption is a primary motivation for terrorist groups, mass casualties are not a driving force. Deliberate contamination of Israeli citrus fruit with mercury in 1978 and similar incidents since then have caused major economic impact. The cyanide contamination of Chilean grapes in the U.S. in 1989 caused losses of $100 million to $300 million and resulted in the bankruptcy of 100+ growers and shippers. The crisis in Belgium over dioxin-contaminated meat and dairy products led to major restrictions in imports and animal feed products, and fed the growing hysteria over imports and genetically modified foods, seen as the erection of nontariff trade barriers by the European Union. The 1997 *Escherichia coli* 0157:H7 infection in ground beef in the U.S. led to a 11-million-kg recall, $50 million to $70 million in direct costs and bankruptcy (CDC, 1997). Similarly, a 1998 incident implicating *Listeria monocytogenes* in deli meats and frankfurters led to a 14-million-kg market withdrawal and tens of millions of dollars in direct costs (CDC, 1997). Incidents such as these have increased regulatory burden on the U.S. industry, increased production costs, and led to new mandatory product testing and environmental testing programs.

Groups with limited resources could perpetrate an attack by employing any number of possible food contamination agents, raising widespread havoc. As seen with the anthrax "mail bombs" in October 2001 and the ricin contamination incidents at the White House (November 2003) and a Senate office building (January 2004), even a limited small-scale terrorist activity can rapidly saturate the emergency response and medical facilities of a community (Sobel et al., 2002). The response to the anthrax mail bombs in Washington, D.C., New York, and Florida in 2001 tied up investigative and response agencies across the nation. Because of enhanced screening and treatment, mail deliveries to Washington, D.C., remained slow for 2 years following the anthrax scares. Precautionary responses to numerous false alarms across the nation in response to the 2001 incidents employed large numbers of police, fire, and hazardous materials response teams. Overreaction to the severity of this agent and its risk has led to bizarre outcomes. For example, it was not possible in 2002 to have even a postcard hand-stamped at the post office at

Edwards Air Force Base, a locale with presumably high security located in the middle of a California desert and accessible only to those with proper security screening or clearances, due to a fear of anthrax contamination to postal employees. Following the 2001 incidents, affected government offices received expensive sanitation treatments with chlorine dioxide and other agents, and then had the opportunity to test decontamination procedures again with hopefully improved efficiency in 2003 and 2004.

To further complicate the responsible handling of an incident, acts of bioterrorism may occur and either not be detected by authorities or not be properly characterized. Failure of state and federal (but not local) authorities to properly characterize an incident was the major failing in the *Salmonella typhimurium* investigation in Oregon in 1984 where over 1000 individuals reported symptoms and 751 confirmed cases of illness resulted. Failure to characterize this incident as a possible criminal act stopped an ongoing criminal investigation, which was not reinitiated until a year later. Fortunately, the cult was not able to conduct attacks using the typhoid fever organism found in their possession during later investigations (WHO, 2002).

There is some concern now that there has been a major paradigm shift as a result of the new Bioterrorism Act that food is presumed to be contaminated until proven otherwise. A spate of unjustified product detentions immediately following implementation of the first round of regulations prior to Christmas 2003 indicates that there are still a number of kinks to be worked out with regard to interpretation as to the scope of the regulation, implementation, and enforcement of provisions under this new act.

The threat of a food-tampering incident involving harmless materials (or no materials) can be as effective as a real attack. Simply claiming that a product has been purposely contaminated with a dangerous material is sufficient to precipitate an extensive product recall with the associated adverse publicity, short-term economic loss, longer-term loss of market share, and the resultant economic impact (Bledsoe and Rasco, 2001a, 2001b). For example, a Class I recall is required when there is " a reasonable probability that the use of or exposure to a violative product will cause adverse health consequences or death" (21 CFR § 7.3(m)(1)). Reasonable probability has been replaced with "credible threat" language in the new Bioterrorism Act, legally smoothing the way for an overreaction to any scare and making a successful hoax more likely.

Perpetrators of terrorist activity targeting the food industry will probably have a variety of different motivations. The most common will be to cause economic damage to a specific company, type of product, or to the industry at large. Targeting food as a political statement or to influence a political outcome, as in The Dalles case, is a lesser possibility. Malicious mischief, copycat crimes, and personal revenge are other likely motives. The type and desire for publicity will depend upon the motive. More detail on terrorist motivation is presented in Chapter 1.

Combating terrorism, in a situation involving food, will most likely involve local efforts as well as a range of policies and programs developed by governmental entities to counter terrorism. This may track the National Strategy for the Physical Protection of Critical Infrastructures and Key Assets issued by the president in February 2003, incorporated in part into Homeland Security Presidential Directives HSPD-7 (Critical Infrastructure Identification, Prioritization, and Protection) and HSPD-8 (National Preparedness) (GAO, 2003, 2004a). This initiative is to foster cooperation between all levels of government and between government and industry and encourage market solutions wherever possible and government intervention when needed. The three strategic objectives of this national strategy are to:

1. Identify and ensure the protection of the most critical assets, systems, and functions in terms of national-level public health and safety governance and economic and national security and public confidence
2. Ensure protection of infrastructures and assets facing specific, imminent threats
3. Pursue collaborative measures and initiatives to ensure the protection of other potential targets that may become more attractive over time

These strategies imply responsibilities that cut across many different governmental and nongovernmental entities and that cross state and international borders. Furthermore, the current national strategies regarding terrorism are not governed by a single, consistent set of requirements, making development of coordinated efforts difficult and implementation complex and inconsistent. For example, the Homeland Security Act of 2002 (Section 801(b)) requires DHS to develop a process for receiving meaningful input from states and localities to assist in the development of a national strategy to combat terrorism and other homeland security activities (GAO, 2004a), but does not establish specific content elements for this, making the determination of the objectives and scope difficult.

Examples of Targets

The types of attacks terrorists have directed against the food industry to date range from false statements or accusations to overt acts designed to destroy property, information and communications systems, crops, animals, and people. Product tampering (real or hoaxes) and vandalism have proven to be particularly productive in terms of perpetrator notoriety and economic damage to targets. Such food terrorism is directed against perceived injustices, and while their actions are not necessarily encompassed within the realm of conventional terrorist activities, the results often are. On a larger scale, attacks against a country's crops and livestock remain a viable aggressive weapon in the strategic planning of many governments, particularly those with reduced conventional weaponry.

Objectives of food terrorism include the desire to severely impact a company and put it out of business by affecting the stock price, product availability, or marketability in a malicious way; a program directed toward the elimination of a specific food, ingredient, or agricultural practice; the prohibiting of the importation of competing crops, research, or development in a particular area; and pressure to erect trade barriers.

Specific targets include primary producers, processors, distributors, retailers, shareholders and investors, consumers, vendors/suppliers, suppliers of other services (e.g., banking and insurance), commercial neighbors (e.g., adjacent tenants or property owners), and researchers (public and private). Thousands of products each year are subject to malicious tampering and accidental contamination, which precipitate a product recall or market withdrawal (Washington State, 2001; Hollingsworth, 2001; Bledsoe and Rasco, 2002). Food, beverages, pharmaceuticals, agricultural chemicals, fertilizers, pest control media, and genetically modified crops are among the products more commonly affected. Activities directed specifically against organizations supporting or directly involved with biotechnology are the greatest threats of late. Most food companies have crisis management plans in place to handle a precautionary recall or market withdrawal. However, a serious incident of intentional contamination is not a situation that is routinely planned for. Unfortunately, this is the type of crisis management planning and response that will be necessary in a world where food has become increasingly politicized. The focus here and in guidance for preparedness planning is directed toward impeding or limiting the damage caused by an attack perpetrated by nongovernmental actors and not an outright act of war (WHO, 2002). Food terrorism directed toward a civilian population (compared to a military target) is also more problematic because of the large number of possible agents and possible foods. Food is, and has been, a common target (Khan et al., 2001).

Current Level of Readiness

Most organizations are ill prepared to deal with tampering incidents let alone other manifestations of bioterrorism. Six of 10 security professionals said their firm was unprepared for a chemical or biological attack, and less than half had upgraded security measures following the start of the Second Gulf War (Laird, 2003), with decisions based upon a cost–benefit analysis. Problems with developing improved security programs are complacency and lack of money to purchase the necessary equipment or services. Even simple measures such as searching bags or packages, not allowing unescorted visitors into a facility, and providing identification for employees are not routine in many sectors.

Most security improvements in a related industry to food, the chemical industry, have included improved fencing, greater control of facility access, greater lighting, and more patrols or computer-aided or video security improvements. More sophisticated drills are becoming common in the petroleum industry. The focus is upon improving physical security, rather than changing operations to switch to the use of less hazardous materials. The fear remains that even if these places "had as much security as Fort Knox, anyone with a plane, a projectile weapon or even a rifle could still do tremendous damage" (Laird, 2003). As a result of 9/11, the Business Roundtable, an association of CEOs from U.S. corporations with a workforce of 10 million, established a secure telephone communication network that links its members with the Department of Homeland Security. Private firms have also established communication links with the FBI, DHS, and local law enforcement, either on their own or through security-based information-sharing networks.

There is a general feeling that companies are not able to respond to increased security demands without support of the government through tax incentives or other types of economic assistance, even though there is a general agreement that the overall risk of terrorist attacks against the private sector is perceived by 85% of security professionals to be higher now than 3 years ago (Laird, 2003; Hall, 2004). However, federal assistance with private-sector programs is unlikely. This increases the likelihood that more businesses will face higher insurance premiums, lose insurance coverage altogether, and voluntarily or involuntarily become self-insured.

Most emergency response programs in place, either by governments or the private sector, still contain gaps in their capacity to prevent or respond to the broadest range of food safety emergencies (WHO, 2002). This is despite major infusions of government funds over the past 2 years in the U.S. to address these concerns. Efforts at coordination between governmental entities at different levels to address these gaps and to solve jurisdictional problems that impede the development of effective plans are currently under way. Hopefully efforts at the national level to handle response to terrorist attacks, including those involving food, will be streamlined through DHS when current executive branch initiatives are fully implemented.

Issues of product liability, insurance coverage, crisis management, and maintaining business viability are of critical concern. A focus here, and in recent conferences, is on analyzing an organization's risk before an incident occurs, utilizing best practices to avoid a tampering or contamination event, formulating and instituting a crisis management and communication plan, conducting a cost–benefit analysis for transferring the risk through insurance coverage, having and exercising product recall programs, litigating a tampering or recall case, and conducting forensic accounting to quantify losses and analyze claims (Bledsoe and Rasco, 2003; ACI, 2000; IFT, 2001, 2002).

High-profile consumer product-tampering instances from the 1980s and 1990s made companies aware of new risks; however, we have unfortunately entered a brave new world of well-organized, internationally based targeting of organizations and products in and related to the food industry. Recent conferences have addressed techniques for monitoring open-space research, covert sensor technology, and crime prevention training (ACI, 2000). According to the FBI, domestic crime targeting biotechnology is the emerging antitechnology crime of the new millennium (FBI, 2000). However, techniques and tools for protecting and monitoring open-space research areas and facilities are limited (FBI, 2000).

Current Government Programs

Government food inspection is already a large program and is growing as a result of an infusion of funds under the Bioterrorism Act. The FDA inspects over 57,000 food processing facilities every 5 years (on average), and the USDA inspects over 6000 meat and poultry slaughter and processing programs daily. Individual state agencies conduct additional annual inspections of roughly 300,000 facilities. These activities cost $1.3 billion per year (GAO, 2003). In 2002, the FDA received $97 million and the USDA Food Safety and Inspection Service (FSIS) $15 million under the Defense Appropriation Act to enhance the security of imported foods (GAO, 2003). Both agencies have used these funds and others to hire additional inspectors, to improve laboratory testing, and to provide training to current personnel, including simulation exercises.

Although 22 states have recently passed legislation increasing the penalties for malicious acts directed at food and agricultural facilities, the effectiveness of these laws is yet to be seen (Anon., 2002). In the current legislative session, numerous bills have been introduced into Congress regarding food security and bioterrorism (see, for example, Public Health Security and Bioterrorism Response Act of 2002). The Food and Drug Administration (FDA) and USDA have introduced guidance documents (FDA, 2002a, 2002b; www.usda.gov; Appendix A, this volume), which most likely will evolve into *de facto* regulations governing food security despite agency claims to the contrary, since inspectors are instructed to discuss (but not to interpret) security guidelines with facility personnel during inspections (GAO, 2003). Because the USDA takes a broader view to food security, it will probably be the first to regulate it.[2] The net effect to the food industry will be increased regulation and operating costs both directly and indirectly as a result of this new bioterrorism bill and recent events, including contamination incidents involving ricin, food-borne outbreaks tied to imported foods (hepatitis A in green onions from Mexico, November 2003), and the first U.S. mad cow (December 2003).

The position of the FDA is that it has broad authority to regulate food safety but not to specifically regulate "physical facility security measures" and that it sees little overlap between food safety and food security, with the possible exception of handling hazardous materials (GAO, 2003; "Good Manufacturing Practices" in 21 CFR Part 110 *et seq.*). The USDA has a more expansive view and believes that it could require processors to implement specific food security measures that are "closely related to the sanitary condition in the processing facility," such as the shipping and receiving of food, requirements for tamper-evident seals, and the protection of water or ice. The USDA claims authority over conditions within the immediate processing environment, but not outside of it (e.g., requiring fences, alarms, or outside lighting). The USDA believes that FSIS can mandate "inside security" guidelines such as controlling access to certain areas, monitoring equipment operation to prevent tampering, and keeping accurate inventories of restricted ingredients and hazardous materials. However, improvements in food security by food processors are not currently being monitored or documented during agency inspections, and inspectors have little or no training in this area (GAO, 2003).

However, food processors are and have been implementing visible security improvements such as improved fencing, access control, and lighting, and making less obvious improvements in tracking missing stock and enhancing mail handling and personnel training and screening. In general, larger plants were found to be implementing more security measures than smaller ones (GAO, 2003). Naturally, facilities are reluctant to discuss specific security measures with federal authorities, as this information could be released under the Freedom of Information Act (FOIA). If security gaps could otherwise be made public, specific vulnerabilities could reach potential terrorists. As of 2002, a large percentage of the food companies (N = 150) in an FDA survey[3] were focused on food security issues and on increasing facilities security (Table 4.1). A similar survey was conducted with regional (circuit) supervisors within the USDA at roughly the same period. In general, security and security awareness was greater at the larger facilities and had a wide range of responses regarding their confidence in food plant security[4]

Table 4.1 FDA Web Survey of FDA Investigators on Food Security Awareness at Food Processing Facilities

Issue	N	Yes (%)	No (%)	No Basis to Judge (%)
Have you read the Food Security Preventive Measures Guidance?	128	90.6	9.4	NA
To what extent has your role changed as a result of the Guidance?	126	2.3 — great extent 18 — moderate extent 39.1 — some extent	34.4	4.7
What new roles or responsibilities apply since the Guidance was issued?	128			
1. Look for presence of security measures during inspection		36.7		
2. Informal conversations with plant operators on security		47.7		
3. Spend more time learning about security		14.8		
4. Informally discuss food security with supervisor		16.4		
Have you observed the following physical security measures?				
1. Access protection, fencing	126	73.4	20.3	4.7
2. Limited access to restricted areas	126	77.3	14.8	6.2
3. Adequate interior and exterior lighting	125	78.1	10.2	9.4
4. Other measures	116	39.1	40.6	10.9
How many of the inspected facilities in the last year have the physical security measures listed above?		3.9 — All 20.3 — Most 45.3 — Some 24.2 — Few 2.3 — None		3.9
Have you observed the following raw material handling or packaging security measures?	127	60.9	29.7	8.6
1. Supervision of off-loading of incoming ingredients, packaging, and labeling				
2. Destruction of outdated or discarded labels	12	21.9	62.5	14.8
3. Investigation of packaging or paperwork irregularities	127	30.5	48.4	20.3
4. Other features	111	18.8	52.3	15.6
How many of the inspected facilities in the last year have the raw material handling or packaging security measures listed above?		2.3 — All 18.8 — Almost all 31.2 — Some 20.3 — Few 15.6 — None		10.9

Table 4.1 FDA Web Survey of FDA Investigators on Food Security Awareness at Food Processing Facilities (Continued)

Issue	N	Yes (%)	No (%)	No Basis to Judge (%)
Have you observed the following inside security features?				
1. Water quality monitoring	128	64.8	30.5	4.7
2. Water source security	128	19.5	68.0	12.5
3. Air quality monitoring	127	16.4	71.9	10.9
4. Air source security	127	4.7	80.5	14.1
5. Secured hazardous chemical storage	127	81.2	13.3	4.7
6. Limited access to hazardous chemicals	128	68.0	26.6	5.5
7. Other features	108	13.3	55.5	15.6
How many of the inspected facilities in the last year have the inside security measures listed above?	126	31.0 — All 31.2 — Almost all 32.8 — Some 21.1 — Few 7.8 — None		2.3
Have you observed the following finished-product security measures?				
1. Investigation of missing stock or other irregularities	128	18.0	59.4	22.7
2. Random inspection of storage facilities, vehicles, or product containers	128	48.4	39.8	11.7
3. Locking and sealing of vehicles, containers, or railcars	128	52.3	32.8	14.8
4. Other measures	104	7.0	53.9	20.3
How many of the inspected facilities in the last year have the finished-product security measures listed above?	128	2.3 — All 17.2 — Almost all 34.4 — Some 17.2 — Few 18.0 — None		10.9
Have you observed the following personnel security measures?				
1. A system of positive ID and recognition	128	60.9	30.2	7.0
2. Limitation of employee access to certain parts of the plant	128	61.7	32.0	6.2
3. Restriction of personal items allowed in plant	128	63.3	25.0	11.7
4. Provision of security training	128	34.4	47.7	18.0
5. Other measures	104	9.4	53.1	18.8

Table 4.1 FDA Web Survey of FDA Investigators on Food Security Awareness at Food Processing Facilities (Continued)

Issue	*N*	Yes (%)	No (%)	No Basis to Judge (%)
How many of the inspected facilities in the last year have the personnel security measures listed above?	127	5.5 — All 18.8 — Almost all 35.2 — Some 28.9 — Few 5.5 — None		5.5
Have you observed the following laboratory security measures? (64% of facilities had in-house labs: *N* is approximate)	82			
1. Restricted access to lab	82	78.9	21.1	
2. Restriction of materials to lab	82	84.3	15.6	
3. Other	69	49.2	35.2	
How many of the inspected facilities in the last year have the laboratory security measures listed above? (numbers are approximate)	80	11 — All 14.7 — Almost all 2.6 — Some 28.0 — Few 19.5 — None		
Have you observed the following general security measures?				
1. Restricted access to critical data systems to those with appropriate clearance	128	60.2	19.5	20.3
2. Restricted entry to facility, especially food handling and storage	128	71.9	22.7	5.5
3. Monitoring of areas vulnerable to tampering	128	37.5	50.0	12.5
4. Encouraging of staff to be alert to signs of tampering with equipment or food	128	38.3	40.6	21.1
5. Other measures	100	3.9	56.2	18.0
How many of the inspected facilities in the last year have the general security measures listed above?	127	4.7 — All 28.9 — Almost all 25.8 — Some 25.0 — Few 7.0 — None		7.8
Keeping in mind the food security measures listed, how satisfied are you that company efforts will ensure that food processed in the U.S. is protected from deliberate contamination?	128	3.9 — Very satisfied 25.0 — Satisfied 34.4 — Neither satisfied nor dissatisfied 26.6 — Dissatisfied 7.0 — Very dissatisfied		3.1

Adapted from GAO. 2003. Food Processing Security. Voluntary Efforts Are Underway, but Federal Agencies Cannot Fully Assess Their Implementation, GAO-03-342. A report to Senators Richard J. Durbin and Tom Harkin. General Accounting Office, Washington, DC.

(GAO, 2003). Roughly 65% felt that the field inspectors should monitor the security measures in the FSIS guidance.

Terrorism Insurance

It came as no surprise that liability insurance covering acts of terrorism virtually disappeared after 9/11. Covered property and casualty losses for 9/11 are projected at $40 billion (GAO, 2004c). This coupled with the emerging trend to exclude coverage for contamination from "biological or other etiologic agents" (Rasco, 2001), increased regulatory requirements, and the desire of insurance companies to maintain high profitability by increasing premium costs (GAO, 2004c) in a poor long-term investment market shifted even more of the financial risk for damage from a terrorist incident to individual private businesses. To make matters worse, reinsurers are covering less (GAO, 2004c) and are monitoring exposure by geographic area, requiring more detailed information from insurers, introducing annual aggregate and event limits, excluding large insurable values, and requiring strict measures to safeguard assets and lives where risks are perceived to be high (GAO, 2004c).

Congress passed the Terrorism Risk Insurance Act (TRIA) in 2002 to help commercial entities and developers obtain insurance by capping *insurer* liability[5] and by providing government compensation for a large share of losses for terrorist acts that meet certain criteria[6] (GAO, 2004b). It is set to expire in 2005 (GAO, 2004c). TRIA does not extend to group life insurance (GAO, 2004c), although it does extend to state residual market insurance entities and state workers' compensation funds. Problems exist because it is not entirely clear as of this writing whether the Department of Treasury will extend the 2002 TRIA requirements that insurers offer terrorism coverage on terms that do not "differ materially" from other coverage after 2005, hindering underwriting and pricing decisions. The Treasury has not fully established a claims processing and payment structure for insurance companies as required under this bill. It has yet to make "the claims payment function operational." Furthermore, the Treasury had not yet decided whether to extend the mandate that insurers make terrorism coverage available for policies issued or renewed in 2005 (GAO, 2004c). The goal of TRIA was to provide a bridge or transitional period for insurance companies to develop ways to cover losses and establish pricing without government involvement, but this has not been achieved (GAO, 2004c).

Although terrorism riders have been available for purchase since 2003, only 10 to 30% of businesses have bought them, with most of the purchasers in the northeastern U.S. (GAO, 2004b), where risks are perceived to be higher (GAO, 2004c). Costs for coverage are more in high-risk locations, including the New York City and the Washington, D.C., areas, which have already experienced attacks; properties in areas with a perceived high risk, possibly near important landmarks; and large commercial properties in major urban centers. The businesses at greatest risk are owners of "trophy properties," owners or developers of high-risk properties in major city centers and those in or near trophy properties seeking to borrow funds or obtain mortgages (GAO, 2004b, 2004c). TRIA has improved the credit ratings for some commercial mortgage-backed securities that had been downgraded following 9/11 because the underlying property lacked or had inadequate terrorism insurance (GAO, 2004c).

Premium costs are high. Naturally, long-standing nuclear, biological, and chemical (NBC) policy exclusions for such events (GAO, 2004c) have been tightened to exclude loss from fires and intentional property damage that could result from a terrorist act. Industrial accidents and chemical skills are encompassed in new NBC exclusions (GAO, 2004c).

In some states, there is also little or no coverage for remediation activities from biological, chemical, or radioactive contamination (including human illnesses), or intentional property destruction by terrorists. These exclusions greatly reduce the category of incident and scope of cover that would qualify for coverage under a TRIA-based terrorism policy.

Prevention Is the First Line of Defense

Prevention, although not completely effective, is the first line of defense (WHO, 2002). The key for preventing food terrorism is to establish and enhance existing food safety management programs and to implement reasonable food security measures. The World Health Organization (WHO) encourages and outlines cooperative efforts between industry and government, which can be effective in countries where the government or governmental employees have a vested interest in the successful operation of business. Such arrangements are quite commonplace. In countries where the relationship between government and the private sector is adversarial, or where the role of government agencies is to exercise police power to protect public health by forcing compliance with food safety regulations (reasonable or not), this cooperative approach is less likely to work.

Regardless of the governmental structure in place, WHO guidance focuses upon strengthening national systems in response to food terrorism, with the advice provided extendible to private-sector entities. An additional emphasis is upon improving and integrating governmental programs for food-borne illness surveillance and communicable disease control systems, emergency preparedness and response systems, and WHO's role in coordinating existing international systems for public health disease surveillance and emergency response, including aspects of food terrorism. Its proposed models and strategies are based upon those already in place for terrorist attacks involving chemical, biological, or radiological agents (including dirty bombs).

The objective here is to provide ideas to businesses to develop strategies for recognizing potential hazards, and then to develop preventive measures that could be incorporated into an existing food safety program to reduce the danger of intentional product contamination. Such a food security plan would use a framework already widely adopted in the food industry and applicable to production, food processing, and food service facilities derived from familiar HACCP principles. Organizational Risk Management (ORM), a program promoted by the U.S. government, is also discussed. In international settings, HACCP-based programs have greater likelihood of successful adoption, are used in the European Community, and are promoted by WHO and the national academies as the most feasible approach to development of a food security plan (GAO, 2003).

Although important security issues, personal safety, preventing the kidnapping or assault of employees or their families, and defenses against armed attacks are not included here. Rather, the focus is directed toward protecting the integrity of the food produced, the systems employed in its production, and facility and employee safety while at work.

Development of a Food Security Plan Based upon HACCP Principles

Each organization is uniquely situated and should develop a sensible, individualized security plan for managing the risk of terrorism. Because different units and locations will most likely have different vulnerabilities and be at differing levels of risk, each location should be evaluated separately. Critical factors for developing a plan will include evaluating specific hazards, determining their relative risk, and evaluating economic realities associated with managing this risk. As mentioned above, there is a strong parallel between developing a preventive strategy for a terrorist attack and the development of preventive measures under Hazard Analysis Critical Control Point (HACCP) plans (see, for example, 21 CFR Part 123). The emphasis here, as with HACCP, is placed upon *preventive* and not *reactive* measures. HACCP is a systematic approach to the identification, evaluation, and control of food safety hazards (Anon., 1997) and is the recommended protocol to follow as per WHO guidelines (WHO, 2002). An example of components in a food security plan is provided beginning on page 187 with factors to take into consideration for production and distribution of a specific food product. A sample of a food security hazard analysis worksheet is provided on page 191, with a plan and plan forms on page 193.

Fundamental to an effective security plan is that it be built upon a foundation that includes and integrates an effective HACCP plan, security and food safety aspects of good manufacturing practices (GMPs) (21 CFR Part 110), workable and effective sanitation standard operating procedures (21 CFR Part 110, 21 CFR § 123.11), and an up-to-date product recall program (21 CFR Part 7). Vigilance in maintaining these already mandated food safety programs, coupled with increased employee awareness, is vital should a suspected or actual contamination occur. Rapid communication regarding an incident to those with responsibility to manage it, recovery and handling of the affected product, and employee and public safety will be critical, and many of these aspects are already covered in some degree within the prerequisite programs listed above.

Similarly, a threat evaluation assessment and management (TEAM) approach can be used. This is a version of organizational risk management (ORM) and it provides a systematic approach to identify and focus food security efforts to address the most critical risks. The six steps in this program are essentially identical to the HACCP-based model described below (DHHS, 2001) and include (1) identify threats, (2) assess the risk, (3) analyze risk control measures, (4) make control decisions, (5) implement risk controls, and (6) supervise and review. (See Appendix A.)

Operational risk management rules are somewhat different than HACCP. These include the following:

1. Accept no unnecessary risk; there must be a commensurate return in terms of real benefits or opportunities. Under "real" HACCP, this holds true; however, under "regulatory" HACCP (e.g., 21 CFR Parts 123, 1240), the development of preventive measures, monitoring, and control procedures for certain risks may be mandated by guidance, whether or not the risk is reasonably likely to occur (the prevailing legal standard for HACCP), or whether or not the risk is applicable to a specific product or in a particular processing operation.
2. Make risk decisions at the appropriate level, as this establishes clear accountability. Include those accountable in the risk decision process. This is also similar to HACCP; however, adoption of plans covered under HACCP regulations has an additional requirement for plans to be formally adopted as corporate policy. For example, a signature of the highest-ranking official at the establishment is required for seafood HACCP programs when the plan is adopted, upon annual reverification, and after modifications to the plan are made.

3. Accept risk when the benefits outweigh the cost. HACCP requires hazards that are reasonably likely to occur (and, more recently, in situations where regulatory agencies mandate control of a risk regardless of the true risk posed). Clearly, ORM involves a cost–benefit analysis where control of hazards in a HACCP program is technically independent of cost. However, under HACCP there is still latitude provided regarding the means selected for controlling a particular hazard, and lower cost options can be selected (and under regulatory HACCP, as long as the regulatory agency agrees that the selected methods are effective).
4. Integrate ORM into planning at all levels. HACCP programs, too, are to be integrated into facilities operations. Because HACCP is focused only upon food safety, the scope of a conventional HACCP program would be less broad than that of a food security program.

ORM uses a matrix to assist with risk assessment. This assessment evaluates the likelihood of a hazard occurring (frequently, likely, occasional, seldom, unlikely) and its potential severity (catastrophic, critical, moderate, negligible) plus an assessment of hazard exposure (time, proximity, volume, repeated) (Figure 4.1). This matrix is more complicated than HACCP, in which there are only two assessments to make: (1) whether a hazard may cause food to be unsafe for human consumption (see, for example, 21 CFR Section 123.3(f)), and (2) whether a hazard is reasonably likely to occur (see, for example, 21 CFR Section 123.6(a)). A hazard that is reasonably likely to occur is one for which a prudent processor would establish controls (id.). Our experience has been that individuals working with the ORM risk assessment matrix get "hung up" with the analysis and have difficulty getting past this step and onto the development of an effective food security plan. The reason is that companies often have limited concrete information upon which to make these risk assessments. Furthermore, the inclusion of the important and realistic category of "unknown," commonly included in risk assessment matrices used by intelligence professionals, has been omitted from ORM. Frankly, we often do not know what the likelihood and severity of many food-borne risks are, and recognizing this fact would help to focus efforts on filling knowledge gaps. Because of the unnecessary complexity with the ORM matrix, we recommend, initially at least, that companies employ a risk assessment based upon the HACCP standards outlined above instead. This will catch the major hazards that should be controlled. The added level of sophistication in the ORM model can be incorporated into later evaluations when better risk assessment data become available.

Analysis of risk control measures under ORM also involves a consideration of cost and then how various risk control options could work together

		PROBABILITY					
		Frequent	Likely	Occasional	Seldom	Unlikely	Unknown
SEVERITY		A	B	C	D	E	F
Catastrophic	I	Extremely high	Extremely high	High	High	Medium	Unknown
Critical	II	Extremely high	High	High	Medium	Low	Unknown
Moderate	III	High	Medium	Medium	Low	Low	Unknown
Negligible	IV	Medium	Low	Low	Low	Low	Unknown
Unknown	V	Unknown	Unknown	Unknown	Unknown	Unknown	Unknown

SEVERITY
Catastrophic — food contamination incident that causes total business failure. Fatalities and numerous serious illnesses or injuries.
Critical — food contamination incident that has caused a major loss of business. Serious illnesses or injuries.
Moderate — food contamination incident that causes some loss of business. No serious illness or injuries.
Negligible — food contamination incident that causes a small loss in business. No serious illnesses or injuries.
Unknown — food contamination incident with a severity that is difficult to predict.

PROBABILITY
Frequent — food contamination incident that occurs often. Specific individuals or the general population is continuously exposed.
Likely — food contamination incident that has occurred several times. Specific individuals or the general population is regularly exposed.
Occasional — food contamination incident that occurs. Exposure is sporadic.
Seldom — food contamination incident that may occur. Exposure is low.
Unlikely — food contamination incident that is not to occur. Exposure is rare.
Unknown — food contamination incident that may or may not occur. Probability of occurrence is difficult to predict. Exposure of targeted individuals or the population at large is difficult to predict.

Figure 4.1 Operational risk assessment matrix with risk levels.

to address a potential hazard. The following are ORM-based risk control options (DHHS, 2001):

1. *Reject* — Refuse to take the risk if the overall costs of the risk exceed its benefits to the business or operation. An example provided is placing a new employee on night shift with minimal or no supervision when the individual has not been properly cleared or vetted. Another example is providing lockers for which only the employee has access.
2. *Avoid* — Avoiding a risk altogether requires canceling or delaying the job or operation and is rarely used. An example would be removing a salad bar from a restaurant to avoid intentional contamination of its contents with pathogenic bacteria.
3. *Delay* — Is it possible to forestall the risk? If time is not critical, then risk could be delayed. During the period of the delay, the risk may go away. For example, a business decision could be made to retain a salad

bar in a restaurant while threat levels are low. However, if threat levels increase, or public health or law enforcement officials notify a business or class of businesses regarding a particular threat, then the business could decide to switch to providing individual servings of salad.

4. *Transfer* — Risk transfer does not change the probability or severity of a hazard, but it may decrease the probability or severity of the risk actually experienced. An example would be a professional society or trade association providing an audited certification program for members of a particular industry segment. Such a program would reduce risk by increasing overall awareness and preparedness. The National Restaurant Association has prepared guidelines and training materials to address this need. Risk transfer could also be accomplished by forming a cooperative to insure against a terrorist loss experienced by a member of the cooperative.
5. *Spread* — Distribute the risk by either increasing the exposure distance or lengthening the time between exposure events. For example, spread deliveries in time and between different suppliers and carriers.
6. *Compensate* — Create a redundant capacity. This includes backup systems for critical equipment, staff, materials, and logistics.
7. *Reduce* — Plan operations and design systems that do not contain hazards by minimizing risk, instituting preventive measures, adding safety and warning devices, exercising a program, and providing training.

Making risk control decisions under ORM requires a determination of what risks to control and how by evaluating the different options available. The overall objective is not to reach the least level of risk, but instead to reach the best level of risk for overall food safety and security (DHHS, 2001).

Implementing risk controls will require an allocation of both human and financial resources. The individuals tasked with implementing a food security plan must know what the specific requirements are and fully involve personnel who will be responsible for functions within the plan. The implementation plan must be clear. Management must be supportive of the plan and provide incentives for improving food safety and security programs.

A possible deficiency with the ORM approach is failure to specifically include provisions to verify effectiveness of the program, to develop a series of control decisions for possible scenarios in advance, and to maintain adequate and systematic records. Our recommendation is that a HACCP-style verification procedure, record monitoring, and corrective actions program be incorporated into an ORM program if this model is selected for development of a food security plan.

Regardless of whether a HACCP-based or ORM model is used, to the extent possible, keep details of any food security plan confidential to limit the possibility of information falling into the wrong hands. Although food security plans are not yet mandated by governmental agencies, guidance has been issued (see appendices), and it will not be long before food security issues start to creep into food safety inspections. There is a legitimate fear that security guidance will morph into *de facto* regulations, as we have seen happen with HACCP over the past 8 years. Food security programs are and should be market driven. These are programs beyond the scope of food safety inspectors, with many components outside their jurisdiction. Many companies have had food security programs in place in the past, prior to governmental concern over these matters. Private sector plans are evolving as companies gain a better understanding of food security risks and how to control them, and as technology develops to improve the ability to manage risk. Let us keep it this way.

Governmental Emergency Response Systems

Governments have developed a myriad of emergency response and communicable disease monitoring and surveillance programs. Depending upon the location, these may or may not be terribly effective. Preparedness programs for response to a food incident should be integrated, when possible, with emergency response programs for more common or conventional disasters, such as fire, flood, earthquake, etc. Planning for these types of events has a long history and is generally effective, depending upon the level of public resources underpinning these programs and the experience emergency responders have with these more conventional incidents. Fortunately, with the possible exception of Israel, there is no government with a program in place to adequately handle a widespread intentional contamination incident involving food; there have simply been too few opportunities to develop expertise in this arena — not necessarily a bad thing.

The public health preparedness models developed for infectious disease outbreaks are not necessarily appropriate for food manufacturing, food distribution, or food service operations because the focus of these programs is by necessity on response and not on preparedness. However, these programs have features that overlap food industry HACCP-style programs.[7] Recent efforts to improve overall food safety infrastructure will go a long way toward handling a food crisis in case such a need arises. The WHO recommends that the following specific points be addressed in public health preparedness programs involving food terrorism:

1. Planning should consider the ability of the surveillance system to detect food safety emergencies, including deliberate contamination.
2. Investigation of a potential outbreak identified through surveillance should include identification of the food and the presence of the responsible agent in the food.
3. An incident response should be made concurrently with all the necessary food safety components unless food as a vehicle for the agent has been ruled out.
4. Preparedness programs should be exercised with the agencies responsible for emergency response to food terrorism being involved. Any new components should be tested for effective response.

In countries where there is substantial government/industry cooperation in building an integrated food safety infrastructure, the WHO guidelines pertaining to resource allocation, needs and risk assessment, evaluation of food system vulnerability, legislation, and regulatory requirements can be applied directly and a logical and well-focused approach to food security developed. However, not all governments have the infrastructure needed to assist industry, particularly small business, in developing effective food safety plans (WHO, 2002); this inability increases the risk of intentional contamination. The underlying presumption here is that it is the role of the government and not the food processor or seller to ultimately make sure that food is safe.

Assuming that this presumption is true, a series of generic guidelines have been proposed for governments to take in assisting private industry (WHO, 2002):

1. Cooperating with industry to develop protocols for assessing the vulnerability of individual food businesses, including assessment of the site, security, and personnel, and potential ways in which food might be maliciously contaminated
2. Ensuring that food safety is addressed and controls are coordinated at all links of the food chain, including traceability and recall
3. Cooperating with industry to strengthen the security of processes, people, and products
4. Providing industry with information on known or possible biological, chemical, and radionuclear agents
5. Cooperating with industry to develop, implement, review, and test crisis management plans
6. Coordinating closely with industry in communicating with the public, including effective media communications

7. In case of an incident, sharing early surveillance information with the food industry to facilitate prompt action to address consumer concerns, contain the threat, and mitigate damages

In the U.S., and in countries where food safety infrastructure is not well developed, the overwhelming burden of ensuring food security has been placed in the hands of the private sector. This situation is not likely to change. In this case, private-sector activities must be proactive, with risk analysis and preventive measures structured in a manner similar to that of programs already in place for handling unintentional contamination incidents. The possibility of an intentional contamination incident needs to be incorporated as an integral part of current food safety considerations, and measures to prevent sabotage should augment, and not replace, other food safety activities (WHO, 2002). Unfortunately for the private sector in the U.S., many aspects of the recent vulnerability assessment conducted by governmental agencies, including those of the USDA and FDA analyzing the terrorist risks facing the food supply, have been classified, meaning that private companies will be kept in the dark regarding the most likely risks to their businesses. This will result in the development of incomplete or erroneous vulnerability self-assessments for food- and agriculture-based businesses. This failure to provide necessary information to the private sector will dilute the efforts of both the private sector and the government to develop and then adopt or promote adequate and appropriate food security planning. It will also mean that the most reasonable and effective preventive measures that could be developed to protect the safety of the food that a company produces and distributes may not be undertaken.

For nations with a weak food safety and public health infrastructure, any vulnerability assessment that is conducted is not likely to be complete since it will be difficult to conduct the necessary studies to ascertain the actual risks to the food supply. Often times, there is little reliable information regarding food production, and with critical sectors within the industry not well understood. For companies operating in situations such as this, or who are importing products from countries with poorly developed public health and food safety infrastructures, the development of effective preventive measures and response planning will be difficult, diffuse, and sporatic.

Evaluating Security Risks and Identifying Hazards

Initially, a company or organization should complete an analysis of its facilities and operations to identify significant hazards, identify the potential exposure to a particular hazard, and evaluate the risks of an occurrence. This analysis, sometimes called a vulnerability assessment, incorporates scientific, economic, political, and social circumstances, measures the extent of the threat when possible, and sets priorities for resource utilization (WHO, 2002). Priorities must be set so that any action taken to deal with a security threat is commensurate with its severity and impact. There are specific factors tied to the public health impact of a food terrorist threat or attack that can be considered by a food operation. Such an assessment will provide the company, if nothing else, an idea of how the emergency response from the public health sector may proceed. Obviously, any incident will have major market impact regardless of the agent used, the terrorist motivation, or the level of preparedness and effectiveness in executing a governmental emergency response preparedness program.[8]

A risk analysis should not be limited to just the production facility or during times of peak operations. The evaluation should cover the entire scope of operations, including:

1. Agricultural production methods and harvest methods of ingredients.
2. Storage and transport of raw materials.
3. Receiving operations.
4. Materials and goods-in-process holding.
5. Security risk presented by other materials used within the facility (process or sanitizer chemicals, pesticides, pressurized gases, water and boiler treatments).
6. Suppliers of various components, their potential vulnerabilities, and degree of preparedness planning.
7. Type of processing used for manufacture of particular food items and risks associated with this.
8. Processing lines and their configuration.
9. Subcontracting facilities, packaging, warehousing, rolling stock, distribution of processed goods, physical plant, wholesale and retail distribution, and retail and consumer use. Evaluation of the risks presented by neighboring facilities is also important (WHO, 2002) and includes power plants, fuel or chemical manufacture, storage or distribution, military, law enforcement or government facilities, transportation and communications operations, and infrastructure.

Any point in the food chain where product changes hands is most vulnerable. Access to all critical areas should be controlled. Factors surrounding a research center, farm, or ancillary site security should also be considered. Any evaluation should examine the raw materials, distribution methods, and handling practices of common carriers or third parties. Water sources and supplies may be of specific concern, particularly if water is used as an ingredient or comes in direct contact with consumable products. In effect, the chain of custody for the product and its components should be monitored from farm to table.

As with HACCP, a team should be used to develop the plan and conduct a vulnerability assessment. This evaluation should include individuals from human resources, marketing, distribution, and sales, as well as those involved with quality control, production, and security functions. In larger organizations, this may actually consist of a series of smaller teams formed within identifiable units. Regardless of the structure of an organization, good leadership and a comprehensive integration of the team recommendations are critical factors, as is buy-in of the resultant program by both management and employees at all levels.

Managing the Risk: Preventive Measures

Since it will probably be impossible to eliminate all hazards, a reasonable procedure must be instituted to manage them. Probably the best strategy is to develop preventive or risk control measures that would reduce or eliminate the most significant hazards, and then work down the list to control less likely or less serious hazards. As part of this, points in an operation that are critical for controlling the identified security risks should be developed. These points may change during the course of a day, seasonally, or with ingredient sourcing, supplier (e.g., domestic or imported product), product type, processing and packaging methods, distribution, and final end user. Following this step, establish a monitoring procedure for these risk control points (similar to the program already in place for monitoring critical control points in a HACCP plan). Along with these monitoring protocols should be corrective actions (again, similar to those in a HACCP program) in case of a breach of security or security failure. A plan for verifying the effectiveness of the preventive and risk control measures in a food security plan should also be included. The use of forms such as the "HACCP Hazard Analysis Worksheet" or the "HACCP Plan Form" (see, for example, FDA, 2001) may be of benefit in some cases. See also the example at the end of this chapter.

Suggested Steps for Developing a Security Plan

Here is an approach for developing a security plan based upon HACCP principles:

1. Develop a comprehensive flowchart(s) depicting an operation from primary production or receiving to consumption by the end user.

2. Examine the flow chart and each element therein to determine whether there are significant food security hazards and evaluate the likelihood of the risk of these hazards.
3. Determine the points in the operation that are critical for managing a specific risk. These could be locations, processes, functions, or times when the operation is at greatest risk.
4. Develop and institute preventive or risk control measures to reduce these hazards to acceptable levels.
5. Where appropriate, establish critical limits or restraints that are not to be violated or breached without a resulting corrective action being initiated and completed.
6. Develop monitoring procedures for each critical point in the security plan. Monitoring is a systematic periodic activity to ensure that critical controls are in place and have not been breached or compromised in any way. These should be in writing. Test to see that the monitoring procedures are working and workable for the operation.
7. Develop a procedure similar to the corrective action program under HACCP to fix security problems or failures that occur if a critical control has been breached or compromised. Ensure that the problems are fixed by rigorously retesting the system and its risk-monitoring procedures. Then revise the security plan to include any changes to the critical controls or monitoring procedures and to reduce the likelihood that a similar breach would happen again. Corrective actions may also include the prompt notification of appropriate authorities and the execution of ancillary steps, such as an evacuation, lockdown, or similar activities.
8. Periodically test or verify the security program to ensure that it works. Verification programs should be written as confidential protocols. Revise written protocols when the operation or any key features of it change. A change in operation procedures, product form, suppliers, distributors, etc., may introduce or remove hazards and require that the plan be revised.
9. Above all, adequate and comprehensive records must be developed. These records should be handled as confidential. They should also be maintained to record monitoring, preventive measures, deviations, corrective actions, and verification activities. Supporting documentation should also be incorporated into the records. These might include outside agency notification protocols, hazardous material information, media protocols, an employee notification plan, response team information, and recall procedures. Supervisory personnel, on a timely basis, must systematically and periodically review these

records. The inclusion of superfluous and unnecessary documentation should be avoided (Bledsoe and Rasco, 2003).

Security Strategies

Surveying a Site

A good digital camera, access to plant plans, and aerial photos are excellent tools to use when developing a security plan. Aerial photos may be obtained specifically for the operation, or access to them may be obtained through the local county assessor or land use office, and sometimes through county extension offices. Aerial photos may also be available off the Web from a number of state, federal, and private sources. Consider using them, particularly since they are also readily available to the terrorists, who are not bashful about employing such technology. Have an up-to-date floor plan available, but accessible from a secure location near the site. Copies of these should kept on file with the local police or fire department and held as confidential.

Specific Suggestions

The key to a successful program is vigilance by management and all employees. Training is critical. A clear standard operating procedure must be developed and followed for day-to-day operations, for handling suspicious incidents or individuals, and for actual attacks. The problems arising from an actual attack may be similar to outcomes already included within an existing crisis management plan for natural disasters, loss of power or communications, etc. If product safety is an issue, recall procedures would need to be followed. As with recall programs, individual farms, companies, or research institutions should periodically use exercises and drills to test whether a security plan is current, workable, and effective.

Unfortunately, cost will often be the controlling factor in development of a food security program since it is impossible physically and financially to guard against every eventuality. Not all of the recommendations included here will be appropriate, practical, and cost-effective for every individual entity. As with HACCP, food security programs will be market driven.

Farming Operations: Agricultural Production and Harvesting

Recent unintentional incidents surrounding the possible contamination of animal feed with prions in Europe, Canada, and the U.S., dioxin contamination of chicken feed in Europe (WHO, 2002), the large outbreak of Avian flu in Southeast Asia in 2004, and the mad cow incidents in 2003 in the U.S. and Canada following the devastation of the U.K. beef industry in the mid-1980s all show the widespread impact that an agricultural incident can have on human and animal health, local and national economies, and consumer confidence. As a result of these incidents, animal feed is under heightened scrutiny, with imported components examined more closely. The growth of genetically modified (GM)-free and organic food and feeds can be directly traced to these food safety incidents, and focuses attention on the needs of these specialty markets. (See Appendix C.)

Ingredient traceability will become increasingly important as new regulations requiring component tracing one step forward and one step back are implemented in the U.S. beginning in 2004. Similar regulations are in place by regulation or market forces in the EU and elsewhere. These traceability requirements have the objective of increasing the ability to conduct a focused and rapid recall of affected foods or food components. Traceability requirements will decrease the practice of commingled components, but will be difficult to implement in many agricultural production settings. It is a very common situation to incorporate products from several smaller farms into a single bulk shipment. Adding clearer lot coding, seals, and a greater degree of packaging will improve security and reduce the likelihood of intentional contamination during feed distribution.

For animal feeding operations, bin locks or other tamper-evident devices should be placed on feed bins, and the security of water delivery systems should be evaluated and appropriate controls instituted.

Food safety issues, in addition to production efficiency, diversification, and value-added production practices, have increased the level of sophistication of many farming operations regardless of their size. Risks for intentional contamination at the farm level include deliberate contamination of feed and water supplies, animals, or crops with toxic chemicals or with human pathogens. Foods that are minimally processed can pose the greatest hazard. Control will be particularly problematic in less developed regions and in low-margin segments of the food industry and in industries with less sophistication.

Greater controls have been in place for farm chemicals in the U.S. since the Oklahoma City bombing in 1996. Measures have also been instituted in recent years to more closely monitor stocks of fuel, ammonia, and ammonia-containing fertilizers because of the risk of theft of these materials for bomb construction and production of methamphetamine. Assistance from local law and drug enforcement authorities to improve monitoring has been generally effective. However, further measures can be taken to protect crops and stock if not already in place. For pesticides, a farmer could consider requiring certifications from seed, feed, livestock, fertilizer, pesticide, and herbicide providers to ensure that materials meet specifications and are not contaminated. In addition, periodically seeking third-party verification is helpful from a food safety and marketing perspective. We recommend that growers avoid stockpiling hazardous materials, keep the amount on site to a minimum, and secure both the stores and applicator equipment.

Growers should develop process controls using HACCP-based systems to ensure that materials processed on the farm comply with current internationally recognized production standards. For example, open-air drying is commonly used for grain, fishery products, fruits, and spices, and is often not well controlled (WHO, 2002).

Product should be monitored systematically from the time of harvest until its safe transportation and storage within a warehouse. To the extent practical, access to croplands and livestock should be controlled and restricted to appropriate personnel. Surveillance equipment is also an option; the cost of such equipment has decreased markedly in recent years. Access to animals at auctions and sales barns should be restricted, and direct contact with animals tightly controlled. Consideration should also be given to compartmentalizing livestock operations, improving hand-washing/sanitation facilities, providing or improving clothes-changing facilities for employees, improving equipment-cleaning operations when animals are to be transported between two locations, and requiring foot and vehicle sanitation dips at critical access locations as ways of controlling the spread of a disease.

Manufacturing or Food Preparation Operations

Ensuring the physical security of a building is critical. This includes the ability to screen, close, and lock any door, window, window well, roof entry, or ventilation inlet or outlet. Metal or metal-clad doors are recommended (WHO, 2002). There should be limited and controlled access to any food storage or processing area. For both food safety and food security considerations, traffic within food preparation areas should be kept to a minimum.

Access to the physical site should be controlled as well. This includes outside boilers, water and waste material handling systems, and HVAC and ventilation systems.

An objective should be to remove any hiding places in food preparation areas, storage areas, janitor's closets, lockers, restrooms, office and reception areas, break rooms, smoking and parking areas, etc., where dangerous materials could be stored for later use. Similarly, elimination of these factors will reduce the temptation for employees to have food, beverages, cigarettes, or contraband within a food production facility.

Surveillance of both the exterior and interior of a facility should be installed if such a system is not already present. The system should be designed so it can be easily and effectively monitored. Alarm systems should be routinely tested to ensure that they function properly.

Food storage and production areas, which should be inspected on at least a daily basis for sanitary and safety issues, can be updated to include a review of food security concerns. Maintaining monitoring records for these activities is important to determine compliance.

An accurate inventory of materials within the operation should be kept and monitored daily. This includes product and raw materials, hazardous components, tools, utensils, and knives.

Naturally, any incoming materials should be inspected to ensure that they are according to specification, with intact packaging, at a proper temperature, clean and with no unusual characteristics. The delivery slips or load manifests should be compared with orders made by the facility for both content and quantity. Ensure that there is enough information on the product and the attendant paperwork to trace materials back to a supplier, supplier production lot (if possible), and location. This is a new requirement under the Bioterrorism Act pertaining to records. Also, employees responsible for receiving and inspecting incoming materials should be provided with training on how to detect adulteration, and instructions on how to segregate affected product, who to notify within the company if adulteration is suspected, and how to contact them after hours. Contact information for public health officials and law enforcement should be available and updated as necessary. If material is suspect, do not move it, if at all possible. Treat the material and surrounding area as if it were a crime scene — leaving the place alone will help to preserve evidence. Likewise, cordoning off the surrounding area will help to preserve evidence. In case of such an incident, employees should be evacuated from the affected area if there is a risk to their safety.

Locations of equipment and utility controls and safety switches, escape routes, and emergency exits should be clearly marked and accessible. Additional guidance is provided in Appendix C and Appendix H.

Facilities Configuration

Facilities should be constructed to control product flow in a rational way that improves operational efficiency and, as an ancillary benefit, food safety and food security. Following the guidance in 21 CFR Part 110 for good manufacturing practices is a reasonable place to start. Operations that are easy to follow, clean, and maintain are those in which contamination (intentional or unintentional) is less likely.

Compartmentalized functions, such as physical segregation of raw and cooked products, are essential. Good lighting should be provided. Access to the building should be controlled in such a manner that it meets the needs of the business, but prevents uncontrolled access from outsiders. Adequate and functioning safety (e.g., sprinklers) and alarm systems are important deterrents and would mitigate damages in case of an arson attack. Recent guidance on plant ventilation, filtration, and air-cleaning systems to protect building environments has been developed (CDC/NIOSH, 2003) and provides good design considerations for new construction. However, the costs for retrofits may not be cost-effective for food businesses.

Facilities Access

Reducing points of access to a facility should be considered (FDA, 2002a). This may include improving the security of or reducing the number of accessible doors, windows, hatches, trucks, railcars, or bulk storage areas. Emergency exit integrity, in appropriate numbers, should be maintained with alarmed "Emergency Use Only" exits where appropriate.

The Water and Air Supplies

Additional preventive measures concerning the safety of the water supply used within a food processing operation should be considered. Evaluating the security of wells, hydrants, and storage and water handling facilities, whether these are on site or controlled by a municipality, are prudent. Even if water is from a municipal source, assuring the integrity of this supply ultimately falls upon the production facility. Normally, the water quality is the responsibility of the operation from the meter on, but questions should be brought to the supplier if liability is an issue. Many water suppliers are notoriously negligent in implementing even the most basic security practices. Unsecured wells, standpipes, reservoirs, and pumping stations are often open to public access. Consider checking water quality more frequently regardless of its source. Locating an alternate source of potable water, providing for additional on-site storage in case of emergency, or providing a backup water purification system may also be desirable (FDA, 2002a, 2002b). As per other FDA requirements, backflow prevention devices must be installed and functioning. Sanitation standard operating procedures (SSOPs) require checks to ensure that there are no cross connections in water supply contacting food or food contact surfaces.

Precautions should also be taken to ensure that air entering the operation is not contaminated. This could include securing access and a routine examination of air intake points for physical integrity (FDA, 2002a).

Suppliers: Letters of Guarantee

Supplier audits coupled with letters of guarantee are effective tools for improving adherence to contract specifications and for maintaining food safety. Food processors should request letters of guarantee from suppliers and require a showing of protected transportation of ingredients. It would be prudent to revisit inspection programs for incoming ingredients, including packaging materials, labels, and supplies used within the production facility and office. Specifically, do not accept unordered ingredients/shipments or product received in opened containers or with damaged packaging. Require tamper-proof packaging or shipping containers as well as numbered seals that can be independently verified with the shipper or vendor.

Also ensure, as part of your recall program, that any specific lot of an ingredient can be tracked from receipt through production to final product and distribution. Work with suppliers and common carriers to ensure that they have instituted appropriate food security programs. Develop an audit program in this area similar to one that may already be in place as part of an existing food safety or food quality preferred supplier program. Periodic inspections of vendors should include an examination of their distribution systems.

Distribution and Transit

Controls during distribution and transit are important and are often outside the control of the producer/manufacturer. Shipping contracts should provide provisions requiring that the carrier secure loads during transit and, if product or container is to be held, that the container be secured by fencing, locks, etc., to prevent unauthorized access to food contents. On-site security, alarm systems, including silent alarms, and other types of intrusion detectors or video monitoring may be appropriate security measures depending upon the circumstances. Transit by common carrier will require an evaluation of current preventive measures and could include an expanded tracking, for example, use of tamper-proof seals on containers coupled with en route monitoring. Modification of current temperature or environmental monitoring systems to improve traceability features is one other consideration. See Chapter 5 on traceability provisions and Appendices B, H, I, and K.

Tamper-evident and tamper-resistant seals unique to any given bulk container should be used. If possible, seal alphanumerics should be communicated electronically, separate from the shipment itself, to the receiver and to the sender, with the numbers and seal integrity verified prior to opening the container and retransmitted to the supplier upon receipt. Off-loading should be conducted under controlled conditions with periodic testing of off-loading security a must. The integrity of finished products (including reconciling the amount received with the amount ordered) should be controlled during storage and distribution (FDA, 2002a). Such monitoring is now required under recordkeeping requirements in regulations promulgated by the FDA under the Bioterrorism and Response Act. Where appropriate, tamper-proof or tamper-evident packaging, at several levels, may be advised. This is easier and cheaper to do now that radio frequency identification (RFID) technologies have advanced to the point of being cost-effective.

Any product returned should be closely examined for integrity and product quality before reshipment is considered.

Wholesale and Retail Distribution

Any product can be tampered with during distribution or at retail by individuals with malicious intent and the right technical knowledge (WHO, 2002; FSIS, 2003). Packaged foods, particularly those in vacuum-sealed containers, glass, heavy plastic, or metal, can be sabotaged by a terrorist with a relatively low degree of sophistication. (See Appendix G.)

Again, controlled access is important, and in retail settings, the vigilance of employees and customers is even more critical than it might be in manufacturing. Many retail establishments have camera systems set up to reduce shoplifting and to provide evidence in case of a robbery or assault. These systems should be reevaluated to determine whether they are effective and whether additions of hardware or monitoring devices would provide deterrents that could increase food security.

Bulk food displays, produce, self-service salad-and-soup bars or delis, and communal food displays, including buffets, are vulnerable and are probably the foods at greatest risk for tampering by customers in a retail store or restaurant. It may be advisable to remove or shrink the size of bulk food displays and replace these as much as possible with packaged foods. Having self-service sections positioned within the store where they can be closely monitored is another way of reducing the likelihood of intentional contamination. Produce sections, at least in newer establishments, are quite prominent and may be easier to monitor than similar sections in older establishments.

Restaurants and retail stores are uniquely vulnerable due to the desire to attract people to enter the business, effectively providing uncontrolled access to the establishment by the general public. These establishments have been targeted for attack through intentional contamination of the food, bombings, arson, and vandalism. For restaurants and retail stores, security can be increased by enforcing existing policies of excluding customers and other nonemployees from food preparation and storage areas. Improving the physical barrier between these locations would improve food security somewhat.

Automatic dispensers for beverages and sauces and vending machines can be tampered with and should be included in food security inspections and examined to determine if risks from intentional contamination can be reduced, for example, by incorporating tamper-evident or tamper-resistant features into the equipment. If not, then monitoring should be increased. Improved monitoring may involve relocating these units to higher-traffic areas or areas under greater employee control.

Employee Screening

Employee and contractor screening have become increasingly important. Where appropriate, a criminal background check can be conducted as a condition of employment. However, our experience is that most individuals engaged in terrorist activities do not have criminal records. Furthermore, many good employees in the food industry have had brushes with the law and should not be barred from employment without consideration of the underlying issues that resulted in a criminal conviction. Food manufacturing and food service businesses serve an important role in society as major avenues for employment of ex-convicts and as sources of job training for individuals on supervised release programs. Discrimination against individuals with criminal records and a variety of other equal protection and due process issues have arisen from concerns over employee background checks, profiling based upon gender, age, religion, and national origin being the most infamous.

Initially there were rumblings that the government would mandate certain types of employee screening for the private-sector employers, but these were quickly dropped. Issues of cost, timeliness, and effectiveness were cited as reasons for not adopting uniform requirements. One proposed model was for a program similar to SEVRIS, which is used for evaluating foreign nationals during visa application processes. Under such a program, a federal agency would expand and refine the current employment eligibility programs to provide national and local agency checks and report the findings to the employer in a timely manner. This would be much more effective than requiring individual employers to conduct such checks but would still be funded by "user fees" and take potentially weeks to complete, much longer than the time restraints food industry employers generally have for hiring employees. Restaurants can have 100% turnover in employees within a quarter, and mandating a national background check for individuals in this sector would be completely unworkable.

Using a credit check may be a more valuable screening tool for potential employees, as it provides an evaluation of an individual's stability within a given community. However, because the food industry commonly provides entry-level positions and is the largest employer of individuals with no prior work experience, credit checks may be of limited use. Young people often have no banking records, credit cards, or rental history and probably have no credit.

The major purpose of any employee screening program should be to check on an individual's veracity. Therefore, any signs of dishonesty in an application, during an interview, or through references should be a greater cause of concern than a criminal record or poor credit history. Reviewing references for potential employees, especially contacting former employers, teachers, and coaches, will provide better insight into an individual's character than record checks. Psychological profile testing is also becoming more common, but it is expensive and probably not feasible for use, at least for entry-level positions.

Consider adopting a policy in which all job applicants apply for positions at a location far removed from the processing facility. As part of this, initial screening and interviews of potential employees and contractors can also occur off site.

A two-person rule should be in effect for both safety and food security reasons. This means that no individual would be left alone during food preparation or handling.

Contractor Screening

Contractors can have relatively open access to the facility (e.g., outside cleaning crews, pest inspectors, construction, maintenance, and repair crews, truckers, security personnel, etc.) and should be held to the same standards as employees.

Employers should also ensure that both employee and (sub)contractor rosters, job, and shift assignments are current, reviewed on a daily basis, and updated as needed. There should be a formalized check-in and check-out system so it is clear who is on site and on duty. It may be prudent for employees/contractors to wear ID while on the job, with the policy that badges are recovered when an individual is no longer on assignment. Such badges can be color coded, or otherwise identified, to indicate to which parts of the plant or operations that the individual has authorized access. These badges should also be periodically collected unannounced, accounted for, and reissued in a different format.

Similarly, there should be adequate surveillance of contractors while on the job, and implementing control measures similar to those in place for employees may also be desirable. Contractors, just like employees, should not be at the work site unless they are scheduled to be there (FDA, 2002a). When possible, schedule new contractors for the shift where the greatest degree of supervision is provided. Avoid having contractors you do not know well and who have not been properly vetted from having access to an unattended or minimally staffed facility.

Personal Items

Under food GMPs, no personal items such as lunches, purses, etc., should be permitted into a food processing area. You may wish to extend this procedure to prohibiting any personal objects from entering the facility. The FDA recommends that employees be provided with mesh lockers with employer-issued locks for storage of personal items on site (FDA, 2002a). A condition of employment would be that the employer may inspect the personal property of any employee at any time.

Compartmentalizing Job Functions

Job functions within a facility should be compartmentalized to the extent practicable. This would mean restricting access to specific areas of a facility to only the individuals who need to be there. This could be done through the use of color-coded badges, uniforms, or other measures. Controlling access is particularly critical for operations processing ready-to-eat food products and has been found to be an effective preventive measure for the spread of *Listeria* sp. and other microbes though a processing facility.

A special note should be made in regards to discharged employees/contractors. Security badges, keys, etc., should be immediately surrendered by the individual. The person should be promptly escorted from the facility and not be permitted to return except as an escorted visitor. Discharges, whether they are simply the result of the end of a season, a reduction in workforce, or a firing for cause, need to be handled carefully and with appropriate sensitivity.

Visitors and Inspectors

Individuals purporting to be inspectors should provide appropriate identification and be vetted by backup procedures. This may simply involve a call to the publicly listed telephone number of the visitor's parent operation. Such individuals should be escorted at all times within the plant. Consider a no-photography policy as a way of improving security and as a means of protecting intellectual property if this policy is not already in effect (Rasco, 1997).

Access to processing areas, including locker and break rooms by visitors (including truckers, delivery people, supplier representatives, customers, applicants for employment, or other visitors) and employees, should be strictly controlled both within the plant and between different areas of the plant. A check-in procedure and issuance of visitor badges should be conducted in a reception area or another location that is not adjacent to the processing area or accessible to the processing area without proceeding through physical barriers. All visitor badges should be accounted for on a daily basis. Some firms will no longer accept visitors on site or visitors who have not made appointments in advance. Where visitors and tours are an important part of public relations or marketing, visitors should be confined to viewing galleries or, at a minimum, be closely monitored and escorted at all times. All individuals with escort authority should be trained and be aware of the importance of their responsibilities.

Keys and Access Cards

Ensure that all keys can be accounted for and that keys each have unique identification numbers. Keys should be marked "do not duplicate." Better yet, consider the use of card-swipe electronic locks that eliminate the need for metal keys. Most of these key card systems allow for improved control over access and maintain a record when individuals have gained entry. Individual access can also be controlled on a time basis, thus permitting entry only during scheduled hours. ID badges commonly serve the dual purpose of also being access cards. Periodic unannounced inventories of keys or cards should be considered as a means of improving security.

Parking

Stricter control of parking at the facility may need to be instituted, including the use of parking permits and vehicle registration. Enclosing the parking area; increasing physical security, "no parking" safe zones, access, and lighting; and instituting a vehicle inspection program are possible options.

Employee Vigilance and Employee Training

Employees should be made aware of their responsibilities to stay alert for and report suspicious activities, objects, and persons at their workplace or at home.

The responsibility for specific security functions should be assigned to qualified individuals and included within job descriptions. One individual within the organization should have primary responsibilities for food security operations.

Food security training programs should be provided to employees with periodic updates that include how to prevent, detect, and respond to a product-tampering incident, terrorist activity or threat, and suspicious or threatening activities or persons. All employees, even those not directly associated with food production or handling, should be trained on food safety and food security programs. This would include training in how to handle hostile employees and individuals within the organization who may threaten to or are suspected of product tampering.

This could be conducted in conjunction with HACCP and recall training or refresher programs. Sales personnel and others, including distributors and retailers, should be made familiar with the products and their packaging and distribution so that they may be able to detect whether a product has been altered or contaminated.

Unusual behavior on the part of employees or contractors, such as staying unusually late or arriving early, reviewing or removing documents or materials not necessary for their assigned work, and removing materials or documents from the facility or seeking information about sensitive subjects, should cause suspicion (WHO, 2002).

Security Checks

Security checks should be conducted on at least a daily basis. All employees and contractors should be trained to be vigilant for the presence of unidentified, unattended, or unauthorized vehicles, the presence of containers in or near the facility, and unauthorized access (even to unsecured areas) by unidentified persons or employees who have no apparent reason to be there. Also, employees should be trained to look for signs of sabotage or tampering of equipment, products, or ingredients; removal or tampering with product or worker safety features of equipment; or attempted unauthorized access to equipment.

Data Security

Ensure that operational, mandatory records and business confidential materials (including product formulations, analytical results, and operational parameters) are backed up with electronic and hard copies stored at a separate location. Restrict access to the computer system and remove access immediately for individuals no longer employed or contracted by the firm. Test and update firewalls and virus detection and cleaning systems. Evaluate the computer system on a regular basis for security.

Mail Handling Procedures

In light of recent developments, it may be prudent to have procedures in place for handling shipments to the facility, including suspicious packages and mail. This could include securing mailrooms and instituting visual or instrument-based package screening. Specific guidance has been provided for identifying suspicious packages and mail by the Centers for Disease Control (2001). These include:

Labeling

Inappropriate or unusual labeling
Excessive postage
Handwritten or poorly typed addresses
Misspellings of common words
Strange return address or no return address
Incorrect titles or title without a name
Not addressed to a specific person
Marked with restrictions such as "personal," "confidential," or "do not x-ray"
Marked with any threatening language
Postmarked from a city or state that does not match return address

Appearance

Powdery substance felt through or appearing on package or envelope
Oily stains, discoloration, or odor
Lopsided or uneven envelope
Excessive packaging material, such as masking tape, string, etc.

Other Suspicious Signs

Excessive weight
Ticking sound
Protruding wires or aluminum foil

Specific recommendations of the CDC for handling a suspicious package are not to open it and not to ship it to the recipient until it has been evaluated and cleared. As a result of the anthrax scares, packages that are suspicious should be handled as biohazards, meaning that the package should not be shaken, emptied, or carried around. Others should not be asked to handle, smell, touch, or taste package contents or to put themselves at risk of inhaling contents. Individuals in the area should be alerted of the concern about a package, and both management and law enforcement should be contacted. If dispersion of possible contaminants such as spores or volatile components within an area is a possibility, then the area should be closed off, including closing doors and windows to contain any airborne materials. If possible, a list of individuals in contact with the package or in the area with it should be written down, in case there is the need for medical evaluation. Also, having such a list will provide public health and law enforcement with sources of information in case of a criminal investigation.

Emergency Evacuation Plans

The multiple simultaneous bombings at 10 commuter stations in Madrid, Spain, on March 11, 2004 (exactly 2 1/2 years after 9/11), killed more than 190 people and injured over a thousand. This is a grim reminder that the most likely terrorist threat we face is from bombings. Corporations have been bombed repeatedly, mostly by political and environmental anarchists of various sorts, with food retailers and restaurants probably at the greatest risk. Fortunately, most businesses have emergency evacuation plans and are required by local law to have these plans in place. These plans should be reviewed for appropriateness in consideration of potential business-specific (e.g., biological) or other forms of terrorist threats. Management should file a copy of the operations' safety and emergency procedures with the local municipal planning department and with emergency response agencies. However, these governmental entities must be required to safeguard these documents and be *prohibited from releasing them to any parties without the knowledge and written consent of management*. An additional option is to have the evacuation plan along with the facility layout in a locked and sealed container outside the facility in case access to it is limited in an emergency. Conduct unannounced tests of evacuation plans for the facility. Train employees how to respond to certain types of emergencies and natural disasters (flood, fire, earthquake, tornado, chemical spill, power outage) and include this training as part of new employee orientation (WHO, 2002). This training will also provide most of the skills needed if the company becomes the target of a terrorist attack. As many individuals as possible should be trained in CPR and basic first aid. Make certain that contractors and visitors know company evacuation procedures and the special hazards at the operation (e.g., high-voltage electricity, steam, toxic and corrosive chemical storage and use). At a minimum, for visitors, ensure that they are escorted through the facility, can understand the instructions of their escort in case of an emergency, and that exits are clearly marked. Make sure that contact information with emergency responders is up to date.

Research and Quality Control Labs

Laboratories should also implement similar safeguards, including controlled access to laboratories, test plots, and the supporting infrastructure. Increased security of, and access to, hazardous materials is advised. GMPs require hazardous materials to be labeled, stored, and handled properly to avoid contamination of food and food contact surfaces (21 CFR § 110.20(b)(2), § 110.35(b)(iv)(2)). Hazardous materials can include, among other things, cleaning materials, solvents, acids, bases, paints, pesticides, lubricants, and water treatment chemicals. Locked access to dangerous biological materials or chemicals could be considered. Think about revisiting inventory control of hazardous materials (including some ingredients) and the safety and security of storage areas, including the use of these materials within the processing area itself. Access to hazardous materials should be limited to only those individuals who need to use them and who have proper training for handling them. Maintaining Material Safety Data Sheets (MSDSs) is mandatory at most locations. Periodic review of chemical inventories should be conducted to remove out-of-date chemicals and to eliminate those no longer necessary from the inventory.

For industrial quality control labs, lab access should be restricted to lab personnel only (FDA, 2002a). Under GMPs, dangerous materials should remain in the lab and not be brought into office or production areas. Assign responsibility for the inventory and control of dangerous materials (e.g., toxic reagents, bacterial cultures, drugs, pesticides, radiological materials) to a specific individual and include this responsibility within his or her job description. Have a plan in place for immediately investigating missing reagents or other potentially dangerous materials. Consider procedures for determining whether a theft has occurred, when and how to contact authorities, and how to maintain an evidentiary chain of custody.

Quality control labs can conduct random product and environmental testing as a preventive measure against contamination within the processing environment. This can include testing different portions than those that are normally sampled, such as sampling different regions of an animal carcass in addition to those proscribed by regulation or collecting samples at different times or different sampling locations. Develop a good working relationship with local or regional food testing and forensic laboratories because their services may be critical if an issue of product or facilities contamination arises.

Costs of Implementation

Implementation of food security plans will require outlays for equipment, materials, and most likely additional personnel (permanent or contracted). To date there are no proposed federal programs to assist the private sector with the costs of preparedness planning or implementation, although hundreds of millions of dollars have been proposed for upgrading and expanding food inspection programs and public health preparedness in government sectors. State and federal legislatures could provide economic support and incentive for these expenditures by implementing an investment tax credit at a modest level of 10%. Tax credits have proven to be a positive motivator for companies and a stimulus to the economy in general. A credit would provide direct tax relief while requiring a 10:1 investment by the tax-paying entity. Funds could be used to improve physical infrastructure or to hire personnel to improve safety at a facility.

Civil Liberties

The new federal and state legislations precipitated by 9/11 and later biological contamination incidents raise constitutional issues, including overreaching provisions on search and seizure, the potential for violations of equal protection safeguards, and loss of procedural due process under the USA Patriot Act and the Public Health Security and Bioterrorism Preparedness and Response Act. At the state level, a Model State Emergency Health Powers Act developed at the request of the CDC and the Bush Administration has been proposed and has been adopted by roughly 20 states to date. This Model Act is structured to legally facilitate preparedness, surveillance, management of property, protection of persons, communications with the public, and to coordinate public health efforts at the state level, which are often fragmented and inconsistent.

The Model Act is to amend often outdated public health laws that have been built up in layers during the 20th century in response to new and differing disease threats. For example, different types of quarantine may be permitted dependent upon the disease (Gostin et al., 2002). This Model Act is based in part upon exercises for smallpox (Dark Winter) and plague (TOPOFF) that simulated biological attacks on the U.S., raising the issue for the need for legal reform. As with the federal legislation above, the potential problem with the Model Act is overreaching government activity and inadequate safeguards to ensure due process and protection of individual rights.

Excessive use of government police power is justified by the drafters of the Model Act, citing "numerous circumstances" that might require "management of property," including closing facilities and destruction of property that is contaminated or dangerous. Unfortunately, as we have seen with the recent bovine spongiform encephalopathy (BSE) incidents, the cost for the property destruction is borne by the private entity and not the public sector. This flaw makes issues with voluntary compliance with surveillance provisions of these acts problematic for businesses. Under provisions of the Model Act, the state public health authority can close, decontaminate, or procure facilities or materials to respond to an emergency and not compensate private parties unless there has been a "taking." An example of a taking might be conversion of a private facility into an emergency hospital. However, there is no compensation for "nuisance abatement" — the situation most likely to arise with a food contamination incident, where the government is authorized under the act to destroy property or close an establishment that poses a serious health threat. Under a nuisance abatement scenario, an affected food business would be forced to destroy product (affected or not), and then to decontaminate, destroy, or replace equipment or facilities before operations could recommence. Most likely, there would be layers of governmental approval (federal, state, and local) required before a facility could reopen, as there would be potential issues tied to animal health and safety, food safety, occupational health and safety, building code, and environmental quality. Criteria for reopening a facility would be determined on an ad hoc basis and are not likely to be consistent between agencies within the different layers of the governmental entities involved.

Civil liberties of individuals victimized by the incident would also be compromised and could include compulsory vaccination, testing, physical exams, treatment, isolation, or quarantine. The drafters of the model act believed that individuals would not resist these intrusions based upon either their individual self-interest or desire for the common good; however, this is not necessarily true for a food contamination incident. Drafters of the Model Act were functioning in the weapons of mass destruction mode, with efforts focused upon containment and remediation of damages from a large-scale attack involving an (exotic) infective agent or toxin. Although use of an exotic agent is possible in food terrorism, it is not the most likely scenario. Intentional contamination with common chemicals (e.g., mercury), pesticides, or other common agents, and with common food-borne pathogens, is more likely, and questions of how to interpret the provisions of the Model Act within the context of a food contamination incident are not clear. Response under the act appears to be tied to the agent used, and not to the public health impact of the contamination incident. Furthermore, in food and agricultural businesses, an individual's concern over his or her legal status may cause that person to flee.

Provisions under the Model Act can be launched by a state's governor when a public health emergency is declared. A public health emergency is defined as:

1. An occurrence or imminent threat of an illness or health condition
2. Caused by bioterrorism or a new or reemerging infectious agent or biological toxin previously controlled (under state law)

3. And poses a high probability of a large number of deaths, a large number of serious or long-term disabilities, or widespread exposure to an infectious or toxic agent that poses a significant risk of substantial future harm to a large number of persons (Gostin et al., 2002)

Although the authors envision an emergency being declared only under extreme conditions, these provisions can be read broadly and constitutional safeguards thwarted. The drafters present four principled limitations[9] to agency action triggering implementation, but these safeguards could be easily circumvented or ignored since the overall goal is to avert a significant public health risk. In states, as with the federal government, much latitude is given under the law to agencies in technical matters, but provisions for judicial review and safeguarding civil liberties, let alone remuneration to injured parties, in general and under this Model Act, are weak at best. In short, the greatest risk to liberty may be that the safeguards are adequate in principle, but insufficient practically in the hysteria of an actual incident.

Government Response

Following the September 11 incidents, the Food and Drug Administration (FDA) contacted major food industry associations requesting that they advise their members to review current procedures and markedly increase vigilance (FDA, 2002a, 2002b).

If a terrorist attack is suspected, the FDA recommends seeking immediate assistance from your local law enforcement and health/hazardous materials handling experts (often the fire department). Additional support can be provided by the Federal Bureau of Investigation (FBI) (National: 202/324-3000), U.S. Department of Agriculture Office of Crisis Planning and Management (877/559-9872, 202/720-5711), the FDA Emergency Operations Office (301/443-1240), and the state emergency management division. Contact information for the relevant safety and law enforcement agencies should be readily available to employees and updated as needed. The FDA recommends that an organization have a capable media spokesman and generic press statements prepared in advance in case of an emergency (FDA, 2002a). In some states, such as Washington, National Guard units may have special training and equipment to respond to chemical or biological terrorist threats.

It is not possible to present a full picture of the bioterrorist threat to food production here, or to present every appropriate defense, let alone to address the full scope of terrorist threats, including cyberterrorism, more conventional acts, arson, vandalism, and economic terrorist acts. Suffice it to say that the threat is real and most likely these incidents will continue and possibly escalate. Individuals, institutions, and companies can become more cognizant of the threat and take steps to reduce the likelihood and impact of any incident. This does not mean that paranoia should reign supreme. These risks, as with others tied to food safety, are manageable. The risks must be kept in perspective. Ensure that common sense prevails. As with HACCP and recall protocols, prior planning, training, and established procedures are essential tools for establishing a successful program.

References

ACI (American Conference Institute). 2000. Product Tampering and Accidental Contamination Conference and Workshop. June 12–14, San Francisco, CA.

Anon. 1997. HACCP: Hazard Analysis and Critical Control Point Training Curriculum. North Carolina SeaGrant Publication UNC-SG-98-07, Raleigh, NC.

Anon. 2002. State Legislative Activity in 2001 Related to Agricultural Biotechnology. Pew Initiative on Food and Biotechnology. http://pewagbiotech.org.

Bledsoe, GE and Rasco, BA. 2001a. Taking the Terror out of Bioterrorism. *Food Quality*, November/December, pp. 33–37.

Bledsoe, GE and Rasco, BA. 2001b. Terrorists at the Table. Part II. Developing an Anti-Terrorism Plan. *Agrichemical and Environmental News*, Cooperative Extension, Washington State University, Tri-Cities, November, No. 187, pp. 5–8.

Bledsoe, GE and Rasco, BA. 2002. Addressing the risk of bioterrorism in food production. *Food Technology* 56:43–47.

Bledsoe, GE and Rasco, BA. 2003. Effective Food Security Plans for Production Agriculture and Food Processing. *Food Protection Trends*, February, pp. 130–141.

CDC. 1997. *Escherichia coli* 0157:H7 infections associated with eating a nationally distributed commercial brand of frozen ground beef patties and hamburgers. Colorado, 1997. *Morbidity and Mortality Weekly Report* 46:77–78.

CDC. 2001. Updated Information about How to Recognize and Handle a Suspicious Package or Envelope. http://www.bt.cdc.gov/documentsapp/Anthrax/10312001.

CDC/NIOSH. 2003. Guidance for Filtration and Air-Cleaning Systems to Protect Building Environments from Airborne Chemical, Biological and Radiological Attacks, DHHS (NIOSH) Publication 2003-136. April, 78 pp. http://www.cdc.gov/niosh/docs/2003-136.

CFR (Code of Federal Regulations). 2000. U.S. Government Printing Office, Washington, DC.

DHHS. 2001. Food Safety and Security: Operational Risk Management Systems Approach. November 29, 2001. Department of Health and Human Services, U.S. Food and Drug Administration, Center for Food Safety and Applied Nutrition.

Drees, C. 2004. U.S. Food Sector May Be Vulnerable to Attack. February 25. Citing Yim, R.A. Combating Terrorism: Evaluation of Selected Characteristics of National Strategies Related to Terrorism, GAO-04-408T. Presented February 3 before the Subcommittee on National Security, Emerging Threats and International Relations, House Committee on Government Reform. http://www.goa.gov/news.itms/d04408t.pdf.

FBI. 2000. FBI Sponsors Genetic Engineering Ecoterrorism Conference in Berkeley, CA. January 26. FBI, National Institute of Justice, Berkeley and Davis Police Departments.

FDA. 2001. *Fish and Fisheries Products Hazards and Control Guides,* 3rd ed. Department of Health and Human Services, Center for Food Safety and Applied Nutrition, Office of Seafood, Rockville, MD.

FDA. 2002a. Guidance for Industry. Food Producers, Processors, Transporters and Retailers: Food Security Preventive Measures Guidance. www.fda.gov.

FDA. 2002b. Guidance for Industry. Importers and Filers. Food Security Preventive Measures Guidance. www.fda.gov.

FSIS. 2002. FSIS Safety and Security Guidelines for the Food Processors. http://www.fsis.usda.gov/OA/topics/securityguide.html.

FSIS. 2003. FSIS Safety and Security Guidelines for the Transportation and Distribution of Meat, Poultry and Egg Products. http://www.fsis.usda.gov/OA/transportguide.html.

GAO. 2003. Food Processing Security. Voluntary Efforts Are Underway, but Federal Agencies Cannot Fully Assess Their Implementation, GAO-03-342. A report to Senators Richard J. Durbin and Tom Harkin. General Accounting Office, Washington, DC.

GAO. 2004a. Combating Terrorism. Evaluation of Selected Characteristics in National Strategies Related to Terrorism, GAO-04-408T. Subcommittee on National Security, Emerging Threats, and International Relations, Committee on Government Reform, House of Representatives. Statement of Randall A. Yim, Managing Director, Homeland Security and Justice Issues. General Accounting Office, Washington, DC.

GAO. 2004b. Terrorism Insurance. Implementation of the Terrorism Risk Insurance Act of 2002, USGAO Report GAO-04-307. Report to the Chairman, Committee on Financial Services, House of Representatives. April.

GAO. 2004c. Terrorism Insurance: Effects of the Terrorism Risk Insurance Act of 2002, USGAO Report GAO-04-806T. Testimony before the Committee on Banking, Housing and Urban Affairs, United States Senate. Statement of Richard J. Hillman, Director, Financial Markets and Community Development. May.

Gisborne, KD. 2004. Farm to Fork. Securing the Food Supply Chain. Risk Strategies Inc., Vancouver, BC. Presentation to the British Columbia Food Protection Association, September 23, 2004, Burnaby, British Columbia.

Gostin, LO, Sapsin, JW, Terret, SP, Burris, S, Mair, JS, Hodge, JG, and Vernick, JS. 2002. The Model State Emergency Health Powers Act. Planning for and response to bioterrorism and naturally occurring infectious diseases. *JAMA* 288:622–628.

Hall, M. 2004. Officials trying to reduce holes in security net. *USA Today.* September 14, 2004. p. 9a.

Hollingsworth, P. 2001. Know a crisis when you see one. *Food Technology* 54:24.

IFT. 2001. Institute of Food Technologists Annual Meeting. New Orleans, LA, June 24–27. (Hot Topic: Bioterrorism: The Threat Reality and Reaction.)

IFT. 2002. Institute of Food Technologists Annual Meeting. Anaheim, CA, June 14–19. (Hot Topic and Forum: Homeland Defense. Food Bioterrorism.)

Khan, AS, Swerdlow, DL, and Juranek, DD. 2001. Precautions against biological and chemical terrorism directed at food and water supplies. *Public Health Report* 116:3–14.

Laird, B. 2003. Corporate America's Security Precautions May Still Be Lax. *USA Today,* April 18, pp. 1, 2.

Rasco, BA. 1997. Protecting Trade Secrets in a HACCP Era. *Food Quality,* September, pp. 16–23.

Rasco, BA. 2001. It's the Water: Legal Issues and Rural H20. *Agrichemical and Environmental News,* Cooperative Extension, Washington State University, Tri-Cities, November, No. 186, pp. 18–21.

Sobel, J, Khan, AS, and Swerdlow, KL. 2002. Threat of biological terrorist attack on the U.S. food supply: the CDC perspective. *Lancet* 359:874–880.

Washington State. 2001. Eco-terrorism. Public Hearing. June 11. Washington State Senate, Senate Judiciary Committee, P.O. Box 40466, Olympia, WA. 60 pp.

WHO. 2002. Food Safety Issues. Terrorist Threats to Food. Guidance for Establishing and Strengthening Prevention and Response Systems. World Health Organization, Food Safety Department, Geneva, Switzerland.

An Example

Production and Retail Distribution of Frozen Pacific Cod Fillets

In this example food security issues surrounding the receipt, processing, storage, and distribution of frozen, wild-harvested ocean fish are described. Each situation is unique and should be evaluated as such.

The operation is as follows:

1. Harvest and primary processing. Fish are harvested using a trawl or longline from the Pacific Ocean off the coast of Alaska, headed, gutted, and filleted on board ship. Fish fillets are layered between sheets of colored plastic inside a fiberboard box that has been placed within a

metal plate freezer frame. There is about 17 pounds of product per box. After filling, the box is closed and the contents frozen in a contact plate freezer. Following freezing, the boxes of frozen product are removed from the freezer frames. Three boxes are placed within a cardboard master carton. The carton is sealed with tape and marked with approximate weight, product identification (product form, size, and species), lot code, harvest location, processing vessel name, and company contact information. Then the master cartons are stacked onto pallets; shrink-wrapped; labeled with product identification (see above), lot code, and company contact information; and placed into a storage freezer. Packaging is tamper evident only to the extent that removal of tape on the master case could be detected.

2. Transit to processing facility. After the ship arrives at port, the contents are loaded into a refrigerated trailer using a forklift or hoist and shipped by truck to the processing facility. Loads are sealed and seal numbers noted. Temperature recorders are included in each shipment.
3. Receiving at processing facility. Product is received at processing facility within 8 hours of the ship being unloaded. Product is transferred from the trucks to the storage freezer in the processing facility using a forklift. Package integrity, quantity shipped, and temperature are noted.
4. Storage at processing facility. Product is held frozen until needed. Product flow is first in, first out.
5. Processing — slacking out. Pallets are removed from the freezer and moved to the production floor to slack out. The number of units from each lot and the lot number and product weight are recorded. Temperatures are taken intermittently, and when product core temperatures are at 28°F, the master cartons are transferred to processing tables.
6. Repackaging. Master cases are opened, and approximately 5 pounds of product is transferred manually into plastic heat-resealable, frozen-product retail bags. Since the product for sale is whole fillets, this will be an approximate weight. Bags are heat sealed. Package is tamper evident to the extent that it is easy to tell that the bag or seal has been torn. Product is weighed and weight label applied. Package information includes a new lot code (containing line, production time, and facility), product country of origin, and company contact information (name, address, phone number, and website/e-mail).
7. Freezing and frozen storage. Ten bags of product are placed into a master case, which in turn is sealed, repalleted, shrink-wrapped, and then refrozen. The master case contains the same information as the retail package. A similar identifying label is wrapped within each pallet.
8. Distribution to retailer. Palletized product is transferred by forklift to a refrigerated container on a truck (common carrier). Truck makes

multiple deliveries within metropolitan area. Retailers receive both full and partial pallets. Product is transferred by either forklift (pallets) or hand truck (partials) from truck to facility. Product is placed into retailer's storage freezer.

9. Retail sales. Personnel from frozen foods department replace stock on sales shelf as necessary. Consumer removes product from freezer case.

The points at which this product is most susceptible to purposeful contamination risk are during production, during the multiple transfers, and at retail sale. A food security plan, fully integrated into the normal food safety programs of a food producer and seller, will provide markedly enhanced protection that continues even beyond the point when the producer properly places the product into transport, e.g., railcar, trailer, container, etc. It would include the supervised sealing of all access to the product, including doors (in some cases inspection doors on vans or reefers), vents, discharge ports, etc. Locks should be used where practical. All records of seal numbers and locations should be independently, electronically provided to the purchaser or its receiving agent.

Many private and common carrier fleets are now equipped with sophisticated, automated trip loggers/recorders that are integrated with the critical elements of the vehicle and Global Positioning Systems (GPS). These systems not only identify individual drivers and monitor vehicle speeds, van temperatures, and engine performance, but they also compare vehicle locations, routes, and times against those scheduled. Some even monitor the physical condition of the driver. Normal operating information, as well as deviations that might indicate hijackings, unauthorized stops, or driver distress, is common, automatically transmitted via satellite communication to the parent company. In many cases, these can also communicate directly with the nearest law enforcement agency.

In some instances, temperature data recorders may also be placed into the cargo or the cargo area and can provide an indication of unauthorized access (by temperature spikes or similar aberrations), in addition to recording normal product temperature profiles. Such temperature monitors are often integrated with automated onboard systems that can remotely notify a shipping company of an unusual condition.

Many modular containers also have integral solid-state devices that monitor and record additional activities related to a particular unit. This may include records of access into the container. Data from such monitoring may be remotely monitored on a continuous or periodic basis, or may simply be recorded for future access by authorized personnel. Regardless of the scope or capability of the monitoring system, this should be a feature incorporated into the security program. It is even entirely possible and quite practical to integrate automated data analysis of the transmitted information into monitoring programs that will provide appropriate alarms and even corrective actions or responses to deviations.

A key to ensuring that shipment integrity has been maintained is inspection at receiving. It is entirely practical and possible for the receiving company to match the data output from even the simplest of the aforementioned devices against schedule profiles. Product that does not meet the critical limits established by the purchasing firm should be rejected, isolated, and the vendor notified immediately. The receiving records and supporting documents should be reviewed in a timely manner by a qualified supervisor for every shipment. Where practical, this review should be backed up by comparative automated data analysis.

At receiving, vendor certification and lot numbers should be matched against those provided by the vendor, normally through the purchasing department of the purchasing company. Volumes and weights should also be compared and matched against purchasing documents. In a similar manner, receiving as well as other personnel at all stages of production should inspect packaging integrity. A copy of the critical receiving information should be independently transmitted to the producer/shipper for comparative verification.

Part of this step would be assurance that the receiving department has the appropriate seal numbers available to them. The driver or other delivery agent should have these data and use them periodically for inspection while transporting the materials. However, for security reasons, the driver should not provide seal data to the receiver. This information should be provided directly from the supplier to the receiver electronically or by fax. Seals and locks should not be removed until immediately prior to unloading. Inspecting agencies that remove the seals en route should replace the seals with new units and then independently transmit the new seal numbers to the intended receiver.

The printout from the truck recorder (often this will be provided electronically by the common carrier from remotely downloaded data) should be examined for indications of unauthorized deviations.

While such measures as those described in this example may appear onerous at first glance, many of the steps are current accounting, quality control, and production records commonly in use. Many of these recommendations are just good business practices that should be employed regardless of a perceived bioterrorist threat.

Production of Frozen Fish Fillets — Hazard Analysis Worksheet

Item, Step or Function	Identify Potential Hazards Introduced, Controlled, or Enhanced at This Step	Are Any Hazards Significant?	What Control Measure(s) Can Be Applied to Prevent the Significant Hazard	Is This Control Measure Critical?
Harvesting and primary processing	Purposeful contamination	Yes	Certification of lot by vendor/supplier guarantee for packaging materials	Yes
			Tamper-evident packaging	Yes
			Periodic testing	No
Transit to processing facility	Purposeful contamination	Yes	All openings, vents, doors, etc., on truck or conveyance locked and sealed by vendor	Yes
			Time and temperature recording included in shipment	Yes
			Automated trip recorder/report	Yes
			Vehicle held in secured facility when unattended	No
			Vehicle secured when driver absent	Yes
Receiving at processing facility	Purposeful contamination	Yes	Verifying paperwork at receipt, check lot, product type, quantity, number of units, origin	Yes
			Check product and packaging integrity	Yes
Storage at processing facility	Purposeful contamination	Yes	Ensure product secure and access limited to authorized individuals only	Yes
Processing — slacking out	Purposeful contamination	No	Product remains in original packaging; process is short and poses little risk	No
Repackaging	Purposeful contamination	Yes	Individual fish handled and placed into packaging in which contamination would not be evident to retail purchaser	Yes
			Product properly coded for traceability at retail level	Yes
Freezing and frozen storage	Purposeful contamination	Yes	Product properly inventoried	No
			Tamper-evident seals present and intact on unitized packs (e.g., pallets)	Yes
			Traceability features incorporated onto unitized product	Yes

Production of Frozen Fish Fillets — Hazard Analysis Worksheet (Continued)

Item, Step or Function	Identify Potential Hazards Introduced, Controlled, or Enhanced at This Step	Are Any Hazards Significant?	What Control Measure(s) Can Be Applied to Prevent the Significant Hazard	Is This Control Measure Critical?
Distribution to retailer	Purposeful contamination	Yes	All openings, vents, doors, etc., on truck or conveyance locked and sealed by vendor	Yes
			Data and temperature recording device(s) included in shipment	Yes
			Automated trip recorder/report	Yes
Retail sale	Purposeful contamination	Yes	Control receipt and storage of inventory	Yes
			Control stock rotation, first in first out	No
			Ensure that all food is properly labeled and has suitable traceability features	Yes
			Do not resell perishable food items found in other store locations or returned by consumers	Yes
			Check package integrity of product on shelf	Yes
			Watch for suspicious activity	Yes
Firm Name:	Fine Fish Co.		Product line/description: Frozen fish fillets, retail pack (plastic)	
Address:	16 Penn Ave. Columbia, MA 01234		Intended use/consumer: Retail sale, direct to consumer	
Prepared by:	R. Tidge			
Date:	March 15, 2004			

Food Security Plan Form

Critical Control Point	Significant Hazards	Critical Limits for each Critical Control Measure	Monitoring				Corrective Actions	Verification	Records
			What	How	Frequency	Who			
Harvesting and primary processing	Purposeful contamination	Vendor certification received with shipment	Certification received	Check against shipping documents	Each lot	Receiving QC	Isolate and investigate; if required records verification is possible, accept, but if not, reject	Daily record review	Receiving record
		Lot ID numbers	Lot identification	Check product ID against shipping documents	Each lot		Isolate and investigate; if cannot adequately ID lots, reject		
		All units have functional tamper-evident packaging; packaging is intact	Package integrity	Visual	Each unit		Reject and isolate; retain product if criminal activity is suspected and contact law enforcement		
Transportation to processing facility	Purposeful contamination	Seals intact and ID numbers match	Seals and ID numbers	Visual and match against shipping documents	Each unit	Shipper	Isolate	Daily record review	Receiving records
		Time/temperature recorder in shipment	Functional recorder present	Visual	Each shipment		Isolate product; install calibrated functional recorder into shipment	Shipping document review with each shipment	
		Vehicle secure when driver is absent	Vehicle secure	Driver log/report	Each shipment		Isolate and investigate; if product may have been tampered with, reject and isolate; retain product if criminal activity is suspected and contact law enforcement	Driver log review with each shipment	

Food Security Plan Form (Continued)

Critical Control Point	Significant Hazards	Critical Limits for each Critical Control Measure	Monitoring: What	Monitoring: How	Monitoring: Frequency	Monitoring: Who	Corrective Actions	Verification	Records
Receiving at processing facility	Purposeful contamination	Time/temperature recorder is in shipment; trip report received	No unexplained deviations	Visual and recorder printout	Each shipment	Receiving QC	Isolate and investigate; if deviations cannot be explained, reject	Daily review of shipping documents and recorder printout	Receiving records
		Product and package integrity	Visual	Visual	Each unit		Document product or packaging defect; isolate and examine product; if suspicious, reject; if criminal activity is suspected, retain product and contact law enforcement		
		Lot numbers match	Lot identification	Check product identifiers against shipping documents	Each lot		Isolate and investigate; if required records verification is possible, accept; if not, reject; if criminal activity is suspected, isolate and retain product and contact law enforcement		
Storage at processing facility	Purposeful contamination	No unauthorized access	Access to stored product	Locked storage with keys issued to authorized individuals only	Weekly	Production supervisor	Keys marked do not duplicate; collect keys from ex-employees and those no longer authorized; if cannot account for all issued keys, reissue new keys	Weekly review of safety inspection and employment records	Safety inspection log; employee or contractor log

Repackaging	Purposeful contamination	Product integrity	Product handled by trusted individuals only	Two-person rule; employee vigilance for suspicious activity	Each lot	Packaging lead	Isolate suspicious product and evaluate; if tampering is possible, reject and destroy; if criminal activity is suspected, isolate and retain product and contact law enforcement	Daily records review	Packaging log
		Package integrity	Package properly sealed	Visual	Each unit	Line worker	Reject, then isolate damaged product		
		Package coded for traceability	Proper package labeling	Visual	Each unit	Line worker	Isolate and relabel uncoded product; if this is not possible, reject and destroy product		
Freezing and frozen storage	Purposeful contamination	Tamper-evident seals present and intact on unitized packs	Seals present and intact	Visual	Each unit	Warehouse supervisor	If product is damaged, isolate and examine; if it may be contaminated, destroy	Daily records review; review of deviation reports for relabeled goods	Freezer log
		Traceability features incorporated onto unitized packs	Traceability features present and functional	Visual	Each unit		Replace tamper-evident packaging if possible; apply new traceability features		

Food Security Plan Form (Continued)

Critical Control Point	Significant Hazards	Critical Limits for each Critical Control Measure	Monitoring: What	Monitoring: How	Monitoring: Frequency	Monitoring: Who	Corrective Actions	Verification	Records
Distribution to retailer	Purposeful contamination	All openings, vents, doors, etc., on truck or conveyance are locked and sealed prior to shipment	Seals present and numbers recorded	Check numbers against vendor documents	Each shipment	Receiving QC	If seal documentation not consistent, reject shipment	Daily review of shipping records	Shipping records
		Data and temperature recorder device(s) included in shipment	Data logger and temperature recorder present and functional	Check for any unexplained deviations			Reject shipments with unexplained deviations		
		Automated trip recorder/report	Trip recorder set	Recover and review report			Isolate shipments for which there is questionable information; reject if suspicious activity is possible; if criminal activity is suspected, retain product and contact law enforcement		
Retail sale	Purposeful contamination	Receipt and storage of product controlled	Product delivered to secure area and unloaded under supervision of retailer	Offload monitored by retail personnel	Each shipment	Receiving QC	Isolate shipment. Do not offload into retail facility until security issues have been adequately addressed. Reject if security cannot be guaranteed and retain product if criminal activity is suspected	Daily review of shipping and receiving records and security inspection reports	Shipping and receiving records. Reports of security inspections

All food is properly labeled and has traceability features	Examine master cases at unload for labeling and traceability features	Visual	Each shipment	Receiving QC	Isolate any food not properly labeled and note defect. Reject. If criminal activity is suspected, retain product and contact law enforcement	Daily review of inventory control and product reports	Inventory reports
	Examine retail units when stocking	Visual	Each unit	Inventory control staff	Isolate unit and note defect if labeling is not clear and traceability or tamper evident features are damaged or absent. If criminal activity is suspected, retain product and contact law enforcement	Daily review of inventory control report	Inventory control reports
Perishable foods are not restocked; returned perishable food items are not resold	Restocked or returned perishable food items	Product recovery location and sale history	Any affected unit	Any employee	Isolate and discard	Review of inventory control and departmental shift reports	Inventory control and security inspection reports
Package integrity for product on shelf	Examine products on shelf during shift and when stocking shelves	Visual	Any affected unit	Department manager	Isolate, report defect, and discard. Retain product if criminal activity is suspected	Review of inventory control and departmental shift reports	Inventory control and security inspection reports
No suspicious activity that could jeopardize food safety	Monitor customers and employee behavior	Visual or via surveillance system	Any individual	All employees	Question suspicious individual. If criminal activity is suspected, retain and contact law enforcement	Daily review security report	Security inspection report

Firm Name: Fine Fish Co.
Address: 16 Penn St.
Prepared by: R. Tidge
Date: March 15, 2004

Product line/description: Frozen fish fillets, retail pack (plastic)
Intended use/consumer: Retail sale, direct to consumer

Notes

1. Possible Animal Diseases and Disease Agents

Animal diseases listed by APHIS (9 CFR Part 121.1(d)	Overlap diseases and agents listed by APHIS and CDC (9 CFR Part 121.3(b))	Plant Diseases (7 CFR Part 331.3)
African horse sickness[a]	Anthrax (*Bacillus anthrasis*) [b]	Citrus greening (*Liberobacter africanus, L. asiaticus*)
African swine fever[a]	*Clostridium* sp. (producing botulinum neurotoxin	Philippine downy mildew (of corn) (*Peronosclerospora philippinensis*)
Akabane[c]	Brucellosis, cattle (*Brucella abortus*)b	Soybean rust (*Phakopsora pachyrhizi*)
Avian influenza (highly pathogenic)[a]	Brucellosis, sheep (*B. melitensis*)b	Plum pox potyvirus
Bluetongue (exotic)[a]	Brucellosis, pig (*B. suis*) [b]	Bacterial wilt , potato brown rot (*Ralstonia solanacearum*, race 3, viovar 2)
Bovine spongiform encephalopathy[b]	Glanders (*Bukholderia mallei*) [b]	Potato wart or potato canker (*Synchytrium endobioticum*)
Camel pox[c]	Melioidosis (*Burkoholderia pseudomallei*)	Brown stripe downy mildew of corn (*Sclerophthora rayssiae* var. *zeae*)
Classical swine fever[a]	*Clostridium perfringens* epsilon toxin	Bacterial leaf streak of rice (*Xanthomonas oryzae* pv. oryzicola)
Contagious bovine pleuropneumonia[a]	Valley fever (Coccidioides imitis)	Citrus variegated chlorosis (*Xylella fastidiosa*)
Contagious caprine pleuropneumonia[b]	Q fever (*Coxiella burnetii*) [cb]	
Foot-and-mouth disease[a]	Eastern equine encephalitis[b]	
Goat pox[a]	Tularemia (*Francisella tularensis*) [b]	
Heartwater (*Cowdria ruminantum*)	Hendra virus (of horses)	
Japanese encephalitis[b]	Nipah virus (of pigs)	
Lumpy skin disease[a]	Rift Valley fever	
Malignant catarrhal fever[b]	Shigatoxin	
Menangle virus[c]	Staphyloccocal enterotoxins	
Newcastle disease (exotic) [a]	T-2 toxin	
Peste des petits ruminants[a]	Venezuelan equine encephalitis[b]	
Rinderpest[a]		
Sheep pox[a]		
Swine vesicular disease[a]		
Vesicular stomatitis[a]		

OIE (Office International des Epizooties) is an arm of the World Health Organization and is a clearinghouse for animal diseases and health. The OIE provides a list of high consequence List A (superscript a in table), List B (superscript b), and other (superscript c) animal diseases (9 CFR Sec. 121.3(d)). Data from Monke, 2004.

2. The GAO recommendations that DHHS and USDA "study their agencies' existing statutes and identify what additional authorities that they may need relating to security measures at food processing facilities" and "seek additional authority from Congress as needed" (GAO, 2003) add credibility to this observation.
3. The average FDA investigator had been on the job 1 to 3 years (18%), 4 to 7 years (32%), 8 to 12 years (27%), and >13 years (23%). They inspect on average annually 1 to 12 facilities (12%), 13 to 24 facilities (22%), 25 to 36 facilities (40%), 37 to 60 facilities (20%), and 61 facilities (7%) (GAO, 2003).
4. Of the supervisors surveyed, 11% were very confident, 36% were somewhat confident, 16% were neither confident nor not confident, 26 % were not very confident, and 10% were not confident at all about food security measures implemented within the industry that they highly regulate.
5. We all share the pain. To recoup costs of offering TRIA, all commercial property casualty policyholders pay a maximum 3% surcharge on annual premiums (GAO, 2004b).
6. A terrorist act must be "certified" by the Department of Treasury in consultation with the Secretary of State and the Attorney General before coverage would apply. Acts meeting these criteria must be perpetrated by an individual acting on behalf of any foreign person or foreign interest (GAO, 2004b). Businesses are self-insured for acts of homegrown crackpots, which would include almost every act of eco- and agroterrorism that has occurred in this country over the past 20 years.

   Aggregate damages must be at least U.S.$5 million for a single event. The deductible for the policyholder is 15% for 2005, and the government pays 90% of any covered loss until the aggregate insured losses for all insurers reach U.S.$100 billion in any calendar year (GAO, 2004b).
7. The key features of a public health preparedness plan are to prepare a public health department to respond to an actual or threatened terrorist event. To meet this objective, a public health department should be capable of the following:
   1. Identify the type of events that might occur in its community
   2. Plan emergency activities in advance to ensure a coordinated response
   3. Build capabilities necessary to respond effectively to an event and its consequences
   4. Identify the type or nature of an event when it happens
   5. Implement the planned response quickly and efficiently
   6. Recover from the incident

   Components of preparedness plans would include:
   1. A surveillance system to detect public health emergencies
   2. Implementation of preparedness planning principles
   3. Testing of preparedness plans for effectiveness
   4. Assessment of vulnerability to a specific threat or incident; capabilities for investigation and verification of the threat or incident and linkage of

the relevant government agencies and other entities that will contribute to management of public health consequences

The key preparedness elements for a terrorist response include:

1. Hazard analysis
2. Emergency response planning
3. Health surveillance and epidemiologic investigation
4. Laboratory diagnosis and characterization
5. Consequence management

See The Public Health Response to Biological and Chemical Terrorism Interim Planning Guidance for State Public Health Officials. http://www.bt.cdc.gov. See also WHO, 2002.

8. Vulnerability assessments should include evaluation of the following factors:
    1. The public health impact, severity of illness or risk of death, and deliberate exposure of the agent to the general public and susceptible subpopulations
    2. Potential for delivery of the agent to large populations
    3. Possibility of mass production or distribution of the agent
    4. Potential for person-to-person transmission
    5. Public perception of the agent, fear, and potential for civil disruption
    6. Special needs for public health preparedness, including stockpiles of drugs, vaccines, etc.
    7. Need for enhanced surveillance or diagnosis capabilities
    8. Possibility of obtaining necessary quantities of agent
    9. Means to do harm and opportunity to carry out a terrorist act
    10. Identification of potential terrorists and willingness to do harm
    11. Availability of effective preparedness plans and capacity for effective response; means to avert harm
9. Agency action should be:
    1. Necessary to avert a significant risk, in the first instance in the judgment of health officials and, ultimately, with reasonable deference, to the satisfaction of a judge
    2. Reasonably well tailored to address the risks in the sense that officials do not overreach or go beyond a necessary and appropriate response
    3. Authorized in a manner allowing public scrutiny and oversight
    4. Correctable in the event of an unreasonable mistake (Gostin et al., 2002).

Security Improvements by Tracking Food

5

> In the war against terrorism, America's vast science and technology base provides us with a key advantage.
>
> **— President George W. Bush, June 6, 2002[1]**

Product traceability is the addition of a feature to the product or packaging that provides information about the identity of the product, its origin, manufacture, or authorized destination. Of major concern to many manufacturers is maintaining the ability to protect the integrity of their product once it leaves their control. A classic example of what can happen occurred in Washington State in the 1980s with the over-the-counter pharmaceutical Tylenol. A man developed an elaborate plan to murder his wife. He purchased four bottles of Tylenol capsules at a local chain drugstore. He then contaminated the capsules in each of the bottles with cyanide and reverse shoplifted three of the poisoned bottles to the same drugstore and saved one bottle for his wife's use. His basic theory behind the crime was that the police would suspect that his wife's death would be considered the random act of a deranged mass murderer and that he would not be suspected. Things did not work out to his benefit. His wife only became ill and recovered. The other bottles were found and the evidence pointed to the husband; he was convicted for attempted murder. This incident led to major changes in over-the-counter drug manufacture and to the production and sale of consumer products in general, most importantly, the development of a myriad of antitampering programs for both bulk and primary packaging of food, drug, and cosmetic items.

Manufacturers require not only controls that are directed to defend against tampering, but also product or package features that can provide effective deterrence, detection, and product tracking. Examples include steps directed toward the prevention of product counterfeiting, shoplifting, maintaining product quality by improved stock rotation practices, inventory control, pricing, and automated retailing. It should be noted, though, that given enough patience and technology, virtually any preventive measure can be circumvented. This does not mean, however, that the task is hopeless and full protection is unachievable. To the contrary, the simple systems in use today provide significant protection and have most likely successfully discouraged many potential perpetrators from contaminating consumer products.

Cost is another issue of concern. The risk of tampering and the net risk reduction to be realized must be weighed against the cost of the protective strategy. As with any defensive or preventive measure, there simply is not enough money to protect against every eventuality. Priorities must be established and efforts concentrated on the most effective and economically feasible measures.

There are, of course, several points in the production and distribution of foodstuffs to which protective measures can be applied, from ingredient formulation to final consumption at the dinner table. The points most commonly address normal production and distribution, including the production of the product itself, and then measures to incorporate identification, tracking, and antitampering features into the packaging at multiple stages, including the primary container, master carton, pallet or super master, and container/trailer or shipping van. Cost-effective protective measures can be applied at each of these levels.

Perhaps the most difficult area of food production with which to instill these features is at the actual product formulations and ingredient usage stage. Purposeful contamination or fraudulent substitutions are very difficult to monitor simply because of the vast quantity and variety of foods in the marketplace. Few governments, let alone distributors and retailers, have the wherewithal to conduct effective monitoring programs, and much of the work would involve detailed or sophisticated chemical analyses, forensic accounting, or both. A tragic example of fraudulent substitution occurred in China in 2003 and continued through March 2004, killing 17 infants with powdered formula (milk substitute) into which several Chinese manufacturers had purposely substituted inferior and less expensive ingredients, causing malnutrition. While this is not technically a bioterrorist act, it illustrates the difficulty of detecting defective and dangerous products in the marketplace even in situations where there are fatalities, and then the problems of investigating these incidents to arrive at the cause.

Product substitutions may not be dangerous. Even so, counterfeiting is a legitimate concern. Protecting legitimately produced products and reducing the introduction of diluted, substituted products of lesser quality, or purposely contaminated materials into commerce is an important concern. International counterfeiting rings target name-brand distilled beverages, mostly expensive scotch, brandies, and cognacs, and this has led to several innovations having potential and effective applications in biodefense. These include the use of electronic devices, radio frequency identifications (RFIDs) embedded into the base of the bottles or other primary packaging materials, holographic and multidimensional graphics, sequential serial numbers for each bottle, and closure seals that are difficult to counterfeit or replace.

As a rule, protective measures are classified as overt, covert, or forensic. Overt security features are readily apparent, visible, and do not require special equipment such as readers or instruments to detect. Covert security features tend not to be readily apparent, while forensic security features normally require unique test methods or instruments. Measures may also be further classified as external or internal. External measures might include antitampering, product control, and authentication steps such as security tape or tear tape, holographic labels, optically variable coatings, specialized printing or printing/coating combinations, coding, tracking, seals, and incorporation of fluorescent fibers and threads into packaging or seal materials (external). Fluorescent fibers and threads may also be added directly to the material (used internally). Similarly, physical, chemical, or genetic taggants provide distinct identification features. Numerous tools are already available for product authentication and tampering and inventory control and are presented in Table 5.1.

Government Strategies

Improving the security and safety of the country's infrastructure and making the nation safer from emerging threats to agriculture and human health is one of the core components of the nation's counterterrorism program. The Gilmore Commission noted in its 2000 report that "the United States has no coherent, functional national strategy for combating terrorism" and recommended that the new administration develop one. Formation of the Department of Homeland Security and the improvements in response programs at the local level have been important advances toward this goal in the past 4 years under the Bush Administration. However, much is left to be done, and as always, much of the burden for protecting critical infrastructure, including the food production and distribution system, will fall into the lap of the private sector.

Numerous government reports on the possible contributions of science and technology to improving response and strategies for dealing with terror-

Table 5.1 Tools for Product Authentication, Antitampering, and Product Control

Product Feature	Authentication — Covert	Authentication — Forensic	Antitampering and Product Control — Tamper Evident	Antitampering and Product Control — Tracking
Product composition	Product analysis	Materials testing Nucleic acid or protein markers, etc.		
Package features	Fluorescent fibers and threads (internal or external)		Labels Tapes (plain, logo, message) Seals/integrity Closure integrity (e.g., presence of vacuum seal)	Labels Coding
Taggants	Physical taggants	Substrate analysis — in product taggants		
Optically active materials	Holographic labeling, tape, tamper-evident packaging	Surface analysis Laser-read holograms — tagged coatings	Holographic labeling, tape, tamper-evident packaging	Product codes incorporating holographic features
Embedded data		Embedded data within print or graphics		Identification or recognition information embedded in graphics
Print features	Micro text	Tagged links Embedded data	Digital watermarks Variable data — ink jet or laser	Coding
Coding	Micro text	Tagged links Embedded data		Batch coding Digital watermarking Numbering Linear bar codes Two-dimensional bar codes PLU lookup stickers RSS Electronic codes — chipless magnetic and electromagnetic, short-range readable (<1 m) Chip tag radio frequency identification (RFID) (active or passive)

ist attacks have been issued over the past 3 years. One of the most influential of these is the 2002 report of the National Research Council (Anon., 2002). The recommendations in this report tied most closely to food security are outlined and expanded upon here.[2] Objectives derived from this report can form the basis for a corporate traceability program for food and agricultural products. These objectives include the following:[1]

1. Collection of the proper intelligence necessary for the development of detection, surveillance, and diagnosis systems in food and agriculture. This would also include the necessary information management systems.
2. Identification of biological or chemical hazards in the environment that could impact food safety.
3. Surveillance and diagnosis of infection- and disease-causing agents in agriculture production environments.
4. Development of programs to prevent, respond to, and recover from, an incident involving crop disease, animal disease, or food-borne illness.
5. The development of monitoring programs, including sensors to check for contaminants in shipments and to monitor food components and packaging, water, and other materials in the production and business environment (chemical, currency, mail, etc.).
6. The development of portable, mobile, and remote sensing technologies for assessment and tracing of contamination events, including technologies to determine whether a site is safe and can be returned to its normal function.
7. Protecting the food supply by extending existing Hazard Analysis Critical Control Point (HACCP) programs to deal with the deliberate contamination of food. A key factor in this food security program will be the development of feasible preventive measures and corrective actions for food and agriculture.
8. Develop criteria for quantifying hazards to define the level of risk for various food and agriculture operations. This information will be useful for developing a "graded security program" and for determining the minimal level of protection necessary to make a facility and food products secure.

The focus in the National Academy of Science (NAS) report is on a multitiered approach, on the food and agriculture side, with an emphasis on biological weapons, although the risk from intentional chemical contaminants is also addressed. Within the NAS report is a recognition that pathogenic microbes and their toxins pose a threat to national security whether they occur naturally or are intentionally introduced (Anon., 2002, p. 84).

The microbes that post the greatest direct threat to human health are different from those posing the greatest threat to the agricultural sector. In production agriculture specifically, the microorganisms posing the greatest risk are present naturally in the environment and may be endemic in parts of the world, but not yet introduced here (see footnote 1, Chapter 4). Unlike the select agents that raise the ugly spectre of weapons of mass destruction, many of the plant and animal pathogens that pose a high risk to agriculture are relatively easy to obtain and could be released with little difficulty. In fact, the effectiveness of current agricultural security requires voluntary compliance on the part of food importers, travelers, and others with current quarantine programs and can be easily circumvented by an individual with malicious intent. For example, foot-and-mouth disease could be introduced by a single individual bringing contaminated soil in on his shoes from an affected part of the world. The more or less voluntary quarantine and destruction of ornamental plants and relinquishment of contraband at the border are primary means of control against the introduction of agriculture disease and pests. This unique situation facing agriculture is often lost in the debate over national security, but is an area where environmental monitoring coupled with improved tracking systems in agricultural production and distribution would be particularly valuable for averting the environmental damage and economic impact that would result from introduction of crop or animal disease.

Food safety is at the intersection of human health and agricultural production. Improving traceability systems by adopting current off-the-shelf technology on a widespread basis, while at the same time ensuring that information management systems are compatible, would have a major impact on improving food security. Improving traceability would reduce the risk to the food supply from intentional contamination. In addition, there would be many collateral benefits that would improve production practices, supply chain management, and product quality and marketability. Feasible and cost-effective technologies are available, and the industry at large is sophisticated enough to adopt many of these relatively new advances into their operations, if the right incentives are provided.

Improvements in traceability become more important as many of our food production operations become more centralized. Food processing is increasingly concentrated, and now four companies slaughter and process 85% of the domestically produced meat (Anon., 2002, p. 93). Centralized feeding operations and large expanses of cropland dedicated to a limited number of cultivars of grain and oilseeds make certain commodities vulnerable to the introduction of an exotic disease. This is a risk, but also an advantage when developing a traceability system that could be instituted from the farm through to retail distribution. The level of sophistication of these operations makes a coordinated traceability system a realistic possibility.

Improvements in traceability are probably even more critical as the demand for relatively low cost food items puts an increasing demand upon food suppliers to import. Demand for year-round availability of perishable products is an area where increasing attention must be paid, as indicated by the fall 2003 outbreak of Hepatitis A from imported green onions or scallions. Although current provisions of the Bioterrorism Act extend to imported foods, the U.S. government is not in the position to take enforcement action against many of these products as a result of conflict of laws and jurisdictional issues. However, the marketplace will be able to require traceability measures that the government cannot, with market demands providing the driving force behind the incorporation of traceability features into the trade in imported foods. Any government-imposed system will tend to be ill suited to the task of increasing market incentives to improve product traceability (Golan et al., 2004).

Traceability Systems Overall

ISO guidelines provide a broad definition of traceability as the ability to trace the history, application, or location of that which is under consideration (ISO, 2000). The primary advantage of traceability programs is the enhancement of management capabilities and expanded market opportunities generated by improving value. The predicted market for traceability systems is U.S.$5 billion to $10 billion, which is not out of line considering that there are at least 150,000 manufacturers of branded products and exporters, which will be one of the first groups of business to adopt improved traceability systems (Boonruang, 2003). Besides food safety issues, other market forces are in effect that will increase verification of product source, an important feature. Over the past 10 years we have seen market premiums for organic, pesticide-free, "green," and non-genetically modified (GM) products. Furthermore, at least 68% of consumers would like specific country-of-origin information on the food products they purchase (McCoy, 2004).

Traceability is objective specific. Even with high-visibility industries, adoption traceability technologies will be motivated by economic incentives rather than governmental regulations (Golan et al., 2004; Appendix K, this volume). The current widespread use of traceability programs complicates the development of a centralized governmental system, many of which are currently operating and vertically integrated into sophisticated food distribution systems. Industry- or product-specific characteristics will create systematic variation in traceability systems and are dependent upon a balance of cost and benefit, and upon a determination of efficient breadth, depth, and precision of the traceability system. Breadth means the amount of information the traceability system records. Depth refers to how far back or forward the system tracks. And precision[3] reflects the degree of assurance the tracking system has to pinpoint a particular characteristic or location of a food product. The hope is that it will be possible to easily interface traceability systems across commodities and with systems being independently developed in important overseas markets, particularly Asia. Traceability systems are established to (Golan et al., 2004):

1. *Improve supply management.* The higher the value of coordination along the supply chain, the greater the benefit, presuming that technological difficulties and the cost of implementing a broader system, which reaches different levels within the supply chain and covers more transactions, are not overwhelming.
2. *Increase safety and quality control by improving trace-back capabilities.* Improves market and legal position in case of a recall, as long as the increase in precision (smaller and more exact tracing of individual units) does not substantially increase the cost of the program.
3. *Provide market differentiation* for food with subtle or undetectable quality differences, assuming new accounting procedures and start-up costs of improved traceability are not excessive.
4. *Market food with credence[4] attributes.* These are characteristics that are difficult for consumers to detect, such as whether a food has been genetically engineered.
5. *Reduce cost of distribution.* The larger the market, the greater the benefit, as long as the larger number of new segregation or identity preservation activities and the degree of product transformation do not create a traceability system that is too complex to manage.
6. *Improve inventory management.*
7. *Reduce cost of product recalls.* The higher the likelihood of safety of quality failures — in terms of the cost of a recall, legal expenses, market impact, trade impediments, and punitive regulatory measures — the greater the benefits of improved traceability.
8. *Expand sales of products* of high value and those with important, but difficult to discern, quality attributes. The higher the premium, the greater the benefits of traceability.
9. *Detect counterfeiting.*

A primary collateral benefit of improved product traceability will be improved quality and process control. Simply knowing more about product characteristics and how these may be affected by growing practices, growing location, harvest time, and time to and at market will lead to improvements to agronomic practices that a company could not otherwise justify funding. The potential for highly targeted recalls through improved tracking of sales data is another clear advantage. High-tech innovations, including holographic labeling, modern inks, printing innovations, packaging graphics, high-tech coatings, multidimensional bar codes, online printing equipment, and electronic tags, now provide a multitude of good opportunities for their incorporation into food security measures at both the overt and covert levels. All can contribute to both antitampering/tamper-evident applications and more covert applications directed toward reducing product counterfeiting or substitution. Advantages and disadvantages of various current systems are presented in Table 5.2.

Tamper-Evident Package Features

Tampering is normally considered to be an act that violates the integrity of a product or packaging for the purpose of adulterating the product or fraudulently altering its documentation. Antitampering measures, or "tamper evidence" measures, are normally designed to provide overt or obvious indications as to the integrity of a product's packaging. A common example is a printed plastic band around the closure of a food package or a container of an over-the-counter drug. The plastic band is normally applied to the container and then subjected to heat that shrinks the plastic tightly to the closure and container interface and so that it cannot be removed without tearing or deformation. Heat-shrinkage plastic has a number of other common antitampering or antitheft applications, such as blister packing for drugs, supplements, or other small articles. For these products, the product to be protected is placed upon a fiber paperboard or rigid plastic sheet, the shrink-wrap material is placed over it, and heat or a combination of heat and vacuum is applied to tightly seal the product to the backing. The inclusion of tear strips and graphics to the shrink material enhances the effectiveness of the antitampering device.

A single-time-use closure, security or tear tape is also quite common and can be used in a myriad of ways. The tape should have graphics on it that would be distorted or otherwise disfigured to indicate whether an individual has removed and then attempted to replace the tape. Company logos, names, safety warnings, etc., can all be incorporated within a tape to provide multiple benefits. The tape can be used on the primary unit packaging, for sealing master cartons, or even strategically applied to wrapped pallets of the product.

There is also a trend toward the increased use of other laser technology or ink jet to apply graphics, unit or lot numbers, bar codes, and other

Table 5.2 Features of Various Traceability Devices

	Use	Advantages	Restrictions
		Product Tags	
Plastic tags	Marking individual animals	Relatively cheap and easy to use; sequential numbering of tags is possible for tracking and inventory control Low tech Can incorporate brand and regulatory information (seafood)	Cannot be read by remote sensing devices Limited amount of information can be placed on a tag Data for tagged animal may be in different locations, held by different users, and in incompatible formats
Labels	Identify product, product code, ingredients, manufacturer and contact information	Can adhere to food product (stickers) Can be used to promote product features, uses, related items Attractive	Cannot be read by remote sensing devices Can be removed, altered, or damaged
Product-specific tape	Sealed container; tamper-evident, tear, or distortion features	The restricted availability of a particular tape provides a tamper-evident feature; removal and resealing is difficult to mask Can provide attractive product identifying features for secondary and tertiary packaging	Provides limited or no traceability features
Heat-shrink seals	Provides tamper-evident seal for container	Widespread, can incorporate graphics and marketing features Difficult to alter if a unique printed seal is used	May be difficult for some consumers to remove; accompanying warning label is recommended
Holographic tags	Product authentication	Can be incorporated into many types of packaging; difficult to counterfeit	Can be expensive

Table 5.2 Features of Various Traceability Devices (Continued)

	Use	Advantages	Restrictions
Digital watermarks	Product authentication	Difficult to alter	Requires special reader
Embedded data	Product authentication	Identification or recognition information embedded in graphics	Requires special reader
		Coding	
Bar codes	Provides product identification and inventory control information	In common use Upgrade from 12- to 14-digit GTINs is relatively straightforward UCC standards	Provides limited information Remote sensing not possible
RSS	Provides machine-readable product information	Can contain substantially more information than current bar codes Multidimensional and matrix features Adaptable to current bar code readers Based upon UCC/EAN system	Use of remote sensing is limited Requires hardware and software upgrades Requires special reader
Radio frequency identification tags (RFIDs)	Programmable and readable; passive or active	Individual packages can have unique identifier with much product-specific information Can be read from a distance of a few feet	Requires hardware and software acquisition Requires special reader Complete set of standards not yet established Reading codes on inner packages on a pallet is difficult Hard to use in cold and high moisture environments
		Taggants	
Taggants	Covert product authentication		Detection technology is sophisticated Must be safe and compatible with food product

information on packaging. Strategically placing such applications at closure junctures or on tape or other sealings can provide an additional benefit of protection against tampering.

Optically Active Packaging Features

Holographic printing, labeling, or seals are also commonly used. Optically variable coatings are receiving increased use in food packaging and can be incorporated into plastics coming into direct contact with the food. Holographics are relatively difficult to reproduce and can provide evidence of authenticity as well as of tampering. The appearance of an optically active coating changes as the container is rotated or placed under different lighting conditions. These are features that are difficult to counterfeit.

Taggants

Taggants are unique chemical features that can be incorporated into a product or a product packaging. For many products, genetic markers are currently used to control the trade and supply of specific cultivars of genetically modified crops and ornamental plants propogated by cuttings. Specific food additives such as dyes, an inert component, or ingredients with a unique spectrophotometric profile can be placed into food to determine its origin and to determine whether a product has been counterfeited or "cut" with inferior materials. Physical taggants are widely used in the paper and fabric for currency and can be adapted for use in food packaging.

Physical Tagging Systems

Uniquely numbered or identification seals are commonly used to discourage tampering and pilfering and to provide product traceability. These tags are often prenumbered and sequential; however, in some cases the numbers are generated electronically, often based upon randomly generated discrete codes, and may be duplicated in the form of bar codes or into an incoherent format in addition to a readable form. Further, multilingual seals or other identification formats may be required for the sale of products in international markets. The integration of a visually readable verification step, such as checking a numbered seal, with a parallel digital electronic verification measure, provides an excellent combination for product traceability. An additional advantage of this type of data duplication is that it can also prevent serious incidents resulting from mislabeling or misuse of products or ingredients.

One example of a serious mishap that could have been prevented by a data duplication system of this sort occurred in the 1990s when several drums of glycol-based antifreeze were purchased from a Chinese manufacturer by a French company. The labeling on the drums was only in Chinese characters, and the French purchaser mistakenly understood the product to be food or pharmaceutical grade. The French company mislabeled the product as food/pharmaceutical-grade glycol for commercial sale. The drums eventually ended up at a Caribbean pharmaceutical manufacture that used the material to produce children's liquid aspirin, which in turn resulted in the death of several children.

Physical tags are ubiquitous in the fresh produce industry for tracing cargo containers in international commerce to the four-digit PLU or lookup stickers on individual oranges and apples. Sales of fresh U.S. produce is a $25 billion industry, with sales equally distributed between retail and food service sales, plus a very small (2%) direct market. Imports constitute about 28% of the value of product consumed, and exports about 16% (Golan et al., 2004). Boxes of produce are labeled with commodity, variety, packer or responsible party, date, quantity, and lot number. Grower cooperatives and advisory boards commonly specify additional tracing information pertaining to product quality, grade, country of origin, pesticide treatment history, and organic certification. Special marketing programs require U.S. Department of Agriculture (USDA) inspection, providing a further record for traceability purposes. Scannable pallet tags are common in the industry, but the Uniform Code Council (UCC) standards are not in wide use. Traceability in the fresh produce industry is greatly influenced by the nature of the individual product and by the fact that products are highly perishable. Features that reflect quality attributes and quality variation have been incorporated into product tags. This is an industry where tagging is critical either in the field or at the packinghouse and which will benefit greatly as reduced-space symbology (RSS) and global trade item number (GTIN) technologies receive widespread adoption (Golan et al., 2004).

Food and animal tracking technologies, including embedded chips, have been in use for decades, widely used in fisheries and ecology programs to study behavior and migration patterns of anadromous fish such as salmon, and migrating populations of large terrestrial and marine wild animals. Various physical tagging systems were developed in the late 1990s to track the country of origin for animal food products, and because of these efforts, tagging has become a more affordable option for higher-value and branded products. Food safety scares tied to bovine spongiform encephalopathy (BSE) and foot-and-mouth disease have created a market demand for country-of-origin labeling and improved traceability features for red meat. Individual cuts at retail markets in the U.K. and elsewhere can be traced back to individual animals with improved systems under development in the U.S. (Pates, 2003) and Europe (Anon., 2004). Simple printed number tags, embedded microchips, and RFIDs (Golan et al., 2004) are in use. These systems can monitor vaccination and health records, breeding characteristics, sales or transfers, and genetic background, including a DNA profile. Tattooing, retina scanning, and iris imaging are other identification methods.

The traceability programs for animals build upon a user's prior experience with animal genetics and carcass quality programs. One producer-owned company in the U.S., BeefOrigins LLC (South Dakota), works with FoodOrigins, a John Deere shared services company that provides an umbrella for tracing systems involving several different food commodities. Currently, FoodOrigins can trace 65% of tomatoes grown in California (Pates, 2003). BeefOrigins incorporates tagging with a data management system in which producers own and control the online data. The market advantage of being able to track animal identification has led to improvements with compliance with country-of-origin requirements, monitoring cattle health, managing animal husbandry practices such as feedlot performance, and carcass grade and other product attributes. Major factors in the animal identification system are health and trace-back capability and, in the near future, feeding and residue levels for individual animals. Investors in this traceability technology see this as adding value, as source-verified animals should demand a higher price, because quality is known in advance.

Current traceability built around the use of plastic animal tags and bar codes lacks critical information that could improve market opportunities and shorten the supply chain. Research including field demonstration projects for sheep (U.K.) and goats (Norway) is under way, examining supply chain management systems from livestock genetic breeding to improve the meat/fat ratio, milk quality, fleece quality, and pharmaceutical uses, such as lanolin and keratin (Anon., 2004). Besides food traceability, other factors incorporated into the DNA and electronic-based tracking program involve tracing and monitoring of animal feeds for toxins, electronic tagging, and the use of natural tracers, such as DNA and smart water.

An interesting barrier to widespread adoption to tagging is the exemption of livestock from implied warranty provisions in common and commercial law because farmers are not considered merchants in the sense that other food sellers are. States such as Kansas have adopted statutes to specificially convey this exemption to livestock operations (Golan et al., 2004). Under this exemption from an implied warranty,[5] a seller of livestock would not be obligated for damages associated with the sale of a defective food product.[6] Normally, an implied warranty exists simply because an individual is in the business of selling a particular product, and thereby guarantees that the product being sold is suitable for its ordinary use (fit for food) and meets the average standard of similar products sold in the trade.

Bar Codes and Microtext

One-dimensional bar codes are ubiquitous in retail following a slow route to widespread adoption (1980s) after market introduction in 1974, when a 10-pack of Wrigley's Juicyfruit gum crossed the first scanner (Golan et al., 2004). Linear bar codes (the 12-digit uniform product code (UPC)) contain a series of numbers reflecting product type, package size, and manufacturer that can be read by a laser scanner. When an item is sold, the sale is recorded in the store's central computer and the inventory corrected. A new 14-digit bar code has been developed by the Uniform Code Council (UCC) called the global trade item number (GTIN) system, expanding bar coding to include more company- and product-specific information.

New reduced-space symbology (RSS) bar codes are replacing current bar code systems. RSS encodes more information onto a smaller space and expands current four-digit price lookup codes on fresh produce, for example, with the UCC/EAN 14-digit GTIN numbering on a small sticker that can be attached to single product items. RSS provides enough information to tie a single piece of fruit or retail package or meat back to the shipping box, pallet, and producer/grower. The system can be integrated with retail store purchasing or club cards to contact consumers in case there is a problem with the product.

Embedded data in print, microdots, or graphics, or as components of a bar code, provide authentication features for products and, with special readers, have features similar to those of RSS. Microtext can also be used to provide a format for embedded data. Embedded data technologies are covert and not readily discernible by the consumer or the potential counterfeiter.

Electronic Tags and Smart Labels

Bar codes are useful but, to compile product or inventory information, require that each unit be individually read by passage through a scanner. This is a labor-intensive process spurring development of remote sensing capabilities, including smart labels. These are computer chips embedded in a package that can then set off a sensor on the product shelf when an item has been removed.

A number of programmable chipless magnetic and electromagnetic tags are available that are readable within a range of 1 m or less. These were an expansion upon magnetic ink technology but were not widely adopted because of the ease with which these tags could be altered in a magnetic field.

The evolution of the smart tag into the RFIDs appears to be the most promising technology.

Radio Frequency Identification Devices

Perhaps one of the best systems that may be employed in protective measures, and one with multiple beneficial uses to a company that reach far beyond simple security, is a system based upon the use of radio frequency identification devices (RFIDs). These devices are receiving dramatically increased applications at all levels of production and distribution, including tagging of animals and container cargo. RFID applications can be overt, covert, or both.

RFIDs are being mandated by major retailers in the U.S. and elsewhere as a result of market and regulatory pressure. Wal-Mart had an initial demand in July 2003 for RFID tagging by January 2004 but has moved this date back to the end of 2004. Gillette was the first major consumer products manufacturer to incorporate RFIDs across a product line and had a plan as of late 2003 to distribute 500 million units of razor blades with tags in 2004. Interest in this technology is widespread, and by the end of 2003, roughly 9% of retailers were experimenting with RFIDs, and the number will be substantially higher by the time this book is released. The RFID provides advantages over the current ear tags for animals and bar coding systems that can provide information over one or two steps in the supply chain. RFIDs can be integrated from agricultural production forward, connecting all links in the supply chain (Savage, 2003). Because information on RFIDs is cumulative and integrative, it shortens the time needed to locate product within the supply chain in the case of a product callback.

RFIDs work in the marketplace and have been successfully incorporated into small expensive units, such as distilled spirits, cosmetics, recordings, and software.

Recommendations are that each case and, if possible, each retail unit of a food product be RFID coded. This technology provides a means to trace when a product has been removed from a shelf, purchased, and returned (and to where). RFIDs can also track the individual purchaser and type of payment, assuming a debit or credit card has been used. There are noncontact retail stores in Japan that have complete remote checkout. In this case, each item has an RFID; when a consumer passes the grocery cart past the RFID reader, each tag is read and a receipt generated without having to unload and then reload a cart (Fitzpatrick, 2004).

There are a number of advantages for employing RFIDs. Firms can use these to obtain a better understanding of many aspects of the business, including how goods flow through the supply chain, stock control, antitheft strategies, buying pattern analysis for individual consumers, various sales units and retail establishments, regions, and food tracking. Information about remaining shelf life could provide the basis for in-store promotions to move product expeditiously. Food traceability may be the initial driver for adoption by food companies, but other collateral benefits will provide the basis for more widespread adoption. RFID technology can reach beyond the point of consumption. For example, recycle programs in Germany use RFIDs to trace who purchased and then who recycled (or who did not) product containers.

New operating systems for RFID technologies such as TRON OS and eTRON are possible hearts for embedded devices such as RFIDs, as its key features are a high degree of standardization and software compatibility. TRON can address some of the security and privacy issues involved with conducting business over open networks. RFID technologies in use in Japan employ TRON for traceability of fresh produce, such as diakon, the source of a major *Escherichia coli* outbreak in 1998.

Information Management Systems

Numerous online traceability services are available along with turnkey software and hardware packages for manufacturers, retailers, and traders. Supply management involves the collection of information on each product from production to point of sale (Golan et al., 2004). In 2000, U.S. companies spent $1.6 trillion on supply-related activities, including transit, storage, and product control throughout the supply chain. Virtually every food product today is tied to a sophisticated and relatively costly distribution system, since food, for the most part, is a small-margin business. Fortunately, almost all food products have some sort of coding associated with them, which makes improvement of supply chain management through improved information management systems possible as technology becomes more user friendly and as uniform standards make widespread adoption a reality. Information management will track to the individual unit or container, lot, or some other identifiable level. The system must be established such that each step can be integrated into a food security plan to provide increased protection. One of the most effective steps currently in use is the authentication of product at all levels of distribution by the simple comparison of product coding. For instance, when a distribution center receives product from a manufacturer, it should compare the product and lot codes to those recorded on the shipping documents. The shipping documents containing this information should not only accompany the product shipment, but also be transmitted electronically directly from one distribution level to the next, as well as be available to all levels for simple and expedient verification. This requires tying the physical and paper/electronic trails together. Because of the importance of this, many of the largest manufacturers and retailers have proprietary systems that they require their suppliers to adopt.

There are many developers of proprietary software for food traceability. The goal of these programs is to automatically capture data as the food moves from production to retail or food service sale while imposing a minimal burden on purchasing staff, chefs, etc. The aggregate data are private, protected, and secure (Anon., 2003), and the programs that are successful will incorporate UCC and EAN standards and be compatible with data management tracking systems in place by major retailers.

The interest in food traceability has created some new market strategies. Concern over GM food labeling and now country of origin is creating a demand for "menu transparency" and the identification of the sources of foods used on restaurant menus. In the U.K., for example, consumers are under the mistaken impression that 70% or more of their food is from Britain, whereas in reality, only about 40% of it is, and this has led some entrepreneurs to promote tracking and sourcing of foods used in their establishments (Anon., 2003).

International Standards

A huge effort is under way to develop compatible and universal international standards for product traceability. Currently the lack of a generally accepted standard will slow adoption of sophisticated traceability systems for food (Savage, 2003). The Health and Consumer Protection Directorate in Europe will become law in January 2005 and, as under the U.S. Bioterrorism Act, will require product traceability (Savage, 2003). However, adoption of uniform standards may mean that the transparency and universal utility of the systems implemented will be limited at best. Standards are being adopted in the marketplace, and this will drive rapid development of uniform product coding. The Uniform Code Council (UCC) and EAN International are leading efforts to facilitate establishment of a uniform set of electronic product codes (EPCs) and an EPC network for RFIDs. Wal-Mart is requiring top suppliers to be EPC compliant. These systems will be in the public domain and will hopefully complement and interface with existing proprietary systems. The UCC is a private nonprofit company promoting multi-industry standards for product identification. This group created bar codes in response to demands of retailers for more rapid checkout processes and improved inventory management (Golan et al., 2004). The EAN, which is a subsidiary of the UCC, and EAN International, a European commercial standard-setting organization, have developed an open integrated system for standardized and automated information across the supply chain to integrate GTINs, along with an industry standard set of 62 product attributes (Golan et al., 2004). The objective of these systems is to integrate in real time information management systems from the producer through to retail so that anyone along the chain can track inputs, production, and inventory using any of a wide variety of product traits. Because of this, European food sector traceability standards will emerge over the next few years, transforming food chain integrity, purchasing, safety, and cost (Anon., 2003) throughout the EU and with its trading partners, but only after there is a set of uniform standards, preferably international standards already in place.

Many trade groups are involved in attempts to develop standardization before RFID technologies and others are widely adopted. There is no industrywide consensus as to the type of data model that should be adopted, and this impasse will mean that traceability will meet the minimum (and pending regulatory standards) of one step forward and one step back. One group, EPCGlobal, is a retailer's organization that is trying to standardize item numbering and limit the information on RFID tags to specified EPCs (Savage, 2003). Unfortunately, this approach limits the ability to develop complete traceability between the firms that handle the food.

Traceability is an area of growing importance and one in which there are rapidly evolving technologies on hardware and information management fronts. Hopefully, international standardization will develop in conjunction with these new technologies, leading to widespread adoption. The collateral benefits to improved quality and the agricultural and production practices that will develop as traceability improves provide additional incentives besides the obvious ones of improved food safety, cost control, and improved supply chain management.

References

Anon. 2002. Executive Summary. In *Making the Nation Safer: The Role of Science and Technology in Countering Terrorism*. Committee on Science and Technology for Countering Terrorism, National Research Council of the National Academies, National Academies Press, Washington, DC.

Anon. 2003. *Caterer and Hotelkeeper*. Reed Business Information, Reed Elsevier, Inc., November 6, p. 20.

Anon. 2004. European Research Grant Could Benefit Dozens of Farms. *Farmers Guardian*, March 12, p. 14.

Boonruang, S. 2003. Technology Can Help Local Food Producers. *Bangkok Post*, November 5.

Fitzpatrick, M. 2004. Microsoft's tiny rival steps up to the plate. *Computing*, March 4, p. 24.

Golan, E, Krissof, B, Kuchler, F, Clavin, L, Nelson, K, and Price, G. 2004. Traceability in the U.S. Food Supply: Economic Theory and Industry Studies, AER 830. Economic Research Service, United States Department of Agriculture, Washington, DC, HD 9005.

ISO. 2000. Quality Management Standards, ISO 900:200. International Organization for Standardization. http://www.iso.ch/iso/en/aboutiso/. As cited in Golan et al. 2004.

McCoy, D. 2004. Byrne in U.S. talks. *Belfast News Letter* (Northern Ireland) First Editions, News, March 27, p. 19.

Pates, M. 2003. South Dakota Company Touts Cattle ID System in Midst of Regulatory Delays. *AgWeek*, December 9. www.agweek.com.

Savage, C. 2003. Forrester Research. New Regulations Open the Way for RFID Tags for Tracing Food. M2 Presswire, M2 Communications Ltd., December 15.

Notes

1. Cited in Anon. 2002. Executive Summary. In *Making the Nation Safer: The Role of Science and Technology in Countering Terrorism*. Committee on Science and Technology for Countering Terrorism, National Research Council of the National Academies, National Academies Press, Washington, DC.
2. Issues surrounding decontamination of food, food contact surfaces and equipment, transport vehicles, or the farm, production, or food distribution environment from a biological agent are difficult since so little is known about decontamination protocols for microbial or chemical agents in these settings. This is a relatively new area of study and an important one in the NAS study.
3. The level of precision affects the type and cost of product differentiation (Golan et al., 2004). If a high degree of accuracy is required, then a stringent system for separating products is needed:

   Segregation separates one crop or batch of food components from others, keeping these apart throughout distribution, and may or may not require a high level of precision, assuming attributes of the different products are easy to detect, e.g., separating white and yellow corn. *Identity preservation* (IP) identifies the source and nature of a crop or food component to ensure that these items are separated from other products and their characteristics maintained, e.g., high-quality product from others, organic, pesticide treated, inspected goods, etc.
4. There are two types of credence attributes (adapted from Golan et al., 2004): *Content attributes* affect the physical or chemical attributes of a product and are difficult for a consumer to perceive, such as level of isoflavones in soy products, calcium in milk or juice, pesticide levels, use of animal drugs, and genetic engineering (for products that have measurable nucleic acid or

expressed protein). *Process attributes* are characteristics tied to production, such as country of origin, fair trade, sustainable practices, "dolphin safe," shade grown, feeding regimes, animal health and welfare issues, genetic modification (but not containing sufficient nucleic acid or protein products for analytical testing), etc., which cannot be detected by analytical testing.

5. The provisions under Washington law for implied warranty of merchantability are presented here and are fairly typical of laws adopted by other states (Rev. Code Wash. (ARCW) § 62A.2-314 (1999)).

   TITLE 62A. UNIFORM COMMERCIAL CODE ARTICLE 2. SALES PART 3. GENERAL OBLIGATION AND CONSTRUCTION OF CONTRACT

   § 62A.2-314. Implied warranty: Merchantability; usage of trade

   (1) Unless excluded or modified (RCW 62A.2-316), a warranty that the goods shall be merchantable is implied in a contract for their sale if the seller is a merchant with respect to goods of that kind. Under this section the serving for value of food or drink to be consumed either on the premises or elsewhere is a sale.
   (2) Goods to be merchantable must be at least such as
      (a) pass without objection in the trade under the contract description; and
      (b) in the case of fungible goods, are of fair average quality within the description; and
      (c) are fit for the ordinary purposes for which such goods are used; and
      (d) run, within the variations permitted by the agreement, of even kind, quality and quantity within each unit and among all units involved; and
      (e) are adequately contained, packaged, and labeled as the agreement may require; and
      (f) conform to the promises or affirmations of fact made on the container or label if any.
   (3) Unless excluded or modified (RCW 62A.2-316) other implied warranties may arise from course of dealing or usage of trade.

6. The Restatement is a compilation of "model laws" based upon trends seen within the law as it evolves. This is an example of a provision from the Restatement related to liability of a food seller for injury to a consumer from a defective or dangerous food product.

   RESTATEMENT OF THE LAW, THIRD, TORTS: PRODUCTS LIABILITY. 1998, American Law Institute RULES AND PRINCIPLES

   CHAPTER 1 — LIABILITY OF COMMERCIAL PRODUCT SELLERS BASED ON PRODUCT DEFECTS AT TIME OF SALE

   TOPIC 2 — LIABILITY RULES APPLICABLE TO SPECIAL PRODUCTS OR PRODUCT MARKETS

   § 7 Liability of Commercial Seller or Distributor for Harm Caused by Defective Food Products

> One engaged in the business of selling or otherwise distributing food products who sells or distributes a food product … is subject to liability for harm to persons or property caused by the defect.
>
> … a harm-causing ingredient of the food product constitutes a defect if a reasonable consumer would not expect the food product to contain it.

Appendix A

Food Safety and Security:
Operational Risk Management Systems Approach

November 2001

Presented by

DHHS
US Food and Drug Administration
Center for Food Safety and Applied Nutrition

INTRODUCTION

The United States Air Force, Office of the Surgeon General is developing guidelines for food safety and security for military personnel. Due to their support for national food safety and security and homeland defense they allowed their document to be used as a model for development of these guidelines. If you need further information, contact Dr Robert Brackett or Mr Louis Carson, FDA, CFSAN 202-260-8920/9653fax.

Our vision: Public Health protection through safe food and water sources.

Our strategy: To both stop and reduce threats before they occur our food safety and security strategy is to:

1. *Identify Food Assets*: Identify our food assets from farm to fork.
2. *Receive Threat Assessments*: Food safety regulatory agencies should develop procedures to receive creditable threats and threat assessments from intelligence personnel (FBI, state Office of Emergency Services, etc.) The threat assessments would be based on availability of agents (biologic, chemical, radiological, physical) and aggressors (terrorists, criminals, subversives etc.).
3. *Conduct Operational Risk Management*: Use ORM to enhance food safety and security by minimizing risk at each step in food production from the farm to the fork. Using ORM we will identify our hazards and conduct risk assessment and risk management for effective food safety and security. The goal is the best food safety and security at the least cost (not at any cost)

FOOD ASSETS

First and foremost, concern is centered on protecting the public, our most important asset by providing them with safe food and water sources. Food and water systems can be very complex and literally stretch around the world. For the purpose of this handbook, we want to identify our national assets.

U.S. Agriculture is a $200 billion business with over $55 billion in exports each year (agriculture has a $1 trillion value and provides 22% of all jobs). The United States is the largest producer of food and agriculture products in the world, and agriculture and food production is the nations largest business. The United States has over 500,000 farms, and over 6,000 meat, poultry and egg product and production establishments.

There are in excess of 57,000 food processors in the United States that provide processed foods to our citizens and exports to the world. These processors include canners, dairy product producers, wineries, and other food and beverage manufacturers and distributors. The United States produces over 50% of the world's processed tomato products, and the majority of the canned peaches, fruit cocktail, and black ripe olives. It also processes millions of tons of garlic, prunes, and strawberries. Retail food facilities (restaurants, grocery stores, and other operations serving/selling foods direct to the consumer) number in excess of 1.2 million.

Unfortunately microbes, toxins, chemicals, and heavy metals can be used to contaminate food sources on the farm, during food processing during transportation or in the restaurant during food preparation. These types of activities can cause extensive morbidity and mortality, and the economic destruction of our food manufacturers and agricultural industries.

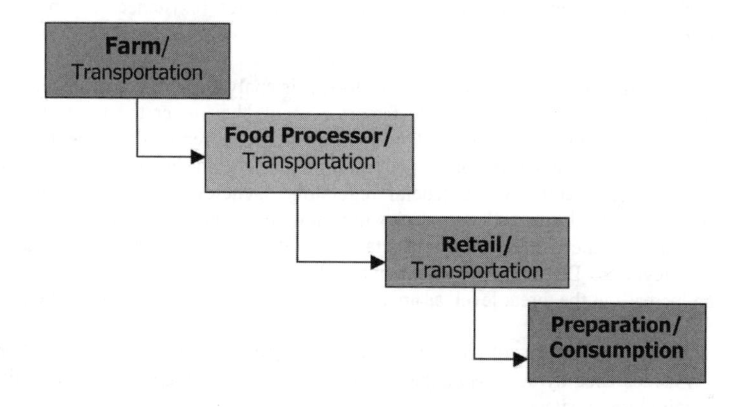

SUMMARY: IDENTIFYING FOOD AND WATER ASSETS

The nations farms, transportation and distribution systems, food processors and retail food establishments are a vital part of our economy and are required for the nations security and health. Local, state and federal food and agriculture regulatory agencies must work together with farmers, ranchers, food processors, food transportation companies and distributors, grocery stores, restaurants and food handlers to address food safety and security from the farm to the fork.

THREAT ASSESMENTS

Food Safety and Security, ORM Systems Approach, November 2001

The threat of terrorism against United States citizens has increased significantly. The nations food supply and agricultural industries could also be subject to this new threat. The nation must develop effective food and agriculture safety and security programs to guard against natural threats and also terrorist attacks. In order to meet our vision of public health protection through safe food and water sources we must stop attacks and also reduce our vulnerability to them before they occur. The United States must be prepared to respond to this new public health and agriculture threat.

Risk analysis of terrorist threats to the public is the responsibility of the FBI. The FBI is the lead law enforcement agency responding to potential terrorist incidents in the nation. The FBI gathers information from numerous sources in an attempt to build a composite picture of threat conditions. This information gathering is shaped by the need to focus on various factors indicating possible terrorist activities (existence, capability, intention, targeting)

As information is gathered on these factors, it is analyzed and a threat level for an area is determined. Basically, the more factors present, the higher the threat level becomes. Incidents of potential food and drug tampering are reported to the FBI. The FBI coordinates investigations of theses crime incidents with the appropriate federal agency (FDA, USDA etc). These federal regulatory agencies have current food safety and security procedures and are reviewing new procedures to communicate creditable increases in the risk of a terrorist attack to state regulatory agencies and the industries they regulate. Do not presume that an attack against food or water could not occur before an increase in the threat level, as an asymmetric attack is very hard to predict.

The three components of an operation against food and water systems are: (1) aggressors (2) tactics used by aggressors and (3) agent used by an aggressor. The following are required for an attack:

1. AGGRESSORS

There are five primary types of aggressor: criminals, protesters, terrorists, subversives, and rogue or disgruntled insiders.

2. TACTICS USED BY AGGRESSORS

A. Exterior attacks occur from outside the facility.
B. Forced entry is made by creating a new opening in the facility in order to gain access
C. Covert entry is accomplished by using false credentials or other means of deception or stealth in order to gain access to food or water systems.
D. Insider compromise involves using someone with legitimate access

3. AGENTS USED BY AGGRESSORS

A. **Biological** agents (bacteria, toxins, viruses, parasites, etc.) can be delivered in the form of liquids, aerosols, or solids
B. **Chemical** agents can be delivered as airborne droplets, liquids, aerosols, or solids. They are categorized as classical chemical warfare agents (nerve, blister, blood and choking agents) and toxic industrial chemicals (e.g., pesticides, rodenticides, and heavy metals).
C. **Radiological** agents are radioactive elements that can be delivered in liquid or solid form.
D. **Physical** agents are materials that could cause adverse health effects if eaten (e.g., bone slivers, glass fragments, and metal filings).

SUMMARY: THREAT ASSESMENT

The DOD January 2001 Proliferation: Threat and Response report for the first time identified that attacks against the U.S. food supply could affect the economic stability of the country and erode military readiness. There are many chemicals, microorganisms and toxins that meet the criteria for effective terrorist weapons for an aggressor who has developed the tactics for an attack against our food production system.

OPERATIONAL RISK MANAGEMENT

All food production procedures involve risk. All operations require decisions that include risk assessment as well as ORM. Supervisors in food production from the farm to the fork, along with every individual, are responsible for identifying potential risk and adjusting or compensating appropriately. Risk should be identified using disciplined, organized, and logical thought-processes that ensure the best food safety and security possible. Good ORM from the farm to retail can provide many benefits to overall food safety and security.

OPERATIONAL RISK MANAGEMENT RULES

Rule 1. Accept no unnecessary risk. Unnecessary risk comes without a commensurate return in terms of real benefits or available opportunities.

Rule 2. Make Risk Decisions at the Appropriate Level. Making risk decisions at the appropriate level establishes clear accountability. Those accountable for the success or failure of the product must be included in the risk decision process

Rule 3. Accept Risk When Benefits Outweigh the costs. All identified benefits should be compared to all identified costs. As an example a lock on a door, lighting and alarms cost less than a 24-hour guard for the door. We accept the risk of entry by an aggressor because we have put in redundant controls and the benefits of the 24-hour guard do not outweigh the additional cost.

Rule 4. Integrate ORM into Planning at all Levels. To effectively apply ORM, managers must dedicate time and resources to incorporate ORM principles into the planning processes. The making of important risk decisions should be preplanned whenever possible.

OPERATIONAL RISK MANAGEMENT IMPLEMENTATION

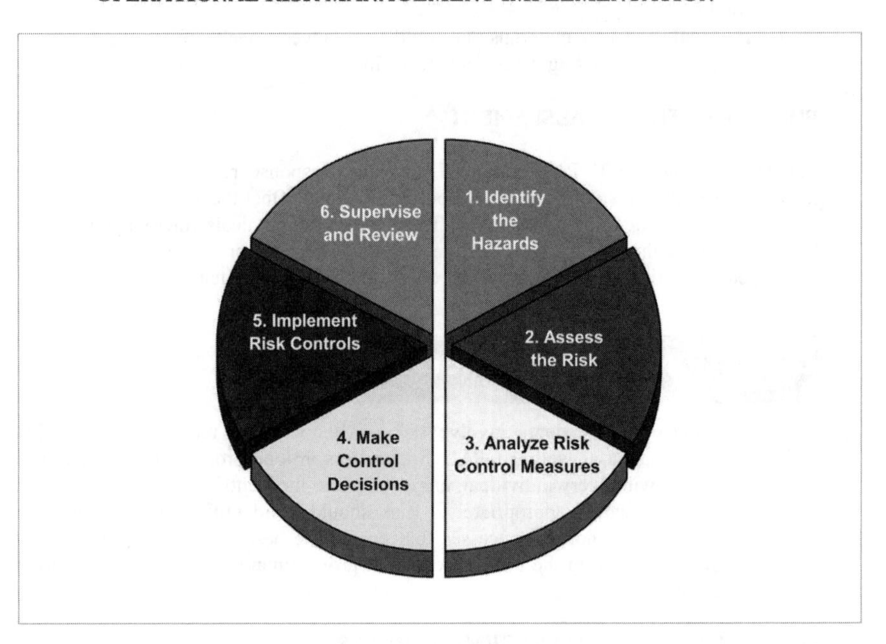

ORM Process Definitions

Mission: The desired outcome (food safety and security)

Management: Directs the food safety and security operation by defining standards, procedures and controls. Management process cited in 80 percent of reported mishaps by the National Safety Council other root (systemic) cause factors for mishaps:

- ✓ *People*: Most common root cause, doesn't know (training) Doesn't care (motivation) can't do (selection
- ✓ *Machines*: Poor design, poor performance, repairs not made, not used as intended, no upkeep or replacement

Food Safety and Security, ORM Systems Approach, November 2001

- ✓ *Plant or Environment forces*: Weak facility design, lighting, noise, Température, ventilation, contamination

Systems Management: ORM is a system-based concept. This means that ORM users understand that operational mistakes and errors have their origin in the design of the system (people, machines, plant/environment and management). Overall system effectiveness is required in order to meet the mission of food safety and security.

Flow Diagram: List of food production events in sequence required to understand the flow of events in food production from farm to fork.

Hazards: A description of a condition with the potential to cause illness, injury or death, property damage or business degradation. Not an indication of its significance to food safety and security.

Hazard Identification: Choose an area or step in food production and conduct an onsite visit. Use the:

- ✓ "What If Tool," conduct a brainstorm session with experts and supervisors on creditable hazards that could result in food contamination. Get input from operational personnel.
- ✓ "Cause and Effect Diagram," Draw a fishbone cause and effect diagram on a worksheet. Evaluate people, procedures, machine and plant/environment causes that could result in a contaminated product.

Risk: A hazard for which we have estimated the severity and probability with which it can impact our food safety and security mission. Supervisors and leaders want to deal with risks, not hazards, because hazards do not have an explicit mission connection.

Risk assessment: Identifying hazards and determining impact on food safety and business or mission (high risk, low risk steps 1 and 2). Remember risk can increase if threat conditions increase. The purpose of risk assessment is to allow us to focus on the worst hazard first.

Severity: Is usually based on worst creditable food safety and security event that can affect the business or mission. Creditable means an event that has some reasonable probability of occurring, not simple an event that conceivable could occur (see attached matrix).

Probability: The probability for a hazard is the closet match with the five levels of probability (see attached matrix). When working with an operation the probability is the cumulative probability of all hazards. Probability is never certain and should be based upon past events, data and expert analysis or group analysis. The probability of a food safety and security mishap goes up in high threat conditions.

Exposure: Is usually captured in probability (the more times we do something the more probable it is) or in severity (the more people exposed the greater is the potential severity). However, there are occasions when it is important to consider exposure in its own right in the final ranking of risks.

Modified risk control matrix: There can be inconsistency in risk assessments because there are at least two dimensions of subjectivity involve in the use of the risk assessment matrix. Interpretations of exposure, severity and probability may be different based on experience. This can be reduced by group discussions and averaging the ratings of several individuals. Remember the goal is to ultimately identify all risks in order of importance in order to prioritize risk control efforts.

Risk management: Analyze food safety and security risks and implement risk control decisions (steps 3, 4 and 5). Remember to conduct a risk assessment after controls are in place to ensure risks are reduced.

Unnecessary risk: Understand what unnecessary risks are and accept risk when the benefits actually outweigh the cost. In assessing risk the three primary causes for unnecessary risks are:

1. Not aware of the risk
2. An incorrect assessment of cost versus benefit
3. Interpreting "Bold Risk Taking" to mean gambling

Risk acceptance: Some degree of risk is a fundamental reality. The goal is the best food safety and security at the least cost, not at any cost.

- Risk management is a process of tradeoffs.
- Risk is a matter of perspective; keep problems in proper perspective.
- Weigh risks and make judgments based on knowledge, experience, and mission requirements.
- There is no best solution. Use good judgment.
- Complete safety is a condition that seldom can be achieved in a practical manner.
- Risk is inherent in all operations; risk can be controlled using ORM

Step 1: Identify the Hazards in Operation

The first step in conducting ORM is to identify the hazards for each activity or step in the process of food preparation. To understand the flow of events in food production, first list the sequence for food preparation. The food-purchasing contracts should be reviewed to ensure the products are from an approved source. This means the food processor and distributor were inspected by a recognized state or federal food safety regulatory agency. To be from an approved food source, the product would also be produced on a farm meeting regulatory requirements for food safety (pesticide use, antibiotic use etc).

The final flow diagram would appear as below:

Produce, milk, eggs and other farm products→ transported in refrigerated truck to food processor→ at food processor→ transported in refrigerated truck to distributor→stored at distributor then transported in refrigerated truck to restaurant→ unloaded at restaurant dock→placed in dry, frozen or refrigerated storage→food preparation in restaurant kitchen → placed in restaurant serving line→ leftovers returned to kitchen

ORM in a Restaurant

An ORM review was conducted on the restaurant identified in the flow diagram above. The following hazards were identified in the during step one of an ORM review of the restaurant identified in the flow diagram above:

Step 1. Identify Hazard	
Operation Activity/Event	**Hazard Identified in Operation**
1. Unloaded at restaurant dock by vendor driver	1.a No identification on drivers and different drivers are used (this person has ability to contaminate bulk food products) b. Trucks are not sealed or locked upon arrival and are making other deliveries c. Unauthorized personnel can enter restaurant through unsecured back door at any time
2. Placed in dry, frozen or refrigerated storage	2a. Storage areas are not monitored at all times and driver has access to entire storage area and kitchen. b. The facility does not have emergency lighting and it serves meals 24 hours a day. d. No tracking system to identify lot numbers on dry food sources served in facility that may have been recalled
3. Food preparation in restaurant kitchen	3a. Exterior door from kitchen area to outside is not secure and can be used to enter kitchen area by aggressor b. No pressure backflow preventor on oven hood cleaner can result in water contamination c. Entry to kitchen area from serving line is not restricted unauthorized personnel can enter kitchen e. Only one backup person if foodhandlers ill. Can result in inadequate staffing and use of unauthorized personnel f. Employees hired from local unemployment agency with only work reference contact. No procedures to conduct complete or random criminal background or drug testing.

	g. Employees locker room provided for changing clothes requires employees to provide own lock. Management cannot enter lockers to check on unauthorized use. j. New employees are hired into graveyard shift does not allow adequate observation
4. Placed in restaurant serving line	4a. Leftovers returned to kitchen could be contaminated b. Individuals have open access to salads and desserts on serving line. Provides ability to contaminate.

Step 2: Assess the Risks

The next step after hazard identification is to assess the risk of each hazard. When we know the various impacts a hazard may have on food safety and security and have an estimate of how likely it is to occur we can now call the hazard a risk. Risk is the probability and severity of loss or adverse impact to the food product from exposure to the hazard. The second aspect of risk assessment is the ranking of risks into a priority order. Remember something that is not a significant risk should still be corrected if it is simple to control.

There are three key aspects of risk. Probability is the estimate of the likelihood that a hazard will cause a loss. Some hazards produce losses frequently, others almost never do. Severity is the estimate of the extent of loss that is likely. The third key aspect is exposure, which is the number of personnel or resources affected by a given event or, over time by repeated events. To place hazards in rank order we must make the best possible estimate of the probability, severity, and exposure of a risk compared to other risks that have been detected.

1. Assess Hazard Probability: Determine the probability that the hazard will cause a negative event (severity). Use the cumulative probability of all causative factors. Probability may be determined through estimates or actual numbers. Use experience, analysis, and evaluation of historical data when possible. Supporting rationale for assigning a probability can be documented for future reference and to acknowledge uncertainty. Probability categories are frequently, likely, occasional, seldom, unlikely.

2. Assess Hazard Severity: Determine the severity of the hazard in terms of its potential impact on people and our mission for food safety and security. Cause and effect diagrams, and "What If" analysis are some of the best tools to assess risk in food production. What is the impact on business? People? Things? (Plant, facility etc). Severity assessment should be based upon the worst possible outcome that can reasonably be expected. Severity categories are catastrophic, critical, moderate, and negligible.

3. Assess Hazard Exposure: Surveys, inspections, observations, and mapping tools can help determine the level of exposure to the hazard and record it. This can be expressed in terms of time, proximity, volume, or repetition. Does it happen often and involve a lot of people. Repeated exposure to a hazard increases the probability of a mishap occurring. Understanding the exposure level can aid in determining the severity or the probability of the event. Additionally, it may serve as a guide for devising control measures to limit exposure.

Using the risk assessment matrix (see attached) the severity and probability for each hazard is identified (see exposure definition above). This process should be conducted individually and then compared with others conducting the same assessment in order to determine the most appropriate risk level. Each risk is labeled with its significance (high, medium, etc.). This allows us to see both the relative priority of the risk and their individual significance.

ORM in a restaurant (continued)

The next step after hazard identification in the restaurant ORM example above would be to conduct a risk assessment of each of the hazards and identify a risk level and ranking for each hazard as identified below:

Step 2. Assess the Risk (example of step 2 from our restaurant review above)

Hazard Identified in Operation	Assess the Risk	Risk Level	Rank
(1) a. No identification on drivers and different drivers are used. b. Trucks are not sealed or locked upon arrival and they are making other deliveries c. Unauthorized personnel can enter restaurant through unsecured back door at any time	1.a. Driver could be an aggressor and contaminate food in bulk form. The severity or potential impact of this hazard is critical due to its large effect on many people and operation (foods could be contaminated in bulk). The Probability is seldom (seldom associated with identified illness outbreaks). b. (Same as one) c. (Same as one)	1a. Medium/11 b. Medium/11 c. Medium/11	4a 4b 4c
2a. Storage areas are not monitored at all times and driver has access to entire storage area and kitchen. b. The facility does not have emergency lighting and it serves meals 24 hours a day. c. No tracking system to identify lot numbers on dry goods served in facility that may have been recalled	2a. Bulk food could be contaminated and the results could be critical. The probability is seldom but it is possible for aggressor to have access. b. Aggressor could cut power to building and stop food service. This could cause moderate damage to operation and loss of food products. The probability is seldom but possible. c. If food is contaminated during production and associated with illness all similar food would have to be destroyed not specific lots and individuals would have to be informed of potential illness associate with product	2a. Medium/11 b. Low/14 c. High/4	4d 5a 1

	even though none may have been served. Could be critical and probability is likely because recalls occur frequently		
3a. Exterior door from kitchen area to outside is not secure and can be used to enter kitchen area by unauthorized personnel b. No pressure backflow preventor on soft drink lines or oven hood cleaner can result in water contamination. c. Entry to kitchen area from serving line is not restricted unauthorized personnel can enter kitchen. d. Only one backup person if foodhandlers ill. Can result in inadequate staffing and use of unauthorized personnel. e. Employees hired from local unemployment agency with only work reference contact. No procedures to conduct complete or random criminal background or drug testing. f.. Employees locker room provided for changing clothes requires employees to provide own lock. Management cannot enter lockers to check on unauthorized use. g. New employees are hired into graveyard shift does not allow adequate observation.	3a. Aggressor could enter from these areas and potentially harm people, food, or machines due to unrestricted access. This could occur occasionally with critical severity. b. Water contamination could occur due to this hazard from a design or maintenance defect. Could potentially expose large group to contaminated ice or water with Critical results. This could be an occasional occurrence with broad exposure. c. An aggressor could enter the kitchen area and potentially result in people, equipment or facility damage with critical results to large group. The probability of occurrence is seldom. d. The individual hired to work temporally may not be trained properly or be an aggressor. This could result in damage to people, equipment and facility with critical results to a large group. The probability of occurrence is unlikely e. The individual hired may be a criminal or aggressor of another type. This individual has access to food, equipment and facility and could cause critical damage. The probability of occurrence is seldom. f. An aggressor could contaminate food resulting in Critical damage to people. The probability of occurrence is unlikely. g. An aggressor hired as a new employee could do critical damage to people, equipment, and facility due to decreased observation. The probability of this occurring is occasional	3a. High /7 b. High/7 c. Medium/11 d. Medium/11 e. Low/15 f. Medium/11 g. High/8	2a 2b 4e 4f 6 4g 3
4a. Leftovers returned to kitchen could be contaminated b. Individuals have open access to salads and desserts on serving line. Provides ability to contaminate	4a. An aggressor could contaminate with moderate results because small number would eat leftover. The probability of occurrence is seldom. b. An aggressor could do critical damage to food by contamination with an agent. The probability of occurrence is seldom. (Has occurred)	4a. Low/14 b. Medium/11	5b 4h

Step 3: Analyze Risk Control Measures

The third step is to analyze risk control measures for the potential hazards that could be introduced into the operation and were identified through the risk assessment step above. Action is taken to investigate specific strategies and tools that reduce mitigate, or eliminate risk. Starting with the highest risk identify as many risk control options as possible for all the risks. Effective control measures reduce or eliminate one of the three components (probability, severity, or exposure) of risk. You should also consider what the risk control costs and how various risk control options work together. The following are risk control options:

- *Reject*-We can and should refuse to take a risk if the overall costs of the risk exceed its benefits to the business or operation. In the example above placing new employees on the midnight shift with no background checks and providing them a locker that only they have access to is a high risk that should be rejected.

- *Avoid*-Avoiding risk altogether requires canceling or delaying the job, or operation, but is an option that is rarely exercised due to operational need. In the example above allowing open public access to the salad bar is a medium risk that has resulted in a bioterrorism incident when salmonella was sprinkled on fruits and veggies, blue-cheese dressing and potato salad at a salad bar. A decision to avoid this hazard by removing the salad bar could be made. As stated above this option is rarely exercised.

- *Delay*-It may be possible to delay a risk. If there is no time deadline or other operational benefit to speedy accomplishment of a risky task, then is often desirable to delay the acceptance of risk. During the delay, the situation may change and the requirement to accept the risk may go away. In the example above a decision may be made to keep the medium risk salad bar but to not use it and go to individual servings when general public threat announcements are made or other concerns with aggressors occur.

- *Transfer*-Risk transfer does not change probability or severity of the hazard, but it may decrease the probability or severity of the risk actually experienced by the individual or organization accomplishing the operation/activity. As a minimum the risk to the original individual or organization is greatly decreased or eliminated because the possible losses or costs are shifted to another entity. An example would be a food commodity group (dairy, produce, etc) developing an audited certification program for food safety and security. The group could work together to reduce risk to individual members by such actions as purchasing insurance to cover losses by members in good standing from food contamination from emerging pathogens (*E coli* 0157:H7 etc) or from potential aggressors (terrorist).

- *Spread*- Risk is commonly spread out by either increasing the exposure distance or by lengthening the time between exposure events. In the example above deliveries to the restaurant by unidentified drivers is a medium risk. Lengthening the time between deliveries (events) could spread the risk of this hazard. The restaurant could move from daily deliveries to twice a week.

- *Compensate*-We can create redundant capability in certain special circumstances. An example is to plan for a back up, and then when a critical piece of equipment or other item is damaged or destroyed we have capabilities available to bring on line to continue the operation. In the example above the low risk (could become larger risk if aggressor tried to remove current staff) created by having only one properly trained back-up food handler could be reduced by working with other restaurants and develop a pool of secure and properly trained food handlers for on call use by all the group.

- *R*educe-The overall goal of ORM is to plan operations or design systems that do not contain hazards. A proven order of precedence for dealing with hazards and reducing the resultant risks is:

  1. **Plan or Design for Minimum Risk.** Design the system to eliminate hazards. Without a hazard there is no probability, severity or exposure. In the example above the back dock for restaurants should be designed with secure delivery doors that trucks can back up to and unload their products securely on the dock and not enter facility. Staff would move products directly from dock to storage with entry doors in dock area. This prevents any aggressor from entering the dock, delivery truck or storage areas.
  2. **Incorporate Safety Devices.** Reduce risk through the use of design features or devices. These devices usually do not effect probability, but reduce severity: an automobile seat belt doesn't prevent a collision but reduces the severity of injuries. In the example above installing a pressure backflow preventor can reduce the high risk for water contamination from the oven cleaner.
  3. **Provide Warning Devices.** Warning devices may be used to detect an undesirable condition and alert personnel. In the example above the high risk of unauthorized potential aggressors entering the facility through the kitchen door to the outside can be reduced by installing a panic bar that prevents the exit from being used except in an emergency such as a fire.
  4. **Develop Procedures and Training.** Where it is impractical to eliminate hazards through design selection or adequately reduce the associated risk with safety and warning devices, procedures and training should be used. In the example above staff can be trained to report suspicious activities and stop or report unauthorized personnel. Procedures could be developed for reward programs for staff actively promoting and supporting food safety and security efforts with new ideas or actions.

In most cases it will not be possible to eliminate risk entirely but it will be possible to significantly reduce it. In making control decisions for the restaurant example above determine the effect of each proposed control on the risk associated with the hazard. The estimated values for severity and probably after implementation of the control measure and the change in overall risk assessment from the risk assessment matrix should be re-reviewed and recorded. Finally, prioritize risk control and for each hazard prioritize those risk controls that will reduce the risk to an acceptable level. The best controls will be consistent with operational objectives and optimize available resources. When implementing risk controls, try to apply risk controls only when the operation is actually at risk. Apply redundant risk controls only when practical and cost effective.

ORM in a restaurant (continued)

The next step after hazard identification and analysis of the risk is to rank the risks and then conduct an analysis of the proposed risk control measures. This risk control analysis and ranking are identified below:

Step 3. Analyze Risk Control Measures			
Hazard Identified in Operation	**Risk Level Rank**	**Risk Control Measures**	**Rank/**
2d. No tracking system to identify lot numbers on dry goods served in facility that may have been recalled	2d High/4	2d. – Maintain on site complete list of where food products and ingredients are purchased, including name, street address, phone and fax numbers. 2d. –Maintain on site records of invoices and bills of lading for all purchases, records of employees work shift, copies of menus and records of invoices and bill for all. 2d. – Maintain on site list containing the names and numbers of primary and secondary contacts of all regulatory agencies. 2d. – Develop written plans for ensuring and maintaining a proper chain of custody, storage and/or destruction of all recalled products. 2d. – Identify in contracts that food products purchased will have commodity codes and expiration dates with written explanations provided to restaurant (for recall purposes).	4. 5. 3. 2. 1.
3a. Exterior door from kitchen area to outside is not secure and can be used to enter kitchen area by unauthorized personnel	3a. High /7	3a. – Provide warning device by installing a panic bar that prevents the exit from being used except in an emergency such as a fire. 3a. – Post the door with signs that identify it is not an exit except in emergencies	1. 2.
3b. No pressure backflow preventor on soft drink lines or oven hood cleaner can result in water contamination	3b. High/7	3b. –Incorporate safety device by installing backflow preventor on oven line 3b. –Install backflow preventor on oven line and soft drink line. 3b Stop using the oven cleaner.	2. 1. 3.
3j. New employees are hired into graveyard shift does not allow adequate observation	3j. High/7	3j. Reject this risk and have new employees complete probation or at least their first 90 days on a day shift with increased oversight.	1.
3g. Employees locker room provided for changing clothes requires employees to provide own lock. Management cannot enter lockers to check on unauthorized use.	3g. High/8	3g. Reject this risk and have management provide locks for all lockers. Management would have master key and authority to enter lockers for periodic safety and security reviews.	1.
(1) a. No identification on drivers and different drivers are used.	1a. Medium/11	1.a. Lengthening the time between deliveries (events) could spread the risk of this hazard. The	2.

Food Safety and Security, ORM Systems Approach, November 2001

		restaurant could move from daily deliveries to twice a week. 1a. Contract with vendors that conduct security checks on employees and require drivers to have picture identification visible on them and provide restaurant with pictures of drivers prior to delivery.	 1.
1b Trucks are not sealed or locked upon arrival and they are making other deliveries	1b. Medium/11	1b. Establish requirement in contract that trucks be sealed or locked at all times. 1b Establish requirement in contract that vendor have locked truck from distributor to restaurant.	1. 2.
3d. Entry to kitchen area from serving line is not restricted unauthorized personnel can enter kitchen	3d. Medium/11	2a Have staff stay in serving line area at all times customers are in facility. 2a Put up security cameras 2a Put up signs that identify entry into kitchen is for authorized personnel only	2. 3. 1.
3f. Employees hired from local unemployment agency with only work reference contact. No procedures to conduct complete or random criminal background or drug testing.	3f. Medium/11	3f Establish procedures at hiring where new employees sigh authorization to submit to drug testing and authorize management to conduct criminal background checks on employees. 3f The same procedure as above, but random drug testing and backgrounds checks are made on all staff.	2. 1.
4b. Individuals have open access to salads and desserts on serving line. Provides ability to contaminate	4b. Medium/11	4b.Avoid this hazard by removing the salad bar 4b. Keep the medium risk salad bar but do not use it and go to individual servings when general public threat announcements are made or other concerns with aggressors occur. Delay the risk during high threat times.	2. 1.
3e. Only one backup person if food handlers ill. Can result in inadequate staffing and use of unauthorized personnel	3e. Low/15	3e. -Compensate by working with other restaurants and develop a pool of secure and properly trained food handlers for on call use by the entire group. 3e. - If food handlers ill reduce menu 3e – If food handlers ill change to self-serving types of food so individual assistance is not required.	1. 2. 3.
2b. The facility does not have emergency lighting and it serves meals 24 hours a day.	2b Low/14	2b Purchase an emergency power generator 2b Reduce hours of operation 2b Purchase emergency battery operated lights for building with extra flashlights	2. 3. 1.
4a. Leftovers returned to kitchen could be contaminated	4a. Low/14	4a Never return leftovers to kitchen 4a. Don't return leftovers to kitchen during times of high threat. 4a Review cooking procedures to reduce leftovers.	3. 2. 1.
4d. Staff reports that food products are taken by customers and then put back.	4d. Low/14	4d. Put up signs that identify food should not be returned 4d Have customer pay when enter facility.	2. 1.

Step 4: Make Control Decisions

In the example above the hazards identified were conditions that could impair our mission to accomplish food safety and security. The hazards were analyzed and we estimated the severity, probability and scope with which it could impact our mission for food safety and security. This evaluation of the hazard establishes a risk level. Then in the above step the risks were ranked from highest to lowest and several proposed controls were identified for each risk. Restaurant management should review the proposed controls with you and then with their staff. After controls have been selected to eliminate hazards or reduce their risk, determine the level of residual risk for the selected course of action.

- Accept the plan as is: Benefits outweigh risks (costs), and total risk is low enough to justify the proposed action if something goes wrong. The decision maker must allocate resources to control risk. Available resources are time, money, personnel, and/or equipment.
- Modify the plan to develop measures to control risk. The plan is valid, but the current concept does not adequately minimize risk. Further work to control the risk is necessary before proceeding.
- Elevate the decision to a higher authority. The risk is too great for the decision-maker to accept, but all measures of controlling risk have been considered. If the operation is to continue, a higher authority must make the decision and accept the risk.

There are several important points to keep in mind when making a risk control decision. Involve the personnel impacted by the risk control to the maximum possible extent in the selection. They can almost always provide ideas to enhance the various options. In addition, be sure to carefully evaluate the impact on the operation of the risk control action. The objective is to choose the option that has the best overall favorable impact on the operation. Be sure to consider all the positive benefits and negative (cost, lower morale, lower productivity) factors associated with the risk decision. It is also important to make risk decisions at the right level. In determining the level, ask who will be accountable if the risk produces a loss. That person should have a voice in the risk decision or actually make it. Once the best possible set of risk control options has been selected, the individual in charge must accept this decision. Remember, the goal is not the least level of risk; it is the best level of risk for overall food safety and security.

There is no single method of preventing or detecting terrorist activity with agents that could be introduced into a food or water system. A multifaceted approach is required to guard against an intentional contamination event. Existing safeguards and safety measures used to protect food and water from naturally occurring contamination are a good starting point. The addition of ORM for food safety and security will be effective deterrents for many terrorist scenarios in our example we accepted the plan as is.

Step 5: Implement Risk Controls

Once the risk control decision is made, resources must be made available to implement the controls. Part of the process of implementing control measures is to inform personnel in the system of the risk management process results and subsequent decisions. Careful documentation of each step in the risk management process facilitates risk communications.

- **Make Implementation Clear**. To make implementation clear fully involve operational personnel, the control measure must be deployed in a method that ensures it will be received positively. Designing in user ownership and describing the benefits of successful implementation. Provide guidance and even conduct small tests on changes before fully implementing'
- **Establish Accountability.** Be clear on accountability; who is responsible for implementation of the risk control. To be successful, management must be behind the control measures put in place. Prior to implementing a control measure, ensure you have approval at the appropriate level. Most failures with risk control are driven by the failure to properly involve personnel impacted by the risk control in the development and implementation of it.
- **Promote Support.** Develop the best possible supporting tools and guides to help in implementation, such as standard operating procedures for safety and security. The easier you make the task the greater the chance for success. Be sure to identify reasonable timelines for implementing. To be fully effective, risk controls must be sustained. This means maintaining the responsibility and accountability for the long haul. Provide management support and positive motivation with incentives for promoting and supporting food safety and security.

Step 6: Supervise and Review

Supervise:

✓ Assure controls are effective and in place
✓ Maintain implementation schedule for controls
✓ Assure needed changes are detected
✓ Correct ineffective risk control

Review:
✓ Obtain feedback from operator involved
✓ Select critical indicators and measure risk (knowledge, attitude, mishaps)
✓ Supervisor conducts work site spot check once a day to review controls
✓ Conduct a brief skill and knowledge quiz on food safety and security control requirements. Goal is 95%.

✓ Review the ongoing cost benefit of control

The sixth step of ORM "Supervise and Review" involves the determination of the effectiveness of risk controls implemented for food or water safety and security. This review process must be systematic and ensure control recommendations have been implemented. In addition, the need for further assessment of an operation due to unanticipated change could result in additional risk management actions. A review by itself is not enough. A feedback system must be established to ensure that the corrective or preventive action taken was effective and that any additional corrective action can be implemented as required. Feedback can be in the form of briefings, lessons learned, or update reports to management are beneficial.

OTHER PREVENTIVE MEASURES

In addition to physical risk control security measures, improvements in routine operations can have a significant impact on reducing threats to food. Routine sanitation can be the most valuable tool to mitigate risk from both intentional and non-intentional biologic contamination.

- Monitor hygiene practices.
- Have foodhandles shower in, and/or change into uniforms without pockets just before arriving in the work area.
- Wash hands thoroughly after any non-food preparation activity.
- Sanitize equipment when necessary to protect against contamination with undesirable microorganisms.
- Take steps necessary to prevent cross contamination with other materials such as wood, glass, and metal.
- Wear latex or other similar gloves when appropriate.
- Wear hairnets, headbands, caps, and beard covers when necessary.
- Do not eat food, chew gum, drink beverages, or use tobacco in areas where food may be exposed or where equipment or utensils are washed.
- Store clothes and other personal belongings in designated areas separate from food or water production and storage areas or utensil/equipment washing areas.
- Store chemicals such as pesticides and cleaning agents away from food or water production areas.
- Leftovers from meals should be monitored to ensure appropriate disposal. This prevents hazardous foods from being re-served or mixed with the next meal.

Minimum Safeguard And Security Measures

In implementing food safety and security systems, the following measures will greatly promote the protection of food or water system.

Eliminate Opportunity for Forced Entry

Defeat of the **forced entry** tactic relies not only on physical barriers, but also detection (electronic sensors, etc.) and interception by a responding force. The purpose of physical barriers is to delay an intruder long enough for a responding force to successfully apprehend the intruder.

Eliminate Potential for Insider Compromise

The basic defeating strategy for **covert entry** or **insider compromise** tactics is to keep people from entering areas that they should not enter. This strategy relies on the use of restricted entry to certain areas, guards, or detection systems. Further, hiring or contracting considerations may be changed based on the risk of an intentional contamination. Other actions to consider:

- Establish locations within the facility where access to assets can be limited to authorized personnel only.
- Minimize the number of entrances to controlled areas.
- Provide locks for entry doors, windows, and roof openings.

Improve Prevention and Control Measures for Food Safety and Security

The following are basic actions recommended for each stage of the food production process. These actions will prevent or mitigate against aggressor attempts to introduce contaminants into the food or water supply of a facility:

Management of Food Security

Security procedures
- Assign responsibility for security

- Reward and hold all staff accountable for being alert to and reporting signs of tampering with product or equipment, other unusual situations, or areas that may be vulnerable to tampering

Procedure for investigating unusual activity

- Immediately investigate all reports of unusual activity
- Document all investigations
- Report all problems to Security Forces

Employees

- Pre-hiring screening for all employees, including seasonal, temporary and contract
- Obtain work references
- Perform criminal background checks
- Place new employees on day shift with increased oversight during probation
- During hiring process obtain authorization to conduct random drug testing

Daily Rosters

- Make them specific to shift
- Know who is and who should be on premises, and where they should be located

Identification

- Issue photo identification badges with identification number; limit employee access to those areas necessary for the employee's position

Restricted access

- Limit access to those areas necessary for the employee's position (e.g. card entry to sensitive areas, cypher locks)

Personal items

- Restrict personal items allowed in establishment
- Prohibit personal items (e.g. lunch containers, purses) in food handling areas
- Reduce the amount of personal belongings brought to the facility. Examples include purses, gym bags, thermoses, drink containers, etc.
- Management should provide locks for locker areas and establish authority (during hiring process etc.) to enter lockers for periodic safety and security reviews. Metal mesh lockers provide additional security because contents are visible.

Training in security procedures

- Provide staff training in food safety and security procedures and inform them to report all unusual activities.
- Place new employees on day shift with increased oversight during probation;

Farm/source

- Promote participation in industry quality assurance programs. Examples include the FDA guidelines for Microbiologic Safety in Produce and shell egg QA production program
- Develop plans for isolation, cleaning and disinfect ion
- Keep records on animals, feed, seed and other products purchased and brought onto the farm
- Restrict entry to farm. For high confinement livestock production could even include employees showering in and out, vehicles being sprayed with disinfectant and other biosecurity precautions
- Conduct work reference checks on all employees
- Illuminate building exteriors and exterior sites where feed and other products are stored

Food Processor

- Improve onsite security programs, such as restricting rights of entry and exit, locking up storage bulk ingredient containers and mounting video surveillance at key internal processing hubs.
- Verify work references for seasonal employees. Conduct random basic criminal and drug checks on all employees
- Develop clearly documented well-rehearsed product recall plans, with crisis management teams that can quickly asses the scope of potential problems and contain them
- Written plans for deciding upon and evaluating the scope of a recall
- List containing the names and numbers of primary and secondary contacts of all regulatory agencies
- Minimize the need for signs or other indicators of food product storage
- Provide metal or metal-clad doors on facilities.
- Eliminate potential hiding places within the facilities where a contaminating agent could be temporarily placed before introduction

Retail Food Service

Raw materials, dry goods and packing

- Use only known, secure, state or locally licensed or permitted sources for all ingredients, compressed gas, packaging, and labels.
- Include in purchase and shipping contracts a requirement that suppliers and transporters practice appropriate food security measures
- Conduct work reference checks on all employees and random criminal background checks and have authority to conduct random drug testing

- Restrict access to food preparation areas to authorized personnel only.

Physical security

- Secure doors, windows, roof openings, vent openings, trailer bodies, railcars, and bulk storage tanks (e.g. locks, seals, sensors, warning devices)
- Use metal or metal-clad doors
- Account for all keys to establishment
- Have security patrols of the facility and video surveillance
- Minimize number of entrances to restricted areas and post areas that unauthorized personnel should not have access to
- Eliminate potential temporary hiding places for intentional contaminants
- Provide adequate lighting both interior and exterior
- Keep parking areas away from storage and water facilities

Storage of hazardous chemicals (e.g. cleaning and sanitizing agents, pesticides)

- Secure storage areas and not in food storage area
- Limit access to storage areas
- Supervise maintenance and sanitation staff
- Keep timely and accurate inventory of hazardous chemicals
- Investigate missing stock or other irregularities immediately

Transportation/Distribution

Suppliers

- Inspect incoming ingredients, compressed gas, packaging, labels, and product returns for signs of tampering or counterfeiting
- Require transportation companies to conduct background checks on drivers and other employees with access to the product (comply with state and local laws in doing this)
- Require locked and sealed vehicles/containers, and require seal numbers to be identified on shipping documents. Verify shipping seals with shipping papers.

Traceability of ingredients, compressed gas, packaging, and, salvage products, rework products, and product returns

- Include in purchasing contracts a requirement that suppliers will have commodity codes and expiration dates with written explanations provided for recalls and other food safety actions
- Use operating procedures that permit subsequent identification of source of ingredients, compressed gas, packaging, labels,
- Keep timely and accurate inventory of ingredients, packaging, labels, Investigate missing stock or other irregularities and report any problems to OSI

Security of Finished Products

- Keep timely and accurate inventory of finished products
- Investigate missing stock or other irregularities and report any problems to local legal authorities
- Include in contracts for shipping (vehicles and vessels) a requirement that they practice appropriate security measures
- Perform random inspection of storage facilities, vehicles,
- Require transportation companies and warehouses to conduct background checks on staff (drivers/warehouse personnel; state and local laws may apply)
- Require locked and sealed vehicles/containers, and identify seal number on shipping documents

Security Plans

Action Plan for tampering or terrorist event

- Include step-by-step SOP for triaging the event
- Include evacuation plan
- Maintain floor and flow plan in secure location and with local fire officials
- Include strategy for continued operation (e.g. at alternate facility)
- Include investigation procedures

Communication protocol

- Have internal, fire, and police emergency phone numbers available
- Identify critical decision-makers
- Identify local, state, and federal government contacts

Computer security

- Restrict access to computer process control systems for food products and critical data systems to those with appropriate clearance (e.g. passwords)

Security of water

- Secure water wells, storage and handling facilities
- Test for potability regularly
- Identify alternate sources of potable water (treat on-site or on-site storage)

OPERATIONAL RISK ASSESMENT MATRIX

Food Safety and Security, ORM Systems Approach, November 2001

			PROBABILITY				
			Frequent	Likely	Occasional	Seldom	Unlikely
			A	B	C	D	E
SEVERITY	Catastrophic	I	Extremely High				
	Critical	II		High			
	Moderate	III		Medium			
	Negligible	IV				Low	
			Risk Levels				

SEVERITY

- *Catastrophic* – Complete business failure due to food product contamination resulting in deaths.
- *Critical* – Major business degradation, due to food product contamination resulting in severe illnesses.
- *Moderate* – Minor business degradation, due to food product contamination resulting in minor illness.
- *Negligible* – Less than minor business degradation, and illnesses

PROBABILITY

- *Frequent* – Occurs often to individual and population is continuously exposed
- *Likely* – Occurs several times and population are exposed regularly
- *Occasional* – Will occur and occurs sporadically in a population
- *Seldom* – May occur and occurs seldom in a population.
- *Unlikely* – So unlikely you can assume it will not occur and occurs very rarely in a population.

Appendix B

FSIS SAFETY AND SECURITY GUIDELINES FOR THE TRANSPORTATION AND DISTRIBUTION OF MEAT, POULTRY, AND EGG PRODUCTS*

Dear Establishment Owner/Operator:

In May 2002, FSIS issued the *FSIS Security Guidelines for Food Processors* to assist Federal- and State-inspected meat, poultry, and egg product plants in identifying ways to strengthen their food security protection. At the time we noted our commitment to providing continued guidance to businesses engaged in the production and distribution of USDA-regulated foods. We have worked with the Food and Drug Administration (FDA) and other agencies to now provide guidance for those handling food products during transportation and storage.

The *FSIS Food Safety and Security Guidelines for the Transportation and Distribution of Meat, Poultry, and Egg Products* are designed to assist small facilities and shippers handling these products. The guidelines provide a list of safety and security measures that may be taken to prevent contamination of meat, poultry, and egg products during loading and unloading, transportation, and in-transit storage. In these voluntary guidelines, we strongly encourage shippers and receivers, as well as transporters of these products, to develop controls for ensuring the condition of the products through all phases of distribution. Such controls are necessary to protect the products from intentional, as well as unintentional, contamination. We recognize that not all of these measures will be appropriate or practical for every facility.

* From Food Safety and Inspection Service, United States Department of Agriculture, Washington, DC 20250-3700, News and Information, Updated August 11, 2003.

Meat, poultry, and egg products are transported by air, sea, and land. Hazards may be present at any point during transportation and distribution, but are most likely at changes between transportation modes and during loading and unloading. Meat, poultry, and egg products frequently are transported multiple times on their way to the consumer and may be exposed to hazards at each step. For example, a product might be transported from a slaughter establishment to a raw-product processing establishment, next to a further processing plant, and then onto distribution sites and retail markets.

The first section of these guidelines provides food safety measures to help prevent the physical, chemical, radiological, or microbiological contamination of meat, poultry, and egg products during transportation and storage. The second section of the guidelines deals specifically with security measures intended to prevent the same forms of contamination due to criminal or terrorist acts. Both sections apply to all points of shipment from the processor to their delivery at the retail store, restaurant, or other facility serving consumers of the products. These guidelines can be applied whether the potential contamination occurs due to an intentional or unintentional act.

For questions or clarification, contact our Technical Service Center at 1-800-233-3935.

Protecting our Nation's food distribution network is essential to the Nation's homeland security. These guidelines are intended to assist the food industry, as well as Federal, State, and local authorities, in that effort.

Sincerely,

Garry L. McKee, Ph.D., M.P.H.
Administrator

SECTION I: FOOD SAFETY DURING TRANSPORTATION AND DISTRIBUTION OF MEAT, POULTRY, AND EGG PRODUCTS

Meat, poultry, and egg products are susceptible to contamination from a wide variety of physical, microbial, chemical, and radiological agents. These products are particularly vulnerable to microbiological hazards because their moisture, pH levels, and high protein content provide ideal environments for the growth of bacteria. Because of these characteristics, the products must be carefully monitored to prevent their exposure.

Food safety protection can be improved by the control of hazards through the use of preventive methods including good sanitation, manufacturing practices, and the Hazard Analysis and Critical Control Point (HACCP) system throughout the food production and distribution chain. Meat, poultry, and egg products must be refrigerated or frozen after processing and before shipment to inhibit spoilage and growth of pathogens. During transportation and storage, the challenge is to maintain proper refrigeration temperatures and to keep the "cold-chain" from breaking during steps such as palletization, staging, loading and unloading of containers, and in storage.

General Guidance

In the United States, most food is transported by truck. However, meat, poultry, and egg products may be transferred to and from other modes of transportation during shipment and held at intermediate warehouses as well as at transfer or handling facilities, such as airports, break-bulk terminals, and rail sidings. Because transportation and storage are vital links in the farm-to-table food chain, effective control measures are essential at each point in the food distribution chain to prevent unintentional contamination.

The following general guidelines address food safety measures that should be taken by shippers from the point of food production through delivery. The guidelines do not cover breeding, feedlot, or any other pre-slaughter live-animal operations or pre-shipment operations at egg-laying farms.

Transportation Safety Plan

- **Identify vulnerable points and develop a comprehensive transportation sanitation and safety plan.**
  - Processors and distributors shipping products should assess and implement measures that will ensure the sanitation and safety of products from initial shipment through delivery to other destinations. A flow diagram from the point-of-origin to final destination, including all shipping modes/routes, can be a helpful assessment tool. (See Attachment for sample flow diagram.) 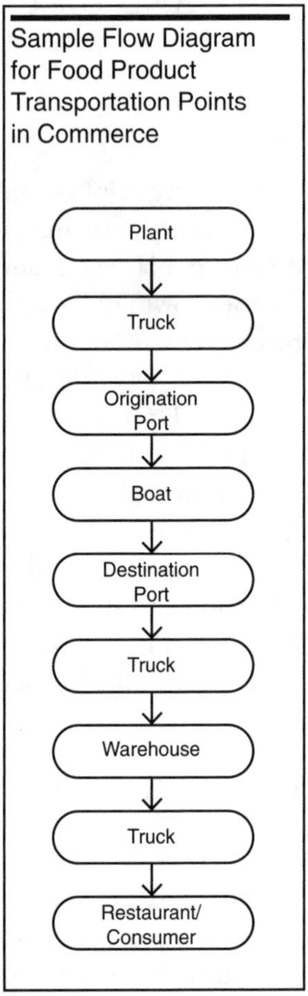
  - Identify all points of vulnerability where there is potential for adulteration or contamination to occur:
    - Identify potential hazards.
    - If control points are identified, then determine the method, frequency, and limit that must be met.
    - Identify if control is possible at the point(s) of hazard and what is the most effective point to exert control.
    - This will determine where and how often monitoring and verification of the limits set should occur and what, if any, corrective and preventive actions should be taken.
    - Define what controls should be put in place to prevent product adulteration or contamination during the transportation and storage process.
  - As an additional check on product condition during and after transportation and storage, processors may want to include special arrangements with receivers to sample and conduct microbiological or other tests on products. The results could be compared with pre-shipment results to determine whether adjustments are needed in transport methods or procedures.
  - Verify that contracted transporters (e.g., air, ground, maritime, rail) and storage/warehouse facilities have a food safety program

Sample Flow Diagram for Food Product Transportation Points in Commerce

    in effect. Consider including specific security measures in contracts and verify that measures are being met.
  - Include procedures for the immediate recall of adulterated products from trade and consumer channels (this applies to processors, transporters, and wholesale and retail distributors).
  - Have a system in place to track your products, including salvage, reworked, and returned products.
- **Train personnel.**
  - Train managers and supervisors involved in the transportation, handling, and storage of food products in food hygiene and sanitation. They should be able to judge potential risks, take appropriate preventive and corrective actions, and ensure effective monitoring and supervision to prevent intentional and unintentional contamination from occurring.
  - Train personnel involved in all phases of transport, handling and storage in personal hygiene, vehicle inspection procedures, and transportation procedures that will ensure the safety of meat, poultry, and egg products.

Storage Food Safety System

- **Design and maintain a storage and warehousing food safety system.**
  - The facility should permit easy access to all areas for cleaning.
  - Adequately insulate the facility and have an adequate temperature control capacity.
  - Prevent access by unauthorized persons through the use of locks and fences, etc.
  - Have an effective, systematic program for preventing environmental contamination and infestation by insects, vermin, etc.

Vehicles Used to Transport Meat, Poultry, and Egg Products

- **Design and construct vehicles to protect product.**
  - Vehicles should be designed and built to make locking and sealing easy, protect the cargo against extremes of heat and cold, and prevent infestation by pests.
  - Vehicle design should permit effective inspection, cleaning, disinfection, and temperature control.

  - Interior surfaces should be made of materials that are suitable for direct food contact. For example, the surfaces may be made with stainless steel or be coated with food-grade epoxy resins.
- **Sanitize and properly maintain vehicles.**
  - Meat, poultry, and egg product transportation vehicles, accessories, and connections should be kept clean and free from dirt, debris, and any other substance or odor that may contaminate the product. They should be disinfected as needed. Cleaning and sanitation procedures should be specified in writing.
  - Different cleaning procedures may be necessary for the different types of meat, poultry, or egg products that are to be transported. The type of product transported and the cleaning procedure used should be recorded. Generally, wash water should be at least 180°F (82°C) and an approved sanitizer may be used to reduce the number of microorganisms and dissolve any fat particles adhering to interior surfaces.
  - Cargo pallets, load securing devices, and loading equipment should be kept clean and free of potential food contaminants and be regularly washed and sanitized.
  - Equipment used in transferring meat, poultry, and egg products, such as hand trucks, conveyors, and forklifts, should be well maintained and kept in a sanitary condition.
  - Secure transport vehicles to prevent tampering when not in use.
- **Use dedicated transport vehicles.**
  - Transport vehicles, containers, and conveyances should be designated and marked "for food use only," and be used only for transporting foods. If feasible, they should be restricted to a single commodity. This reduces the risk of cross contamination from previous cargoes.

Pre-Loading

- **Loading and unloading areas should be configured, cleaned, disinfected (where appropriate), and properly maintained to prevent product contamination.**
  - Loading or unloading facilities should be designed to permit easy access to all areas for cleaning.
  - Facilities should be adequately insulated and have an adequate temperature control capacity.
  - Facilities should have an effective, systematic program for preventing environmental contamination and infestation by insects, vermin, etc.

- **Examine vehicles before loading.**
  - Trailer or truck body should be sufficiently insulated and be in good repair with no holes in the body that might allow heat, dust, or other adulterants to enter the cargo area.
  - Check for residues of previous cargoes.
  - Check for residues from cleaning and sanitizing compounds.
  - The cooling unit must be in good repair and operating. Both truck drivers and plant personnel should check the functioning of the trailer refrigeration unit.
  - Trailers and trucks should be pre-cooled for at least 1 hour before loading to remove residual heat from the insulation and inner lining of the trailer as well as from the air of the trailer. For pre-cooling, the doors should be closed and the temperature setting of the unit should be no higher than 26°F. (Note, however, that poultry products labeled "fresh" must be shipped at temperatures higher than 26°F, usually between 26°F and 32°F.)
  - Inspect trailers prior to loading to determine that the air chutes, if used, are properly in place and that the ribbed floors are unclogged so that adequate air circulation can occur.
  - Examine trailer doors and seals to ensure that they can be secured and that there will be no air leaks.
  - When shipping a mixed load of products, such as both frozen and refrigerated products, it may be necessary to use a trailer with compartments that accommodate different temperature or other handling requirements.
- **Stage loads to facilitate proper stowage and minimize exposure during loading and unloading.**
  - Proper staging of loads is especially important when there are loads of products with different temperature requirements, or different delivery destinations.
  - Dock foremen should document that all freight is 40°F or lower before loading. Freight should not be allowed to remain on the loading dock in warm weather in order to prevent the product temperature from rising above 40°F.
  - *Note:* Federal regulations require processed poultry to be packaged and shipped at a temperature no higher than 40°F.

Loading

- **Protect products from exposure to environmental contaminants such as microbes, dust, moisture, or other physical contamination.**

- **Maintain the "cold chain" to ensure meat, poultry, and egg products are kept at appropriate temperatures continuously throughout all phases of transport.**
  - Meat, poultry, and egg products must be kept refrigerated and protected from temperature changes. All persons involved in the transportation, storage, and handling of these products are responsible for keeping them at appropriate temperatures and preventing any break in the cold chain.
  - Maintain the appropriate temperature of the pre-cooled product by minimizing the time of loading or unloading, conducting the loading and unloading in an appropriately chilled environment, and reducing the amount of surface contact of the product with floors and walls of the storage areas or loading equipment.
  - Appropriately packaged meat, poultry, or pasteurized egg products can be stacked, provided that air circulation is sufficient to maintain the temperature of the products during shipment.
  - Product should be at the desired transit temperature before loading. The boxes and pallets should be secured within the vehicle and pallets should be center-loaded off the walls of the vehicle.
  - Seal vehicles shipping egg products from one official plant to another for pasteurization, re-pasteurization, or heat treatment. (A certificate stating that the products are not pasteurized or that they have tested positive for Salmonella should accompany applicable shipments.)
- **Use appropriate loading procedures and equipment.**
  - Use spacers on sidewalls and at the ends of trailers as well as pallets on the floor so that proper air circulation can be maintained.
  - Keep loading time as short as possible to prevent temperature changes (increases or decreases) that could threaten the safety or quality of food products.
  - Close doors immediately after the truck/trailer has pulled away from dock.
- **Use special care with mixed or partial loads.**
  - Partial and mixed loads increase the frequency and duration of open doors, leading to a greater possibility of temperature fluctuations and exposure to tampering.
  - Other factors affecting temperature include the time of loading and unloading, the number of stops, the total length of the haul from origin to destination, and the outside temperature.
  - During periods of warm weather, loading or unloading should be done in the evening or early morning to minimize the likelihood of products warming.

In-Transit

- **Establish procedures to periodically check integrity of the load during transit.**
  - Check for leakage of heating or cooling fluid onto food products.
  - Monitor the temperature and function of the refrigeration unit at least every 4 hours. If there is a unit malfunction, the problem should be corrected by an authorized refrigeration mechanic before the temperature of the load rises.
  - Check for breakdown of temperature control.
  - Use time-temperature recording, indicator, or integrator devices, if they are available, to monitor the condition of cargo. Check the devices every 4 hours.
- **Establish procedures to ensure product safety during interim storage.**
  - Maintain logbook documenting product condition upon arrival and during storage.
  - Ensure proper temperatures are maintained during storage of meat, poultry, and egg products.

Unloading

- **Carefully examine incoming products.**
  - Product should be inspected and sorted before being accepted at any point during transportation.
  - Develop and implement methods to check and document condition of product and packaging upon receipt at destination. Examine checks of time-temperature recording, indicating or integrator devices or, by prior arrangement with the shipper, test to determine if bacterial growth has occurred after the product was packaged and shipped.
  - Include procedures for the safe handling and disposal of contaminated products. Identify where and how to separate contaminated products.
  - Establish policy and procedures for rejection of packages and products that are not acceptable, can't be verified against the delivery roster, or contain unacceptable changes to shipping documents. Have a monitoring plan and record-keeping system in place to document steps taken.
  - Do not accept products known to be, or suspected of being, adulterated.

- **Move product from the loading dock into cold storage immediately to minimize product exposure to heat and contaminants.**

SECTION II: FOOD SECURITY DURING TRANSPORTATION AND DISTRIBUTION OF MEAT, POULTRY, AND EGG PRODUCTS

The tragic events of September 11, 2001, forever changed our world. They proved to us that the unthinkable could become reality, and that threats to our Nation's food supply are very plausible from those who want to harm us through any possible means. Since the terrorist attacks on America, security – including food security — has been the highest priority at both the Federal and State levels.

Ensuring safe food within the processing plant, during transportation, in storage, and at retail is a vital function to protect public health. We must now look at all possible threats, examine the risks, and take action to prevent any intentional attack on the food supply.

General Guidance

Meat, poultry, and egg products are susceptible to intentional contamination from a wide variety of physical, chemical, biological, and radiological agents. Everyone in the food distribution system is responsible for ensuring that these products are safe, wholesome, and unadulterated. Therefore, as part of this system, those responsible for transportation and delivery should implement every possible security measure to ensure the integrity of the products throughout the supply chain.

There are many potential benefits of having an effective security plan in place such as:

- Protects public health and assets;
- Increases public and customer confidence, including trading partners;
- Provides value-added component to product;
- Deters theft and tampering;
- Creates production and distribution efficiencies;
- Maintains greater control over product through supply chain; and
- Possibly reduces insurance premiums and freight rates.

The guidelines below provide a list of security measures to be considered by processing plants, shipping companies, and warehouse facilities to minimize the risk of tampering or other criminal action for each segment of the food-delivery system.

Security Plan

Assess Vulnerabilities

- Identify a food protection management team and assign a leader to verify required actions are implemented and effective.
- Develop a comprehensive transportation security plan and assess vulnerabilities using a recognized threat/risk/vulnerability model such as Operation Risk Management (ORM) and Systematic Assessment of Facility Risk (SAFR). A flow diagram from your point-of-origin to final destination, including all shipping modes/routes, can be a helpful tool in your assessment.
- In your security plan, identify all points of vulnerability where there is the potential for intentional adulteration or contamination to occur during the transportation and distribution process:
  - Identify potential biological, chemical, and physical hazards.
  - Identify if control is possible at the point(s) of hazard and what is the most effective point to exert control.
  - If control points are identified, then determine the method, frequency, and limit that must be met.
  - This will determine where and how often monitoring and verification of the established limits should occur and what, if any, corrective and preventive actions should be taken.

Develop and Implement Procedures

- Implement identified security measures at each point to ensure the protection of products from the time of shipment through delivery to each destination.
- The plan should include a system to identify and track your product at any time during transportation and distribution such as the use of tamper-resistant seals corresponding to specific shipments and their documentation.
- Verify that contracted transporters (e.g., air, ground, maritime, rail) and storage/warehouse facilities have a security program in effect.

Consider including specific security measures in contracts and verify that measures are being met.
- Include procedures for the immediate recall of adulterated products from trade and consumer channels.
- Have a system in place to track salvaged, reworked, and returned products.
- Include procedures for handling threats to and actual cases of product tampering.
- Establish an evacuation plan for the facility.
- Include procedures for the safe handling and disposal of contaminated products. Identify where and how to separate suspected products.
- Develop and implement methods to check and document condition of product and packaging upon receipt at destination.
- Establish policy and procedures for rejection of packages and products that are not acceptable, can't be verified against the delivery roster, or contain unacceptable changes to shipping documents. Have a monitoring strategy and recordkeeping system in place to document steps taken.
- Establish policy and procedures for allowing rail crew, truckers, etc., to enter the facility and monitor their activities while on the property.
- Food security plans should be kept in a secure location and shared on a "need-to-know" basis.

Emergency Operations

- Regularly update a list of local, State, and Federal emergency contacts, local Homeland Security contacts, and local public health official contacts.
- Develop procedures for notification of appropriate authorities if an event occurs.
- Identify all entry and exit points available to emergency personnel in the plan.
- Develop a strategy for communicating with the media, including identifying a spokesperson, drafting press statement templates, or referring media to trade association or corporate headquarters.

Train and Test

- Train each team member in all provisions of the plan.
- Conduct drills regularly to test and verify the effectiveness of the plan. Continually review policies and procedures in the plan. The food-protection management team leader should coordinate these activities.

Screen and Educate Employees

- Screen all potential employees, to the extent possible, by conducting background and criminal checks appropriate to their positions, and verifying references (including contract, temporary, custodial, seasonal, and security personnel). When this is not practical, such personnel should be under constant supervision and their access to sensitive areas of the facility restricted.
- Consider participating in the Immigration and Naturalization Service (INS) pilot program for screening (#1-888-464-4218).
- All employees should be trained on how to prevent, detect, and respond to threats or terrorist actions so they can recognize threats to security and respond if necessary.
- Promote ongoing security consciousness and the importance of security procedures.
- Personnel involved in the transport, handling, and storage of meat, poultry, and egg products should be trained in procedures that will ensure the security of these products (e.g., train dock and security personnel on documentation requirements for incoming and outgoing shipments).
- Train appropriate personnel in security procedures for incoming mail, supplies, and equipment deliveries. Mail handlers should be trained to recognize and handle suspicious mail using U.S Postal Service guidelines.
- Ensure employees know emergency procedures and contact information.
- Encourage employees to report any suspicious activities such as signs of possible product tampering or break in the food security system. Have a tracking system in place for these reports and follow-up activities.

Secure the Facility

Access

- Maintain a positive ID system for employees. Require identification and escort visitors at all times in your facility.
- Collect company-issued IDs, keys and change lock combinations when a staff member is no longer employed by the company.
- Ensure clear identification of personnel to their specific functions (e.g., colored hats or aprons, ID cards).

- Restrict types of personal items allowed in the establishment, especially firearms or other weapons.
- Secure and restrict access to facilities, transportation trucks, trailers, or containers, locker rooms, and all storage areas with alarms, cameras, locks and fences or other appropriate measures, to prevent access by unauthorized persons.
- All visitors should be escorted while on the premises. Establish procedures for handling unauthorized persons in a restricted access area.
- Control access to food products by unauthorized persons by limiting access to food delivery, storage, food ingredient, and chemical storage areas.
- Restrict access to computer data systems. Protect them using firewalls, virus detection systems and secure passwords, changing them routinely.
- Restrict access to outside water tanks, water supplies, ice machines, and conveying water pipes.
- Restrict access to central controls for heating, ventilation, and air conditioning (HVAC), electricity, gas, and steam systems to prevent contamination from entering the air distribution systems.

Shipping/Receiving

- Consider developing a checklist for shipping and receiving procedures (this can also help identify anomalies).
- Loading docks should be secured to prevent unauthorized deliveries.
- All deliveries should be scheduled and truck drivers should show proper identification upon arrival.
- Shipping documents should contain product information, name of carrier(s), driver information, and seal numbers.
- Establishments should require that incoming shipments be sealed with tamper-proof, numbered seals, and that the seal numbers are shown on the shipping documents for verification prior to entry to the facility.
- Shipping documents with suspicious alterations should be thoroughly investigated. Product should be held and segregated during investigation process.
- Properly secure transportation trucks, trailers, or containers:
  - Doors should not be left open when picking up a load from a warehouse.
  - Ensure shipping trucks, trailers, and containers are secured after loading is complete.
  - Lock transportation trucks, trailers, and containers when not in use, during meal breaks and at night.

- Apply seals to all containers being shipped and maintain a seal log. Have a system in place to verify seal numbers and the integrity of the seals throughout the distribution process.
- Ensure security procedures are in effect for interim storage at in-transit warehouses.

Facility

- Designate limited and specific entry and exit points for people and trucks.
- Secure all access and exit doors, vent openings, windows, outside refrigeration and storage units, trailer bodies and bulk storage tanks.
- Ensure adequate interior and exterior lighting at the facility.
- Parking areas for visitors should be situated away from the main facility, if practical. Vehicles of employees and visitors should be clearly marked (e.g., placards, decals). This is intended to identify vehicles authorized to be on the premises and deter bombing attempts.
- Hazardous chemical storage areas or rooms should be secured and located away from food preparation and storage areas. In addition, they should be constructed and safely vented in accordance with national or local building codes.
- Incoming mail should be handled in an area of the facility separate from food handling, storage, or preparation areas.
- Install backflow devices on all water supply equipment.

Monitor Operations

Employees

- Maintain a daily shift roster to easily identify persons who are/should be on the premises and indicate that they are in their appropriate location.
- Provide appropriate level of supervision to all staff, including food handlers, cleaning and maintenance staff, and computer support staff.
- Monitor employees for unusual behavior (e.g., staying unusually late, arriving unusually early, taking pictures of the establishment, or removing company documents from the facility).

Shipping/Receiving

- Purchase all food ingredients, food products, and packaging materials only from known, reputable suppliers. Require Letters of Guaranty, if possible.

- Require locked or sealed trucks, trailers, or containers for deliveries. Maintain an inbound load verification logbook. Verify inbound trucks for seal numbers and integrity and load manifest. Document seal numbers and the truck or trailer number.
- Hold unscheduled deliveries outside the premises pending verification of shipper and cargo. Do not accept deliveries from, or release product to, unknown shippers using only cell phone numbers or known shippers with unknown phone/fax numbers or e-mail addresses.
- Supervise off-loading of incoming products, ingredients, packaging, labels, and product returns. Only a supervisor or other agent of the owner should break seals and sign off in the trucker's logbook.
- Have system in place to ensure integrity of product when seal will need to be broken prior to delivery due to multiple deliveries or for inspection by government officials.
- Verification of the last company seal put on a truck should be available throughout the delivery chain.
- Examine incoming products and their containers for evidence of tampering or adulteration:
  - Determine a random or other sound plan for checking incoming product;
  - The warehouse supervisor should note on the bill of lading any problems with the condition of the product, packaging, labels, and seals;
  - Do not accept products known or suspected of being adulterated;
  - Check food for unusual odor or appearance.
- Processors may want to arrange with receivers to sample and conduct microbiological or other tests on products.
- This would require an in-house testing plan prior to shipment.
- The results should be compared with pre-shipment results to determine whether adjustments are needed in transport methods or procedures.
- Establish chain-of-command procedures providing for the proper handling of samples.
- Samples should be clearly marked and kept in a secure area.
- Ensure all trucks leaving the facility are sealed.
- Maintain a logbook of seal assignments.

Storage/Water

- Maintain an accurate inventory of food and chemical products and check daily to allow detection of unexplained additions to, or withdrawals from, existing stocks. Include information about the

sources and date of shipment. All discrepancies should be investigated immediately.
- Perform random inspection of storage facilities (including temporary storage trailers or containers), trucks, trailers, containers, and vessels regularly. Keep a log of results. Designate an individual to conduct the inspection and have a record-keeping system in place.
- Regularly test water and ice supply to ensure it is safe to use.
- Inspect water storage and conveying lines inside and outside of the facility regularly for tampering or irregularities.

Respond

- Be aware of and report any suspicious activity to appropriate authorities (e.g., unscheduled maintenance, deliveries, or visitors should be considered suspicious).
- Processors, transportation managers, and wholesale and retail distributors should ensure traceability and recall of products.
- Ensure procedures are in place to accomplish a complete, rapid recall, and removal from the market of any shipment of meat, poultry, and egg products in the event products are found to present a hazard to public health.
- Keep detailed production records, including packaging lot or code numbers and where finished product was stored or served.
- *Trace Forward* — Shippers (including operators of federally inspected meat, poultry, and egg processing establishments) and carriers should have systems in place for quickly and effectively locating products that have been distributed to wholesalers and retailers.
- *Trace Backward* — Retailers, wholesalers, carriers and others who have received products from federally inspected meat, poultry, or egg processing establishments should be able to identify the source of the products quickly and efficiently.
- Investigate threats or reports of suspicious activity swiftly and aggressively.
- In the event of a food security emergency, first contact your local law-enforcement authority.

Additional Guidance for Specific Modes of Transportation

Of the approximately 200.5 billion metric tons of food shipped internationally each year, 60 percent goes by sea, 35 percent by land, and 5 percent by air. Domestically, most food products move via ground transportation (truck and rail). Thus, it is critical that everyone involved in the food delivery system understands his or her role and responsibility to ensure the security of meat, poultry, and egg products to the end point or consignee. Recognizing the inter-modal nature of this system, a multi-layered approach to protecting food is essential.

General Guidance for All Modes

Make certain that contracted shippers and consignees have security measures in place to ensure product integrity and traceability and verify that they are meeting contractual security obligations. Security measures should include:

- Physical boundaries of the facility/terminal are secure;
- Background checks are conducted for all potential employees by shipping, trucking, and drayage companies;
- A positive identification system is in place for all employees. (Recommend requiring participation in the Transportation Worker Identification Card (TWIC) program which is coordinated by the Transportation Security Administration);
- A security training and awareness program for all employees on how to prevent, detect, and report suspicious activity is conducted;
- A system is in place to track movement of products and truck, trailer, and containers/vessels (e.g., Global Positioning System);
- Maintains record-keeping system to document chain-of-custody, which will aid in tracing product;
- Uses a system (e.g., X-ray scanners) to detect tampering and radiological, biological, and chemical agents in shipping containers;
- Policies and procedures are in place for the handling of suspicious product; and
- Ensures all containers are properly secured at all times when held in storage yards.

Aviation

Although fewer meat, poultry, and egg products are transported by air than by other modes, it is still critical to ensure the security of these products when this mode of transport is utilized.

- Check all trucks entering a terminal facility.
- Trucks carrying meat, poultry, and egg products should have seal logbooks and the seals should be examined and numbers verified.
- Inspect containers arriving at a terminal for loading before admitting them to the terminal.

- Immediately report suspicious or inconsistent servicing of a container to terminal security.
- Design internal and external packaging so customers will be able to determine if the product was tampered with and can immediately notify you. Provide instructions and contact information with shipment.

Truck

Approximately 21 million trucks transport products across the United States every day. Keeping containers secure is a huge undertaking as there can be many opportunities for tampering.

- Develop and implement procedures for drivers to ensure security of the truck, trailer, or container when stopping for meals, gas, and repairs.
- Transportation trucks, trailers, and containers should be designed and built to make locking and sealing easy and should permit effective inspection.
- Examine trailer doors and seals to ensure that the trailer can be secured.
- Keep empty trailers locked at all times.
- Check product load periodically during transit to ensure its integrity has not been compromised (e.g., use weigh station stops as an opportunity to check condition of products).
- Processors, distributors, and transporters should have action plans for emergencies, such as breakdowns or reporting criminal activity. The plans should include notification of the relevant Federal, State, and local authorities.
- Drivers should be trained to take appropriate precautions while en route (e.g., do not pick up hitchhikers, do not discuss the nature of cargo at stops, be aware of surroundings, lock truck, trailer, or container when unattended and avoid low-lit areas).
- Prevent unauthorized access to delivery truck, trailer, or containers. Require drivers to secure truck, trailer, or containers while en route, including while on break, at restaurants, at overnight stays, etc.
- Drivers should report unusual circumstances, such as being followed, to appropriate authorities.
- Develop procedures to be followed when reefer boxes or trailers are found unlocked.
- Deter diversion or hijacking of cargo by keeping track of trucks. Ensure time logs for trips are maintained and provide trucks with communication and tracking equipment.

- Hold drivers accountable for ensuring security measures are taken to prevent contamination of meat, poultry, and egg products while under their control.

Maritime

Ports are vulnerable due to their size, accessibility by water and land, location in metropolitan areas, and the quantity of products moving through them. Approximately 80% of U.S. imports arrive via American seaports, yet U.S. Customs physically inspects only a fraction of all containers; the remainder are electronically screened. Therefore, enhanced security measures are necessary for products shipped by sea.

- Check all trucks entering a terminal facility. Trucks carrying meat, poultry, and egg products should be sealed, drivers should have seal logbooks, and the seals should be verified.
- Seals should be removed in the presence of terminal personnel so they can verify the seal number and its integrity.
- Immediately report suspicious or inconsistent servicing of a container to terminal security.
- Supervise opening of ship hatches.
- When unloading product from sea-going vessels, inspect seals for evidence of tampering. Have documentation system in place.
- Document cutting of seals (e.g., when seal is cut for inspection by government official).
- Shipping line agents should provide importers and customs brokers with a record of vessel discharge and checks at discharge and in transit.
- Establish policy and procedures to download reefer electronic information during inspection (this will also allow identification of anomalies);
- Have reporting system in place when the discharging of any product looks suspicious or the product shows evidence of tampering.
- The terminal facility should be locked during meal breaks and at night.
- Facility doors should be closed immediately after the truck/trailer has pulled away from dock.

*Importers and Exporters may want to consider participation in government initiatives pertaining to maritime shipment of products such as:

- Customs-Trade Partners Against Terrorism (C-TPAT)
- Operation Safe Commerce
- Container Security Initiative
- Sea Carrier Initiative Agreement

Rail

Rail transportation is an integral part of the domestic food distribution system, therefore, it is important to recognize that unsecured containers can be easy targets for tampering and address this vulnerability.

- Use boxcars dedicated for food products.
- Employ measures to secure loaded and empty containers from tampering when being stored at the trainyard for any length of time.
- Locks/seals on boxcars should be inspected at pull and place.
- Review shipping documents upon arrival at the trainyard and before the train engineer leaves.
- Inspect integrity of seals upon arrival and before departure of the load.

- If you have questions or need clarification about the guidelines, contact the FSIS Technical Service Center at: 1-800-233-3935.
- Obtain additional copies of the guidelines at: http://www.fsis.usda.gov or call 202-720-9113.
- Further information on the safe and secure transportation of food is available from:
    - USDA Agricultural Marketing Service: http://www.ams.usda.gov/tmd/tsb/
    - Transportation Security Administration: http://www.tsa.dot.gov
    - Food and Drug Administration: http://vm.cfsan.fda.gov/~dms/secguid6.html
    - Federal Aviation Administration: http://www.faa.gov
    - U.S. Postal Service: http://www.usps.com/cpim/ftp/pubs/pub166/welcome.htm
    - U.S. Customs: http://www.customs.ustreas.gov
    - American Association of Railroads: http://www.aar.org
    - American Trucking Association: http://www.trucking.org
    - National Cargo Security Council: http://www.cargosecurity.com/ncsc/
    - The World Health Organization: http://www.who.int/fsf

*Consider researching government Internet sites to obtain funding resources (e.g., grants and loans) to enhance your security program.

*Note: To read and print a **PDF** file, you must have the Adobe® Acrobat® Reader installed on your PC. You can download a version suitable for your system, free of charge, from the Adobe Home Page. Adobe also provides tools and information to help make Adobe PDF files accessible to users with visual disabilities at http://access.adobe.com.

The U.S. Department of Agriculture (USDA) prohibits discrimination in all its programs and activities on the basis of race, color, national origin, sex, religion, age, disability, political beliefs, sexual orientation, or marital or family status. (Not all prohibited bases apply to all programs.) Persons with disabilities who require alternative means for communication of program information (Braille, large print, audiotape, etc.) should contact USDA's TARGET Center at 202-720-2600 (voice and TTY).

To file a complaint of discrimination, write USDA, Director, Office of Civil Rights, Room 326-W, Whitten Building, 14th and Independence Avenue, SW, Washington, D.C. 20250-9410 or call 202-720-5964 (voice or TTY). USDA is an equal opportunity provider and employer.

For Further Information Contact:

- Media Inquiries: (202) 720-9113
- Congressional Inquiries: (202) 720-3897
- Constituent Inquiries: (202) 720-9113
- Consumer Inquiries: Call the USDA Meat and Poultry Hotline at 1-888-MPHotline; TTY: 1-800-256-7072.

FSIS Home Page | USDA Home Page www.usda.gov

Appendix C

FSIS SECURITY GUIDELINES FOR FOOD PROCESSORS*

Dear Establishment Owner/Operator:

The Food Safety and Inspection Service (FSIS) has prepared the enclosed, FSIS Security Guidelines for Food Processors, to assist Federal and State inspected plants that produce meat, poultry and egg products in identifying ways to strengthen their biosecurity protection. FSIS recognizes that inspected plants may also be aware of, and are adopting, guidelines from other government and private sector organizations and agencies. However, businesses or plants that do not have access to specialized security-planning advice should find these guidelines particularly useful as they develop and improve their food security plans.

These guidelines were developed to meet the particular needs of meat, poultry and egg processing plants and to be easily understood and readily adaptable by plant officials. While the guidelines are voluntary and plants may choose to adopt measures suggested by many different sources, it is vital that all food businesses take steps to ensure the security of their operations.

FSIS intends to provide these guidelines to our field employees who will assist in directing plants that seek further clarification or advice. However, inspectors will not mandate adoption of any guideline.

FSIS intends to continue working to enhance guidance to businesses engaged in the production and distribution of USDA-regulated food and to work with Food and Drug Administration (FDA) and other agencies to

* From *FSIS Security Guidelines for Food Processors,* Food Safety and Inspection Service, United States Department of Agriculture, Washington, D.C. 20250-3700, News and Information, May 2002.

provide guidance for transportation, storage and handling. Guidelines for inspected establishments are a first step, but we recognize the need for protections from the farm to the consumer's table. We invite your comments as we work to strengthen these steps. Homeland Security for our food and agricultural sector requires a commitment by all parties — Federal, State, local and private. We trust that these guidelines will be useful in giving specific focus to the commitment that we all share.

If you have any questions or comments, please contact our Technical Service Center at 1-800-233-3935.

Sincerely,

Linda Swacina
Assistant Administrator for Staff Services

FSIS SECURITY GUIDELINES FOR FOOD PROCESSORS

Food Security Plan Management

- A food security management team and a food security management coordinator should be identified for each plant or company. Each member should be assigned clear responsibilities.
- A food security plan using established risk management principles should be developed and implemented. The plan should include procedures for handling threats and actual cases of product tampering and an evacuation plan for each facility.
- Corrective action taken in all cases of product tampering should ensure that adulterated or potentially injurious products do not enter commerce.
- The plan should include the immediate recall of adulterated products from trade and consumer channels. Safe handling and disposal of products contaminated with chemical or biological agents should also be included in the plan.
- A relationship should be established with appropriate analytical laboratories for possible assistance in the investigation of product-tampering cases.
- Procedures for notifying appropriate law enforcement and public health officials when a food security threat is received, or when evidence of actual product tampering is observed, should be detailed in the plan.

- Specially designated entry points for emergency personnel should be identified in the plan.
- Current local, State and Federal Government Homeland Security contacts and public health officials should be listed in the plan. This list should be updated regularly.
- Members of the food security management team should be trained in all provisions of the plan. Drills should be conducted periodically. The plan should be periodically reviewed and revised as needed.
- Food security inspections of the facility should be conducted regularly by plant officials to verify key provisions of the plan.
- All employees should be encouraged to report any sign of possible product tampering or break in the food security system. Consider implementing an award system or establishing performance standards related to food security consciousness.
- All threats and incidents of intentional product tampering should be immediately investigated and reported to the local law enforcement officials and the FSIS/State Inspector-in-Charge.
- Liaison with local Homeland Security officials and other law enforcement officials should be pre-established by the food security management team.

Outside Security

- Plant boundaries should be secured to prevent unauthorized entry. "No Trespassing" signs should be posted.
- Integrity of the plant perimeter should be monitored for signs of suspicious activity or unauthorized entry.
- Outside lighting should be sufficient to allow detection of unusual activities.
- All access points into the establishment should be secured by guards, alarms, cameras or other security hardware consistent with national and local fire and safety codes.
- Emergency exits should be alarmed and have self-locking doors that can be opened only from the inside.
- Doors, windows, roof openings, vent openings, trailer bodies, railcars and bulk storage tanks should be secured (e.g., locks, seals, sensors) at all times.
- Outside storage tanks for hazardous materials and potable water supply should be protected from, and monitored for, unauthorized access.
- An updated list of plant personnel with open or restricted access to the establishment should be maintained at the security office.

- Entry into establishments should be controlled by requiring positive identification (e.g., picture IDs, sign-in and sign-out at security or reception, etc.).
- Incoming and outgoing vehicles (both private and commercial) should be inspected for unusual cargo or activity.
- Parking areas for visitors or guests should be situated at a safe distance from the main facility. Vehicles of authorized visitors, guests and employees should be clearly marked (placards, decals, etc.).
- Truck deliveries should be verified against a roster of scheduled deliveries. Unscheduled deliveries should be held outside the plant premises, if possible, pending verification of shipper and cargo.

Inside Security

General Inside Security

- Restricted areas inside the plant should be clearly marked and secured.
- Access to central controls for airflow, water systems, electricity and gas should be restricted and controlled.
- Updated plant layout schematics should be available at strategic and secured locations in the plant.
- Airflow systems should include a provision for immediate isolation of contaminated areas or rooms.
- Emergency alert systems should be fully operational and tested, and locations of controls should be clearly marked.
- Access to in-plant laboratory facilities should be strictly controlled.
- Comprehensive and validated security and disposal procedures should be in place, particularly for the control of reagents, hazardous materials and live cultures of pathogenic bacteria.
- Visitors, guests and other non-plant employees (contractors, salespeople, truck drivers, etc.) should be restricted to non-product areas unless accompanied by an authorized plant representative.
- Computer data systems should be protected using passwords, network firewalls and effective and current virus detection systems.

Slaughter and Processing Security

- Procedures should be in place to monitor the operation of pieces of equipment (blenders, choppers, poultry chill tanks, etc.) to prevent product tampering.

- A program should be in place to ensure the timely identification, segregation and security of all products involved in the event of deliberate product contamination.
- A validated procedure should be in place to ensure the trace-back and trace-forward of all raw materials and finished products.
- Projected and actual use of restricted ingredients should be verified at the end of each day, preferably by someone other than the employee who logs the ingredient.
- Returned goods should be examined for evidence of possible tampering before salvage or use in rework. Records should be kept on the use of all returned goods in rework.
- The integrity of packaging materials of all spices and restricted ingredients (including premixes prepared in the plant) should be verified before use.
- Accurate inventory of finished products should be maintained to allow detection of unexplained additions to or withdrawals from existing stock.
- Access to product production or holding areas should be restricted to plant employees and FSIS inspection personnel only.
- Plants should use a system that ensures clear identification of personnel to their specific functions (e.g., colored garb).
- An updated daily or shift roster of plant personnel should be maintained and distributed to plant supervisors.

Storage Security

- Controlled access should be maintained for all product and ingredient storage areas. An access log may be maintained.
- Security inspection of all storage facilities (including temporary storage vehicles) should be performed regularly, and the results logged.
- A daily inventory of hazardous chemicals or other products should be made, and all discrepancies should be investigated immediately.
- Hazardous chemical storage areas or rooms should be secured and isolated from other parts of the plant. In addition, they should be constructed and safely vented in accordance with national or local building codes.

Shipping and Receiving Security

- All outgoing shipments should be sealed with tamper-proof, numbered seals that are included on the shipping documents.

- Establishments should require that incoming shipments be sealed with tamper-proof, numbered seals, and that the seal numbers be shown on the shipping documents for verification prior to entry to the plant.
- Shipping documents with suspicious alterations should be thoroughly investigated.
- All trailers on the premises should be locked and sealed when not being loaded or unloaded.
- A policy for off-hour deliveries should be established to ensure prior notice of such deliveries and require the presence of an authorized individual to verify and receive the shipment.
- Packaging integrity of all incoming shipments should be examined at the receiving dock for evidence of tampering.
- Advance notification (by phone, e-mail, fax) should be required from suppliers for all incoming deliveries. Notification should include pertinent details about the shipment, including the name of the driver.
- The FSIS Inspector-in-Charge should be notified immediately when animals with unusual behavior and symptoms are received at the plant.
- Loading docks should be secured to avoid unverified or unauthorized deliveries.
- The integrity of food security measures should be a significant consideration in the selection of suppliers of meat and non-meat ingredients, compressed gas, packaging materials and labels.

Water and Ice Supply Security

- Outside access to wells, potable water tanks and ice-making equipment should be secured from unauthorized entry.
- In-plant ice-making equipment and ice storage facilities should have controlled access.
- Potable and non-potable water lines in food processing areas should be inspected periodically for possible tampering.
- The plant should arrange for immediate notification by local health officials in the event the potability of the public water supply is compromised.

Mail Handling Security

- Mail handling activity should be done in a separate room or facility, away from in-plant food production/processing operations, if possible.

- Mail handlers should be trained to recognize and handle suspicious pieces of mail using U.S. Post Office guidelines.

Personnel Security

- A system of positive identification/recognition of all plant employees should be in place.
- Procedures should be established for controlled entry of employees into the plant during both working and non-working hours.
- New hires (seasonal, temporary, permanent, and contract workers) should be subjected to background checks before hiring.
- Orientation training on security procedures should be given to all plant employees.
- The plant should establish and enforce a policy on what personal items may and may not be allowed inside the plant and within production areas.

#

The U.S. Department of Agriculture (USDA) prohibits discrimination in all its programs and activities on the basis of race, color, national origin, sex, religion, age, disability, political beliefs, sexual orientation, or marital or family status. (Not all prohibited bases apply to all programs.) Persons with disabilities who require alternative means for communication of program information (Braille, large print, audiotape, etc.) should contact USDA's TARGET Center at 202-720-2600 (voice and TDD).

To file a complaint of discrimination, write USDA, Director, Office of Civil Rights, Room 326-W, Whitten Building, 14th and Independence Avenue, SW, Washington, D.C. 20250-9410 or call 202-720-5964 (voice or TDD). USDA is an equal opportunity provider and employer.

- In the event of a biosecurity-related emergency, first contact your local law-enforcement authority.
- If you have questions or need clarification about the guidelines, contact the FSIS Technical Service Center at: 1-800-233-3935.
- For additional copies of the guidelines, go to: www.fsis.usda.gov

Full version of the guidelines is available in PDF format. To read and print a PDF file, you must have the Adobe® Acrobat® Reader installed on your PC. You can download a version suitable for your system, free of charge, from the Adobe Home Page. Adobe also provides tools and information to help make Adobe PDF files accessible to users with visual disabilities at http://access.adobe.com.

For Further Information Contact:

- Media Inquiries: (202) 720-9113
- Congressional Inquiries: (202) 720-3897
- Constituent Inquiries: (202) 720-9113
- Consumer Inquiries: Call the USDA Meat and Poultry Hotline at 1-888-MPHotline; TTY: 1-800-256-7072.

FSIS Home Page | USDA Home Page www.usda.gov

Appendix D

EMERGENCY PREPAREDNESS COMPETENCIES (ANNOTATED)

Public Health Professionals*

Professional occupations require knowledge in a field of science or learning characteristically acquired through education or training equivalent to a bachelor's degree or higher degree with major study in or pertinent to the specialized field. The work of a professional occupation requires the exercise of discretion, judgment, and personal responsibility for the application of an organized body of knowledge that is constantly studied to make new discoveries and interpretations, and to improve data, materials, and methods. (Adapted from the U.S. Office of Personnel Management.)

Note: Competencies marked with an asterisk (*) may also apply to public health leaders/administrators, depending on his or her role in the public health agency.

In order for the public health system to meet performance standards in emergency preparedness, the public health professional must be competent to:

1. *Describe* the public health role in emergency response in a range of emergencies that might arise. (For example, "This department provides surveillance, investigation, and public information in disease outbreaks and collaborates with other agencies in biological, environmental, and weather emergencies.")

* From Center for Health Policy, Columbia University School of Nursing, April 2001.

- Describe the role of the local health department and the individual's own profession and unit in relation to other governmental and health organizations in the event of one or more types of public health emergencies.

2. *Describe* the chain of command in emergency response. This description should be appropriate to the individual's level/placement in the organization.
3. *Identify and locate* the agency emergency response plan (or the pertinent portion of the plan).
   - Demonstrate knowledge of the local health department's emergency response plan.
   - Locate a wallet card, short form, or checklist, if available for the agency.
   - Assure that emergency response contact information is current for any organizations or individuals likely to be contacted in an emergency.
4. *Describe* his or her functional role(s) in emergency response and demonstrate his or her role(s) in regular drills. This includes demonstrating knowledge of the agency's emergency response plan.
   - Describe a method to use the appropriate mechanisms for contacting public safety and emergency response organizations.
   - Describe his or her own specific role in an emergency response.
   - Describe how his or her level of response relates to other resources at the local, regional, and federal organization levels.
   - *Demonstrate readiness to apply professional skills to a range of emergency situations during regular drills as appropriate to the professional role and position.
   - *Demonstrate a procedure for accessing surveillance data relevant to the locality or jurisdiction.
   - *Use surveillance data to identify, investigate, and monitor health events in the locality or jurisdiction.
   - *Interpret surveillance data to identify unusual events or emergencies.
   - *Describe how to access science-based protocols to guide an immediate investigation and risk/threat assessment of communicable disease outbreaks, suspected environmental health hazards, or biological agent threats.
   - *Demonstrate the selection and use of appropriate personal protective equipment.
   - *Identify and access written protocols for an emergency program of contact and source tracing when a suspected exposure occurs.

- *Identify steps for accessing laboratory resources capable of supporting the investigation of adverse health events.
- *Identify the scientific bases and strength of evidence to evaluate the potential impact of a possible emergency.
- *Describe a method to develop and adapt approaches to an emergency situation that takes into account cultural differences in the affected community.
- *Describe a method to regularly evaluate laws and regulations to enforce in response to specific emergency conditions.
- *Describe a method to identify those laws and regulations to enforce in response to specific emergency conditions.
- *Recognize when to enforce laws and regulations in response to specific emergency situations.
- *Describe a method to initiate appropriate enforcement of laws and regulations in response to emergency situations.
- *Participate in research to improve recognition and management of emergencies that have public health impact.
- *Share research technical expertise with those conducting research and emergency preparedness and response.

5. *Demonstrate* correct use of all communication equipment used for emergency communication (phone, fax, radio, etc.).
6. *Describe* communication role(s) in emergency response:
   - Within agency
   - Media
   - General public
   - Personal (family, neighbors)

   Within agency

   —Identify persons from whom directions might be received, and to whom information is communicated.

   —Describe a method to inform officials within the agency about the potential public health impact of actions affecting community preparedness for emergencies and explain policy options appropriate to the emergency.

   Media

   —For most professionals this is "Refer media calls to the press office #."

   —The role of a professional with leadership responsibilities may include appropriately communicating the public health aspects of emergency situations to print, radio, or television media serving the community.

   General public

   —In most cases this is "Refer inquiries to the hot line #."

Personal (family, neighbors)

—Have/activate a family emergency plan, if needed.

—Explain how a public health professional is viewed as a representative of the agency when communicating with family, friends, and others in the community.

- *Maintain regular communication with partner professionals in other agencies involved in emergency response as appropriate to the professional role and position.
- *Maintain regular and periodic contact with a network of consultants, technical assistants, and community-based assets to collect and analyze health data.
- *Communicate the public health aspects of emergencies to other emergency planning agencies.
- *Communicate risk and appropriate medical management accurately and effectively to health professionals who may diagnose or treat those possibly exposed or injured in an emergency.
- *Describe a method to maintain a current directory of public safety and emergency response organizations.
- *Describe a method to maintain regular and periodic contact with local and regional public safety and emergency response agencies.
- *Describe a method to build an effective community-wide response to an emergency using leadership, team building, negotiations, and conflict resolution.
- *Describe a method to manage an effective community-wide response to an emergency using leadership, team building, negotiations, and conflict resolution.
- *Identify emergency medical services capacity to meet the needs of those who may be injured, exposed, or need care in a public health emergency.
- *Describe a method to coordinate roles and responsibilities for emergency medical care providers meeting the needs of those who may be injured, exposed, or need care in a public health emergency.
- *Seek input from partners in conducting the evaluation of an emergency response.

7. *Identify* limits to own knowledge/skill/authority and identify key system resources for referring matters that exceed these limits.
8. *Apply* creative problem solving and flexible thinking to unusual challenges within his or her functional responsibilities and evaluate effectiveness of all actions taken.
9. *Recognize* deviations from the norm that might indicate an emergency and describe appropriate action. (For example, communicate clearly within the chain of command.)

- List examples of possible deviations in the individual's work setting, such as a cluster of unusual lab reports, a series of atypical phone calls, a suspicious delivery, etc., that might indicate a need for emergency response.

10. *Participate* in continuing education to maintain up-to-date knowledge in areas relevant to emergency response (for example, emerging infectious diseases, hazardous materials, diagnostic tests, etc.).
    - Communicate public health emergency response issues for inclusion in research.

Suggestions for Improving Food Security*

The Washington State Department of Health has prepared the following suggestions in response to inquiries about enhancing the security of the food supply at the retail level. This information is being provided on an advisory basis and is based on the HACCP (Hazard Analysis and Critical Control Point) principles of reviewing processes, identifying critical points and potential hazards, and addressing possible problems.

1. Review your business procedures and physical facility, shipping and distribution systems. Identify and list all relevant areas where you may be vulnerable to a potential sabotage or terrorist attack. Identify and outline control measures for each of these areas.
2. Ensure that your facility has adequate security measures for the physical facility and for employees, including:
    - Restricted access to the facility, ingredient, product, and chemical storage areas;
    - Checks of all restrooms, janitor closets, and storage areas on a regular basis for concealed packages or other anomalies;
    - Training for employees on security measures; and
    - Allowing only designated employees in sensitive areas.

    You may also want to consider monitoring or video surveillance systems for display areas that do not have a system in place already.
3. Have procedures in place to ensure the composition and integrity of all products and materials brought into the facility. Some tips follow:
    - Know your supply source and ensure that the supplier meets your requirements for control of goods and access to products during production, shipping, and distribution;

* From Washington State Department of Health, *Notice to Food Retailers.*

- Inspect all incoming materials and products for signs of tampering or other anomalies;
- Compare delivery slips/load manifests with orders made by your facility; and
- Ensure that you have the ability to trace back materials to the supplier.

4. Create an emergency response team in your store and develop an action plan in the case of sabotage or terrorist event, or any other type of emergency. Ensure that this plan includes a procedure to perform trace back and suppliers trace out, and includes lists with 24-hour contact information. Test your plan to make sure that it works and then follow through with any necessary adjustments, employee training, or equipment purchases.
5. Ensure that all hazardous chemicals and pesticides are stored in a secure area away from food storage and food display areas. Have safety sheets available for all chemicals on site.
6. Keep a written record of all products returned to the store for refund. Record the individual's name, address, etc., and the reason for the return. Do not place the product back on the shelf or store it with product destined for the shelf.
7. In the event that there is a suspicion of sabotage or terrorism:
   - Call 911 or contact the local police and indicate the concern;
   - Where possible, do not move or touch the product/equipment/material;
   - Cordon off the area to limit access to the potential hazard; and
   - Move any employees who were in the vicinity or may have been affected to a secure area away from other people.

Recommendations on dealing with suspicious packages or mail are available from the CDC website at:

http://www.bt.cdc.gov/DocumentsApp/Anthrax/10122001Handle/10122001Handle.asp

General information on bioterrorism-related issues and links to other websites are available at the Department of Health website at:

http://www.doh.wa.gov/bioterr/BioTerGenInfo.htm

Appendix E

TERRORIST THREATS TO FOOD — GUIDELINES FOR ESTABLISHING AND STRENGTHENING PREVENTION AND RESPONSE SYSTEMS*

Executive Summary

The malicious contamination of food for terrorist purposes is a real and current threat, and deliberate contamination of food at one location could have global public health implications. This document responds to increasing concern in Member States that chemical, biological or radionuclear agents might be used deliberately to harm civilian populations and that food might be a vehicle for disseminating such agents. The Fifty-fifth World Health Assembly (May 2002) also expressed serious concern about such threats and requested the Organization to provide tools and support to Member States to increase the capacity of national health systems to respond.

Outbreaks of both unintentional and deliberate food borne diseases can be managed by the same mechanisms. Sensible precautions, coupled with strong surveillance and response capacity, constitute the most efficient and effective way of countering all such emergencies, including food terrorism. This document provides guidance to Member States for integrating consideration of deliberate acts of food sabotage into existing programs for controlling the production of safe food. It also provides guidance on strengthening existing communicable disease control systems to ensure that surveillance, preparedness and response systems are sufficiently sensitive to meet the threat of any food safety emergency. Establishment and strengthening of such systems and programs will both increase Member States' capac-

* From Food Safety Department, World Health Organization, e-mail: foodsafety@who.int; http://www.who.int/fsf.

ity to reduce the increasing burden of food borne illness and help them to address the threat of food terrorism. The activities undertaken by Member States must be proportional to the size of the threat, and resources must be allocated on a priority basis.

Prevention, although never completely effective, is the first line of defense. The key to preventing food terrorism is establishment and enhancement of existing food safety management programs and implementation of reasonable security measures. Prevention is best achieved through a cooperative effort between government and industry, given that the primary means for minimizing food risks lie with the food industry. This document provides guidance for working with industry, and specific measures for consideration by the industry are provided.

Member States require alert, preparedness and response systems that are capable of minimizing any risks to public health from real or threatened food terrorism. This document provides policy advice on strengthening existing emergency alert and response systems by improving links with all the relevant agencies and with the food industry. This multi-stakeholder approach will strengthen disease outbreak surveillance, investigation capacity, preparedness planning, effective communication and response.

The role of the World Health Organization (WHO) is to provide advice on strengthening of national systems to respond to food terrorism. WHO is also in a unique position to coordinate existing international systems for public health disease surveillance and emergency response, which could be expanded to include considerations of food terrorism. This document complements other guides and advice developed by WHO, the Food and Agriculture Organization of the United Nations (FAO) and other international agencies related to the threat of terrorist acts with chemical, biological or radionuclear agents.

For more information or purchase refer to http://www.who.int/bookorders/index.htm.

Appendix F

THE PUBLIC HEALTH RESPONSE TO BIOLOGICAL AND CHEMICAL TERRORISM*

Ten Essential Services for Public Health

To respond effectively to terrorism, states should have the capacity to:

1. Monitor health status to rapidly detect and identify an event due to hazardous biological, chemical, or radiological agents (e.g., community health profile before an event, vital statistics, and baseline health status of the community);
2. Diagnose and investigate infectious disease and environmental health problems and health hazards in the community specific to detecting and identifying an emergency event due to a hazardous biological, chemical, or radiological agent (e.g., effective epidemiologic surveillance systems, laboratory support necessary for determining a biological, chemical, or radiological event in a time-sensitive manner);
3. Inform, educate, and empower people about specific health issues pertaining to a threat or emergency event due to the release of hazardous biological, chemical, or radiological agent (e.g., health communication effectiveness in implementing a rapid and effective response);
4. Mobilize state and local partnerships to rapidly identify and solve health problems before, during, and after an event due to a hazardous biological, chemical, or radiological agent, including issues related to

* From *The Public Health Response to Biological and Chemical Terrorism,* Interim Planning Guidance for State Public Health Officials, http://www. bt.cdc.gov/Documents/Planning/PlanningGuidance.PDF.

the National Pharmaceutical Stockpile (e.g., demonstrate an effective knowledge of all key partners involved in effectively responding to an emergency event, including terrorism);

5. Develop policies and plans that support individual and community health efforts in preparing for and responding to emergencies due to hazardous biological, chemical, or radiological agents (e.g., demonstration of practical, realistic, and effective emergency response plans);
6. Enforce laws and regulations that protect health and ensure safety in case of an emergency or threat due to a hazardous biological, chemical, or radiological agent (e.g., enforcement of sanitary codes to ensure safety of the environment during a terrorism event);
7. Link people to needed personal health services in the course of a threat or event due to a hazardous biological, chemical, or radiological agent (e.g., services that increase access to health care in a timely and effective manner);
8. Assure a competent and trained public and personal health-care workforce for rapid response to a threat or event due to a hazardous biological, chemical, or radiological agent (e.g., education and training for all public health-care providers in effective response to an emergency event or threat);
9. Evaluate effectiveness, accessibility, and quality of personal and population-based health services available to respond to a threat or event due to a hazardous biological, chemical, or radiological agent (e.g., continuous evaluation of public health programs which respond effectively to a public health emergency); and
10. Participate in research for new insights and innovative solutions to health problems resulting from exposure to a hazardous biological or chemical agent (e.g., links with academic institutions and capacity for epidemiologic and economic analyses of a chemical or bioterrorism event).

Key Elements of a Public Health Preparedness Program

For public health department officials to effectively prepare their departments to respond to an actual or threatened terrorism event, the departments must be capable of the following:

- Identifying the types of events that might occur in their communities.
- Planning emergency activities in advance to ensure a coordinated response to the consequences of credible events.

- Building capabilities necessary to respond effectively to the consequences of those events.
- Identifying the type or nature of an event when it happens.
- Implementing the planned response quickly and efficiently.
- Recovering from the incident.

To meet these capabilities, a health department should develop the following Key Preparedness Elements for Terrorism Response:

Key Preparedness Elements

1. Hazard Analysis
2. Emergency Response Planning
3. Health Surveillance and Epidemiologic Investigation
4. Laboratory Diagnosis and Characterization
5. Consequence Management

Enhanced Collaboration among Public Health Partners

The first step toward information sharing is the effective collaboration among members of the public health community. To accomplish this task, it is necessary to identify which agencies and organizations must be integrated. For surveillance purposes, the public health system is much more than state and/or local health departments. At the very least, the following organizations should coordinate information and share public health-related data:

Surveillance Partners

- State health department
- Emergency Medical Services
- Social services agencies
- Hospitals
- Clinics and physicians
- Epidemiologists
- Medical examiner/coroner
- Laboratories
- County/health departments
- Dispatch/911
- Volunteer organizations
- Mental health professionals

- Poison centers
- Pharmacists
- Veterinary services
- (Others)

Coordination among these agencies and organizations can be enhanced through activities such as the following:

- Identify and distribute points of contact and communications information to critical response partners.
- Provide education about public health surveillance, disease reporting, epidemiology, and response activities related to bioterrorism to public health response partners.
- Collaborate on educational activities on topics related to bioterrorism preparedness for the general public or general medical community.
- Provide or promote in-service training or grand rounds for the medical community.
- Develop and implement collaborative surveillance projects by utilizing traditional and nontraditional data sources.

Epidemiologic Clues That May Signal a Covert Bioterrorism Attack

- Large number of ill persons with similar disease or syndrome.
- Large number of unexplained disease, syndrome, or deaths.
- Unusual illness in a population.
- Higher morbidity and mortality than expected with a common disease or syndrome.
- Failure of a common disease to respond to usual therapy.
- Single case of disease caused by an uncommon agent.
- Multiple unusual or unexplained disease entities coexisting in the same patient without other explanation.
- Disease with an unusual geographic or seasonal distribution.
- Multiple atypical presentations of disease agents.
- Similar genetic type among agents isolated from temporally or spatially distinct sources.
- Unusual, atypical, genetically engineered, or antiquated strain of agent.
- Endemic disease with unexplained increase in incidence.
- Simultaneous clusters of similar illness in noncontiguous areas, domestic or foreign.

- Atypical aerosol, food, or water transmission.
- Ill people presenting near the same time.
- Deaths or illness among animals that precedes or accompanies illness or death in humans.
- No illness in people not exposed to common ventilation systems, but illness among those people in proximity to the systems.

Appendix G

GUIDANCE FOR INDUSTRY

RETAIL FOOD STORES AND FOOD SERVICE ESTABLISHMENTS: FOOD SECURITY PREVENTIVE MEASURES GUIDANCE*

Draft Guidance

This draft guidance represents the Agency's current thinking on the kinds of measures that retail food store and food service establishment operators may take to minimize the risk that food under their control will be subject to tampering or other malicious, criminal, or terrorist actions. It does not create or confer any rights for or on any person and does not operate to bind FDA or the public.

Purpose and Scope

This draft guidance is designed as an aid to operators of retail food stores and food service establishments (for example, bakeries, bars, bed-and-breakfast operations, cafeterias, camps, child and adult day care providers, church kitchens, commissaries, community fund raisers, convenience stores, fairs, food banks, grocery stores, interstate conveyances, meal services for home-bound persons, mobile food carts, restaurants, and vending machine operators). This is a very diverse set of establishments, which includes both very large and very small entities.

* From U.S. Department of Health and Human Services, U.S. Food and Drug Administration, Center for Food Safety and Applied Nutrition, March 21, 2003.

This draft guidance identifies the kinds of preventive measures they may take to minimize the risk that food under their control will be subject to tampering or other malicious, criminal, or terrorist actions. Operators of retail food stores and food service establishments are encouraged to review their current procedures and controls in light of the potential for tampering or other malicious, criminal, or terrorist actions and make appropriate improvements.

This draft guidance is designed to focus operator's attention sequentially on each segment of the food delivery system that is within their control, to minimize the risk of tampering or other malicious, criminal, or terrorist action at each segment. To be successful, implementing enhanced preventive measures requires the commitment of management and staff. Accordingly, FDA recommends that both management and staff participate in the development and review of such measures.

Limitations

Not all of the guidance contained in this document may be appropriate or practical for every retail food store or food service establishment, particularly smaller facilities. FDA recommends that operators review the guidance in each section that relates to a component of their operation, and assess which preventive measures are suitable. Example approaches are provided for many of the preventive measures listed in this document. These examples should not be regarded as minimum standards. Nor should the examples provided be considered an inclusive list of all potential approaches to achieving the goal of the preventive measure. FDA recommends that operators consider the goal of the preventive measure, assess whether the goal is relevant to their operation, and, if it is, design an approach that is both efficient and effective to accomplish the goal under their conditions of operation.

Structure

This draft guidance is divided into five sections that relate to individual components of a retail food store or food service establishment operation: management; human element — staff; human element — public; facility; and operations.

Related Guidance

FDA has published two companion guidance documents on food security, entitled, "Food Producers, Processors, and Transporters: Food Security Preventive Measures Guidance" and "Importers and Filers: Food Security Preventive Measures Guidance" to cover the farm-to-table spectrum of food production. Both documents are available at: http://www.access.gpo.gov/su_docs/aces/aces140.html.

Additional Resources

A process called Operational Risk Management (ORM) may help prioritize the preventive measures that are most likely to have the greatest impact on reducing the risk of tampering or other malicious, criminal, or terrorist actions against food. Information on ORM is available in the Federal Aviation Administration (FAA) System Safety Handbook, U.S. Department of Transportation, FAA, December 30, 2000, Chapter 15, Operational Risk Management. The handbook is available at: http://www.asy.faa.gov/Risk/SSHandbook/Chap15_1200.PDF.

The U.S. Department of Transportation, Research and Special Programs Administration has published an advisory notice of voluntary measures to enhance the security of hazardous materials shipments. It is available at: http://frwebgate.access.gpo.gov/cgi-bin/getdoc.cgi?dbname=2002_register&docid=02-3636-filed.pdf. The notice provides guidance to shippers and carriers on personnel, facility and en route security issues.

The U.S. Postal Service has prepared guidance for identifying and handling suspicious mail. It is available at: http://www.usps.com/news/2001/press/mailsecurity/postcard.htm.

The Federal Anti-Tampering Act (18 USC 1365) makes it a federal crime to tamper with or taint a consumer product, or to attempt, threaten or conspire to tamper with or taint a consumer product, or make a false statement about having tampered with or tainted a consumer product. Conviction can lead to penalties of up to $100,000 in fines and up to life imprisonment. The Act is available at: http://www.fda.gov/opacom/laws/fedatact.htm.

The National Infrastructure Protection Center (NIPC) serves as the federal government's focal point for threat assessment, warning, investigation, and response for threats or attacks against U.S. critical infrastructure. The NIPC has identified the food system as one of the eight critical infrastructures, and has established a public-private partnership with the food industry, called the Food Industry Information and Analysis Center (Food Industry ISAC). The NIPC provides the Food Industry ISAC with access, information and analysis, enabling the food industry to report, identify, and reduce its vulnerabilities to malicious attacks, and to recover from such attacks as quickly as possible. In particular, the NIPC identifies credible threats and crafts specific warning messages to the food industry. Further information is available at http://www.nipc.gov/ and http://www.foodisac.org/.

Finally, FDA encourages trade associations to evaluate the preventive measures contained in this guidance document and adapt them to their specific products and operations and to supplement this guidance with additional preventive measures when appropriate. FDA welcomes dialogue on the content of sector specific guidance with appropriate trade associations.

Retail Food Store and Food Service Establishment Operations

Management

FDA recommends that retail food store and food service establishment operators consider:

- Preparing for the possibility of tampering or other malicious, criminal, or terrorist events
  - assigning responsibility for security to knowledgeable individual(s)
  - conducting an initial assessment of food security procedures and operations, which we recommend be kept confidential
  - having a crisis management strategy to prepare for and respond to tampering and other malicious, criminal, or terrorist actions, both threats and actual events, including identifying, segregating and securing affected products
  - planning for emergency evacuation, including preventing security breaches during evacuation
  - becoming familiar with the emergency response system in the community
  - making management aware of 24-hour contact information for local, state, and federal police/fire/rescue/health/homeland security agencies
  - making staff aware of who in management they should alert about potential security problems (24-hour contacts)
  - promoting food security awareness to encourage all staff to be alert to any signs of tampering or malicious, criminal, or terrorist actions or areas that may be vulnerable to such actions, and to report any findings to identified management (for example, providing training, instituting a system of rewards, building security into job performance standards)
  - having an internal communication system to inform and update staff about relevant security issues
  - having a strategy for communicating with the public (for example, identifying a media spokesperson, preparing generic press statements and background information, and coordinating press statements with appropriate authorities)
- Supervision
  - providing an appropriate level of supervision to all staff, including cleaning and maintenance staff, contract workers, data entry and computer support staff, and especially, new staff (for example, supervisor on duty, periodic unannounced visits by supervisor, daily visits by supervisor, two staff on duty at same time, moni-

tored video cameras, off line review of video tapes, one way and two way windows, customer feedback to supervisor of unusual or suspicious behavior by staff)
  - conducting routine security checks of the premises, including utilities and critical computer data systems (at a frequency appropriate to the operation) for signs of tampering or malicious, criminal, or terrorist actions, or areas that may be vulnerable to such actions
- Investigation of suspicious activity
  - investigating threats or information about signs of tampering or other malicious, criminal, or terrorist actions
  - alerting appropriate law enforcement and public health authorities about any threats of or suspected tampering or other malicious, criminal, or terrorist actions
- Evaluation program
  - evaluating the lessons learned from past tampering or other malicious, criminal, or terrorist actions and threats
  - reviewing and verifying, at least annually, the effectiveness of the security management program (for example, using knowledgeable in-house or third party staff to conduct tampering or other malicious, criminal, or terrorist action exercises and to challenge computer security systems), revising accordingly (using third party or in-house security expert, where possible), revising the program accordingly, and keeping this information confidential
  - performing random food security inspections of all appropriate areas of the facility (including receiving and storage areas, where applicable) using knowledgeable in-house or third party staff, and keeping this information confidential
  - verifying that security contractors are doing an appropriate job, when applicable

Human Element — Staff

Under Federal law, retail food store and food service establishment operators are required to verify the employment eligibility of all new hires, in accordance with the requirements of the Immigration and Nationality Act, by completing the INS Employment Eligibility Verification Form (INS Form I-9). Completion of Form I-9 for new hires is required by 8 USC 1324a and nondiscrimination provisions governing the verification process are set forth at 1324b.

FDA recommends that retail food store and food service establishment operators consider:

- Screening (pre-hiring, at hiring, post-hiring)

    - examining the background of all staff (including seasonal, temporary, contract, and volunteer staff, whether hired directly or through a recruitment firm) as appropriate to their position, considering candidates' access to sensitive areas of the facility and the degree to which they will be supervised and other relevant factors (for example, obtaining and verifying work references, addresses, and phone numbers, participating in one of the pilot programs managed by the Immigration and Naturalization Service and the Social Security Administration [These programs provide electronic confirmation of employment eligibility for newly hired employees. For more information call the INS SAVE Program toll free at 1-888-464-4218, fax a request for information to (202) 514-9981, or write to US/INS, SAVE Program, 425 I Street, NW, ULLICO-4th Floor, Washington, D.C. 20536. These pilot programs may not be available in all states], having a criminal background check performed by local law enforcement or by a contract service provider [Remember to first consult any state or local laws that may apply to the performance of such checks])
- Note: screening procedures should be applied equally to all staff, regardless of race, national origin, religion, and citizenship or immigration status.
- Daily work assignments
    - knowing who is and who should be on premises, and where they should be located, for each shift
    - keeping information updated
- Identification
    - establishing a system of positive identification and recognition (for example, issuing uniforms, name tags, or photo identification badges with individual control numbers, color coded by area of authorized access), when appropriate
    - collecting the uniforms, name tag, or identification badge when a staff member is no longer associated with the establishment
- Restricted access
    - identifying staff that require unlimited access to all areas of the facility
    - reassessing levels of access for all staff periodically
    - limiting staff access to non-public areas so staff enter only those areas necessary for their job functions and only during appropriate work hours (for example, using key cards or keyed or cipher locks for entry to sensitive areas, color coded uniforms [remember to consult any relevant federal, state or local fire or occupational safety codes before making any changes])

    - changing combinations, rekeying locks and/or collecting the retired key card when a staff member who is in possession of these is no longer associated with the establishment, and additionally as needed to maintain security
- Personal items
    - restricting the type of personal items allowed in non-public areas of the establishment
    - allowing in the non-public areas of the establishment only those personal use medicines that are necessary for the health of staff (other than those being stored or displayed for retail sale) and ensuring that these personal use medicines are properly labeled and stored away from stored food and food preparation areas
    - preventing staff from bringing personal items (for example, lunch containers, purses) into non-public food preparation or storage areas
    - providing for regular inspection of contents of staff lockers (for example, providing metal mesh lockers, company issued locks), bags, packages, and vehicles when on company property (Remember to first consult any federal, state, or local laws that may relate to such inspections)
- Training in food security procedures
    - incorporating food security awareness, including information on how to prevent, detect, and respond to tampering or other malicious, criminal, or terrorist actions or threats, into training programs for staff, including seasonal, temporary, contract, and volunteer staff
    - providing periodic reminders of the importance of security procedures (for example, scheduling meetings, providing brochures, payroll stuffers)
    - encouraging staff support (for example, involving staff in food security planning and the food security awareness program, demonstrating the importance of security procedures to the staff)
    - encouraging staff support (for example, involving staff in food security planning and the food security awareness program, demonstrating the importance of security procedures to the staff)
- Unusual behavior
    - watching for unusual or suspicious behavior by staff (for example, staff who, without an identifiable purpose, stay unusually late after the end of their shift, arrive unusually early, access files/information/areas of the facility outside of the areas of their responsibility; remove documents from the facility; ask questions on sensitive subjects; bring cameras to work)

- Staff health
  - being alert for atypical staff health conditions that staff may voluntarily report and absences that could be an early indicator of tampering or other malicious, criminal, or terrorist actions (for example, an unusual number of staff who work in the same part of the facility reporting similar symptoms within a short time frame), and reporting such conditions to local health authorities

Human Element — Public

FDA recommends that retail food store and food service establishment operators consider:

- Customers
  - preventing access to food preparation and storage and dishwashing areas in the non-public areas of the establishment, including loading docks
  - monitoring public areas, including entrances to public restrooms (for example, using security guards, monitored video cameras, one-way and two-way windows, placement of employee workstations for optimum visibility) for unusual or suspicious activity (for example, a customer returning a product to the shelf that he/she brought into the store, spending an unusual amount of time in one area of the store)
  - monitoring the serving or display of foods in self-service areas (for example, salad bars, condiments, open bulk containers, produce display areas, doughnut/bagel cases)
- Other visitors (for example, contractors, sales representatives, delivery drivers, couriers, pest control representatives, third-party auditors, regulators, reporters, tours)
  - restricting entry to the non-public areas of the establishment (for example, checking visitors in and out before entering the non-public areas, requiring proof of identity, issuing visitors badges that are collected upon departure, accompanying visitors)
  - ensuring that there is a valid reason for all visits to the non-public areas of the establishment before providing access to the facility — beware of unsolicited visitors
  - verifying the identity of unknown visitors to the non-public areas of the establishment
  - inspecting incoming and outgoing packages and briefcases in the non-public areas of the establishment for suspicious, inappropriate or unusual items, to the extent practical

Facility

FDA recommends that retail food store and food service establishment operators consider:

- Physical security
  - protecting non-public perimeter access with fencing or other deterrent, when appropriate
  - securing doors (including freight loading doors, when not in use and not being monitored, and emergency exits), windows, roof openings/hatches, vent openings, ventilation systems, utility rooms, ice manufacturing and storage rooms, loft areas and trailer bodies, and bulk storage tanks for liquids, solids and compressed gases to the extent possible (for example, using locks, "jimmy plates," seals, alarms, intrusion detection sensors, guards, monitored video surveillance [remember to consult any relevant federal, state or local fire or occupational safety codes before making any changes])
  - using metal or metal-clad exterior doors to the extent possible when the facility is not in operation, except where visibility from public thoroughfares is an intended deterrent (remember to consult any relevant federal, state or local fire or occupational safety codes before making any changes)
  - minimizing the number of entrances to non-public areas (remember to consult any relevant federal, state or local fire or occupational safety codes before making any changes)
  - accounting for all keys to establishment (for example, assigning responsibility for issuing, tracking, and retrieving keys)
  - monitoring the security of the premises using appropriate methods (for example, using security patrols [uniformed and/or plain-clothed], monitored video surveillance)
  - minimizing, to the extent practical, places in public areas that an intruder could remain unseen after work hours
  - minimizing, to the extent practical, places in non-public areas that can be used to temporarily hide intentional contaminants (for example, minimizing nooks and crannies, false ceilings)
  - providing adequate interior and exterior lighting, including emergency lighting, where appropriate, to facilitate detection of suspicious or unusual activity
  - implementing a system of controlling vehicles authorized to park in the non-public parking areas (for example, using placards, decals, key cards, keyed or cipher locks, issuing passes for specific areas and times to visitors' vehicles)

    - keeping customer, employee and visitor parking areas separated from entrances to non-public areas, where practical
- Storage and use of poisonous and toxic chemicals (for example, cleaning and sanitizing agents, pesticides) in non-public areas
    - limiting poisonous and toxic chemicals in the establishment to those that are required for the operation and maintenance of the facility and those that are being stored or displayed for retail sale
    - storing poisonous and toxic chemicals as far away from food handling and food storage areas as practical
    - limiting access to and securing storage areas for poisonous or toxic chemicals that are not being held for retail sale (for example, using keyed or cipher locks, key cards, seals, alarms, intrusion detection sensors, guards, monitored video surveillance [remember to consult any relevant federal, state or local fire codes before making any changes])
    - ensuring that poisonous and toxic chemicals are properly labeled
    - using pesticides in accordance with the Federal Insecticide, Fungicide, and Rodenticide Act (for example, maintaining rodent bait that is in use in covered, tamper-resistant bait stations)
    - knowing what poisonous and toxic chemicals should be on the premises and keeping track of them
    - investigating missing stock or other irregularities outside a normal range of variation and alerting local enforcement and public health agencies about unresolved problems, when appropriate

Operations

FDA recommends that retail food store and food service establishment operators consider:

- Incoming products
    - using only known and appropriately licensed or permitted (where applicable) sources for all incoming products
    - informing suppliers, distributors and transporters about FDA's food security guidance, "Food producers, processors, and transporters: Food security preventive measures guidance" and "Importers and filers: Food security preventive measures guidance," available at: http://www.access.gpo.gov/su_docs/aces/aces140.html.
    - taking steps to ensure that delivery vehicles are appropriately secured
    - requesting that transporters have the capability to verify the location of the load at any time, when practical

  - establishing delivery schedules, not accepting unexplained, unscheduled deliveries or drivers, and investigating delayed or missed shipments
  - supervising off-loading of incoming materials, including off hour deliveries
  - reconciling the product and amount received with the product and amount ordered and the product and amount listed on the invoice and shipping documents, taking into account any sampling performed prior to receipt
  - investigating shipping documents with suspicious alterations
  - inspecting incoming products and product returns for signs of tampering, contamination or damage (for example, abnormal powders, liquids, stains, or odors, evidence of resealing, compromised tamper-evident packaging) or "counterfeiting" (for example, inappropriate or mismatched product identity, labeling, product lot coding or specifications, absence of tamper-evident packaging when the label contains a tamper-evident notice), when appropriate
  - rejecting suspect food
  - alerting appropriate law enforcement and public health authorities about evidence of tampering, "counterfeiting," or other malicious, criminal, or terrorist action
- Storage
  - having a system for receiving, storing and handling distressed, damaged, and returned products, and products left at checkout counters, that minimizes their potential for being compromised (for example, obtaining the reason for return and requiring proof of identity of the individual returning the product, examining returned or abandoned items for signs of tampering, not reselling returned or abandoned products)
  - keeping track of incoming products, materials in use, salvage products, and returned products
  - investigating missing or extra stock or other irregularities outside a normal range of variability and reporting unresolved problems to appropriate law enforcement and public health authorities, when appropriate
  - minimizing reuse of containers, shipping packages, cartons, etc., where practical
- Food service and retail display
  - displaying poisonous and toxic chemicals for retail sale in a location where they can be easily monitored (for example, visible by staff at their work stations, windows, video monitoring)

- periodically checking products displayed for retail sale for evidence of tampering or other malicious, criminal, or terrorist action (for example, checking for off-condition appearance [for example, stained, leaking, damaged packages, missing or mismatched labels], proper stock rotation, evidence of resealing, condition of tamper-evident packaging, where applicable, presence of empty food packaging or other debris on the shelving), to the extent practical
- monitoring self-service areas (for example, salad bars, condiments, open bulk containers, produce display areas, doughnut/bagel cases) for evidence of tampering or other malicious, criminal, or terrorist action

- Security of water and utilities
  - limiting to the extent practical access to controls for airflow, water, electricity, and refrigeration
  - securing non-municipal water wells, hydrants, storage and handling facilities
  - ensuring that water systems and trucks are equipped with backflow prevention
  - chlorinating non-municipal water systems and monitoring chlorination equipment and chlorine levels
  - testing non-municipal sources for potability regularly, as well as randomly, and being alert to changes in the profile of the results
  - staying attentive to the potential for media alerts about public water provider problems, when applicable
  - identifying alternate sources of potable water for use during emergency situations where normal water systems have been compromised (for example, bottled water, trucking from an approved source, treating on-site or maintaining on-site storage)
- Mail/packages
  - implementing procedures to ensure the security of incoming mail and packages
- Access to computer systems
  - restricting access to critical computer data systems to those with appropriate clearance (for example, using passwords, firewalls)
  - eliminating computer access when a staff member is no longer associated with the establishment
  - establishing a system of traceability of computer transactions
  - reviewing the adequacy of virus protection systems and procedures for backing up critical computer based data systems
  - validating the computer security system

Emergency Point of Contact

U.S. Food and Drug Administration
5600 Fishers Lane
Rockville, MD 20857
301-443-1240

If a retail food store or food service establishment operator suspects that any of his/her products that are regulated by the FDA have been subject to tampering, "counterfeiting," or other malicious, criminal, or terrorist action, FDA recommends that he/she notify the FDA 24-hour emergency number at 301-443-1240 or call their local FDA District Office. FDA recommends that the operator also notify local law enforcement and public health authorities.

FDA District Office telephone numbers are listed at:

http://www.fda.gov/ora/inspect_ref/iom/iomoradir.html

Appendix H

GUIDANCE FOR INDUSTRY

FOOD PRODUCERS, PROCESSORS, AND TRANSPORTERS: FOOD SECURITY PREVENTIVE MEASURES GUIDANCE*

This guidance represents the Agency's current thinking on the kinds of measures that food establishments may take to minimize the risk that food under their control will be subject to tampering or other malicious, criminal, or terrorist actions. It does not create or confer any rights for or on any person and does not operate to bind FDA or the public.

Purpose and Scope

This guidance is designed as an aid to operators of food establishments (firms that produce, process, store, repack, relabel, distribute, or transport food or food ingredients). This is a very diverse set of establishments, which includes both very large and very small entities.

This guidance identifies the kinds of preventive measures operators of food establishments may take to minimize the risk that food under their control will be subject to tampering or other malicious, criminal, or terrorist actions. It is relevant to all sectors of the food system, including farms, aquaculture facilities, fishing vessels, producers, transportation operations, processing facilities, packing facilities, and warehouses. It is not intended as guidance for retail food stores or food service establishments.

* From U.S. Department of Health and Human Services, U.S. Food and Drug Administration, Center for Food Safety and Applied Nutrition, March 21, 2003.

Operators of food establishments are encouraged to review their current procedures and controls in light of the potential for tampering or other malicious, criminal, or terrorist actions and make appropriate improvements. FDA recommends that the review include consideration of the role that unit and distribution packaging might have in a food security program. This guidance is designed to focus operator's attention sequentially on each segment of the farm-to-table system that is within their control, to minimize the risk of tampering or other malicious, criminal, or terrorist action at each segment. To be successful, implementing enhanced preventive measures requires the commitment of management and staff. Accordingly, FDA recommends that both management and staff participate in the development and review of such measures.

Limitations

Not all of the guidance contained in this document may be appropriate or practical for every food establishment, particularly smaller facilities and distributors. FDA recommends that operators review the guidance in each section that relates to a component of their operation, and assess which preventive measures are suitable. Example approaches are provided for many of the preventive measures listed in this document. These examples should not be regarded as minimum standards. Nor should the examples provided be considered an inclusive list of all potential approaches to achieving the goal of the preventive measure. FDA recommends that operators consider the goal of the preventive measure, assess whether the goal is relevant to their operation, and, if it is, design an approach that is both efficient and effective to accomplish the goal under their conditions of operation.

Structure

This guidance is divided into five sections that relate to individual components of a food establishment operation: management; human element — staff; human element — public; facility; and operations.

Related Guidance

FDA has published a companion guidance document on food security entitled, "Importers and filers: Food security preventive measures guidance" to cover the farm-to-table spectrum of food production. This document is available at: http://www.access.gpo.gov/su_docs/aces/aces140.html.

Additional Resources*

A process called Operational Risk Management (ORM) may help prioritize the preventive measures that are most likely to have the greatest impact on reducing the risk of tampering or other malicious, criminal, or terrorist actions against food. Information on ORM is available in the Federal Aviation Administration (FAA) System Safety Handbook, U.S. Department of Transportation, FAA, December 30, 2000, Chapter 15, Operational Risk Management. The handbook is available at: http://www.asy.faa.gov/Risk/SSHandbook/Chap15_1200.PDF.

The U.S. Department of Transportation, Research and Special Programs Administration has published an advisory notice of voluntary measures to enhance the security of hazardous materials shipments. It is available at http://frwebgate.access.gpo.gov/cgi-bin/getdoc.cgi?dbname=2002_register&docid=02-3636-filed.pdf. The notice provides guidance to shippers and carriers on personnel, facility and en route security issues.

The U.S. Postal Service has prepared guidance for identifying and handling suspicious mail. It is available at: http://www.usps.com/news/2001/press/mailsecurity/postcard.htm.

The Federal Anti-Tampering Act (18 USC 1365) makes it a federal crime to tamper with or taint a consumer product, or to attempt, threaten or conspire to tamper with or taint a consumer product, or make a false statement about having tampered with or tainted a consumer product. Conviction can lead to penalties of up to $100,000 in fines and up to life imprisonment. The Act is available at: http://www.fda.gov/opacom/laws/fedatact.htm.

The National Infrastructure Protection Center (NIPC) serves as the federal government's focal point for threat assessment, warning, investigation, and response for threats or attacks against U.S. critical infrastructure. The NIPC has identified the food system as one of the eight critical infrastructures, and has established a public-private partnership with the food industry, called the Food Industry Information and Analysis Center (Food Industry ISAC). The NIPC provides the Food Industry ISAC with access, information and analysis, enabling the food industry to report, identify, and reduce its vulnerabilities to malicious attacks, and to recover from such attacks as quickly as possible. In particular, the NIPC identifies credible threats and crafts specific warning messages to the food industry. Further information is available at http://www.nipc.gov/ and http://www.foodisac.org/.

Finally, some trade associations have developed food security guidance that is appropriately focused for that specific industry. For example, the International Dairy Food Association has developed a food security guidance document as an aid to the dairy industry.

FDA encourages other trade associations to evaluate the preventive measures contained in this FDA guidance document and adapt them to their specific products and operations and to supplement this guidance with additional preventive measures when appropriate. FDA welcomes dialogue on the content of sector specific guidance with appropriate trade associations.

Food Establishment Operations

Management

FDA recommends that food establishment operators consider:

- Preparing for the possibility of tampering or other malicious, criminal, or terrorist actions
  - assigning responsibility for security to knowledgeable individual(s)
  - conducting an initial assessment of food security procedures and operations, which we recommend be kept confidential
  - having a security management strategy to prepare for and respond to tampering and other malicious, criminal, or terrorist actions, both threats and actual events, including identifying, segregating and securing affected product
  - planning for emergency evacuation, including preventing security breaches during evacuation
  - maintaining any floor or flow plan in a secure, off-site location
  - becoming familiar with the emergency response system in the community
  - making management aware of 24-hour contact information for local, state, and federal police/fire/rescue/health/homeland security agencies
  - making staff aware of who in management they should alert about potential security problems (24-hour contacts)
  - promoting food security awareness to encourage all staff to be alert to any signs of tampering or other malicious, criminal, or terrorist actions or areas that may be vulnerable to such actions, and reporting any findings to identified management (for example, providing training, instituting a system of rewards, building security into job performance standards)
  - having an internal communication system to inform and update staff about relevant security issues
  - having a strategy for communicating with the public (for example, identifying a media spokesperson, preparing generic press statements and background information, and coordinating press statements with appropriate authorities)

- Supervision
  - providing an appropriate level of supervision to all staff, including cleaning and maintenance staff, contract workers, data entry and computer support staff, and especially, new staff
  - conducting routine security checks of the premises, including automated manufacturing lines, utilities and critical computer data systems (at a frequency appropriate to the operation) for signs of tampering or malicious, criminal, or terrorist actions or areas that may be vulnerable to such actions
- Recall strategy
  - identifying the person responsible, and a backup person
  - providing for proper handling and disposition of recalled product
  - identifying customer contacts, addresses and phone numbers
- Investigation of suspicious activity
  - investigating threats or information about signs of tampering or other malicious, criminal, or terrorist actions
  - alerting appropriate law enforcement and public health authorities about any threats of or suspected tampering or other malicious, criminal, or terrorist actions
- Evaluation program
  - evaluating the lessons learned from past tampering or other malicious, criminal, or terrorist actions and threats
  - reviewing and verifying, at least annually, the effectiveness of the security management program (for example, using knowledgeable in-house or third party staff to conduct tampering or other malicious, criminal, or terrorist action exercises and mock recalls and to challenge computer security systems), revising the program accordingly, and keeping this information confidential
  - performing random food security inspections of all appropriate areas of the facility (including receiving and warehousing, where applicable) using knowledgeable in-house or third party staff, and keeping this information confidential
  - verifying that security contractors are doing an appropriate job, when applicable

Human Element — Staff

Under Federal law, food establishment operators are required to verify the employment eligibility of all new hires, in accordance with the requirements of the Immigration and Nationality Act, by completing the INS Employment Eligibility Verification Form (INS Form I-9). Completion of Form I-9 for new hires is required by 8 USC 1324a and nondiscrimination provisions governing the verification process are set forth at 8 USC 1324b.

FDA recommends that food establishment operators consider:

- Screening (pre-hiring, at hiring, post-hiring)
  - examining the background of all staff (including seasonal, temporary, contract, and volunteer staff, whether hired directly or through a recruitment firm) as appropriate to their position, considering candidates' access to sensitive areas of the facility and the degree to which they will be supervised and other relevant factors (for example, obtaining and verifying work references, addresses, and phone numbers, participating in one of the pilot programs managed by the Immigration and Naturalization Service and the Social Security Administration [These programs provide electronic confirmation of employment eligibility for newly hired employees. For more information call the INS SAVE Program toll free at 1-888-464-4218, fax a request for information to (202) 514-9981, or write to US/INS, SAVE Program, 425 I Street, NW, ULLICO-4th Floor, Washington, D.C. 20536. These pilot programs may not be available in all states], having a criminal background check performed by local law enforcement or by a contract service provider [Remember to first consult any state or local laws that may apply to the performance of such checks])
- Note: screening procedures should be applied equally to all staff, regardless of race, national origin, religion, and citizenship or immigration status.
- Daily work assignments
  - knowing who is and who should be on premises, and where they should be located, for each shift
  - keeping information updated
- Identification
  - establishing a system of positive identification and recognition that is appropriate to the nature of the workforce (for example, issuing uniforms, name tags, or photo identification badges with individual control numbers, color coded by area of authorized access), when appropriate
  - collecting the uniforms, name tag, or identification badge when a staff member is no longer associated with the establishment
- Restricted access
  - identifying staff that require unlimited access to all areas of the facility
  - reassessing levels of access for all staff periodically
  - limiting access so staff enter only those areas necessary for their job functions and only during appropriate work hours (for example, using key cards or keyed or cipher locks for entry to sensitive

  areas, color coded uniforms [remember to consult any relevant federal, state or local fire or occupational safety codes before making any changes])
  - changing combinations, rekeying locks and/or collecting the retired key card when a staff member who is in possession of these is no longer associated with the establishment, and additionally as needed to maintain security
- Personal items
  - restricting the type of personal items allowed in establishment
  - allowing in the establishment only those personal use medicines that are necessary for the health of staff and ensuring that these personal use medicines are properly labeled and stored away from food handling or storage areas
  - preventing staff from bringing personal items (for example, lunch containers, purses) into food handling or storage areas
  - providing for regular inspection of contents of staff lockers (for example, providing metal mesh lockers, company issued locks), bags, packages, and vehicles when on company property (Remember to first consult any federal, state, or local laws that may relate to such inspections)
- Training in food security procedures
  - incorporating food security awareness, including information on how to prevent, detect, and respond to tampering or other malicious, criminal, or terrorist actions or threats, into training programs for staff, including seasonal, temporary, contract, and volunteer staff
  - providing periodic reminders of the importance of security procedures (for example, scheduling meetings, providing brochures or payroll stuffers)
  - encouraging staff support (for example, involving staff in food security planning and the food security awareness program, demonstrating the importance of security procedures to the staff)
- Unusual behavior
  - watching for unusual or suspicious behavior by staff (for example, staff who, without an identifiable purpose, stay unusually late after the end of their shift, arrive unusually early, access files/information/areas of the facility outside of the areas of their responsibility; remove documents from the facility; ask questions on sensitive subjects; bring cameras to work)
- Staff health
  - being alert for atypical staff health conditions that staff may voluntarily report and absences that could be an early indicator of

tampering or other malicious, criminal, or terrorist actions (for example, an unusual number of staff who work in the same part of the facility reporting similar symptoms within a short time frame), and reporting such conditions to local health authorities

Human Element — Public

FDA recommends that food establishment operators consider:

- Visitors (for example, contractors, supplier representatives, delivery drivers, customers, couriers, pest control representatives, third-party auditors, regulators, reporters, tours)
  - inspecting incoming and outgoing vehicles, packages and briefcases for suspicious, inappropriate or unusual items or activity, to the extent practical
  - restricting entry to the establishment (for example, checking visitors in and out at security or reception, requiring proof of identity, issuing visitors badges that are collected upon departure, accompanying visitors)
  - ensuring that there is a valid reason for the visit before providing access to the facility — beware of unsolicited visitors
  - verifying the identity of unknown visitors
  - restricting access to food handling and storage areas (for example, accompanying visitors, unless they are otherwise specifically authorized)
  - restricting access to locker room

Facility

FDA recommends that food establishment operators consider:

- Physical security
  - protecting perimeter access with fencing or other deterrent, when appropriate
  - securing doors (including freight loading doors, when not in use and not being monitored, and emergency exits), windows, roof openings/hatches, vent openings, ventilation systems, utility rooms, ice manufacturing and storage rooms, loft areas, trailer bodies, tanker trucks, railcars, and bulk storage tanks for liquids, solids, and compressed gases, to the extent possible (for example, using locks, "jimmy plates," seals, alarms, intrusion detection sensors, guards, monitored video surveillance [remember to consult

any relevant federal, state or local fire or occupational safety codes before making any changes])

- using metal or metal-clad exterior doors to the extent possible when the facility is not in operation, except where visibility from public thoroughfares is an intended deterrent (remember to consult any relevant federal, state or local fire or occupational safety codes before making any changes)
- minimizing the number of entrances to restricted areas (remember to consult any relevant federal, state or local fire or occupational safety codes before making any changes)
- securing bulk unloading equipment (for example, augers, pipes, conveyor belts, and hoses) when not in use and inspecting the equipment before use
- accounting for all keys to establishment (for example, assigning responsibility for issuing, tracking, and retrieving keys)
- monitoring the security of the premises using appropriate methods (for example, using security patrols [uniformed and/or plain-clothed], video surveillance)
- minimizing, to the extent practical, places that can be used to temporarily hide intentional contaminants (for example, minimizing nooks and crannies, false ceilings)
- providing adequate interior and exterior lighting, including emergency lighting, where appropriate, to facilitate detection of suspicious or unusual activities
- implementing a system of controlling vehicles authorized to park on the premises (for example, using placards, decals, key cards, keyed or cipher locks, issuing passes for specific areas and times to visitors' vehicles)
- keeping parking areas separated from entrances to food storage and processing areas and utilities, where practical

- Laboratory safety
  - restricting access to the laboratory (for example, using key cards or keyed or cipher locks [remember to consult any relevant federal, state or local fire or occupational safety codes before making any changes])
  - restricting laboratory materials to the laboratory, except as needed for sampling or other appropriate activities
  - restricting access (for example, using locks, seals, alarms, key cards, keyed or cipher locks) to sensitive materials (for example, reagents and bacterial, drug, and toxin positive controls)
  - assigning responsibility for integrity of positive controls to a qualified individual

  - knowing what reagents and positive controls should be on the premises and keeping track of them
  - investigating missing reagents or positive controls or other irregularities outside a normal range of variability immediately, and alerting appropriate law enforcement and public health authorities about unresolved problems, when appropriate
  - disposing of unneeded reagents and positive controls in a manner that minimizes the risk that they can be used as a contaminant
- Storage and use of poisonous and toxic chemicals (for example, cleaning and sanitizing agents, pesticides)
  - limiting poisonous and toxic chemicals in the establishment to those that are required for the operation and maintenance of the facility and those that are being held for sale
  - storing poisonous and toxic chemicals as far away from food handling and storage areas as practical
  - limiting access to and securing storage areas for poisonous and toxic chemicals that are not being held for sale (for example, using keyed or cipher locks, key cards, seals, alarms, intrusion detection sensors, guards, monitored video surveillance [remember to consult any relevant federal, state or local fire codes that may apply before making any changes])
  - ensuring that poisonous and toxic chemicals are properly labeled
  - using pesticides in accordance with the Federal Insecticide, Fungicide, and Rodenticide Act (for example, maintaining rodent bait that is in use in covered, tamper-resistant bait stations)
  - knowing what poisonous and toxic chemicals should be on the premises and keeping track of them
  - investigating missing stock or other irregularities outside a normal range of variation and alerting appropriate law enforcement and public health authorities about unresolved problems, when appropriate

Operations

FDA recommends that food establishment operators consider:

- Incoming materials and contract operations
  - using only known, appropriately licensed or permitted (where applicable) contract manufacturing and packaging operators and sources for all incoming materials, including ingredients, compressed gas, packaging, labels, and materials for research and development
  - taking reasonable steps to ensure that suppliers, contract operators and transporters practice appropriate food security measures (for

example, auditing, where practical, for compliance with food security measures that are contained in purchase and shipping contracts or letters of credit, or using a vendor approval program)

- authenticating labeling and packaging configuration and product coding/expiration dating systems (where applicable) for incoming materials in advance of receipt of shipment, especially for new products
- requesting locked and/or sealed vehicles/containers/railcars, and, if sealed, obtaining the seal number from the supplier and verifying upon receipt, making arrangements to maintain the chain of custody when a seal is broken for inspection by a governmental agency or as a result of multiple deliveries
- requesting that the transporter have the capability to verify the location of the load at any time, when practical
- establishing delivery schedules, not accepting unexplained, unscheduled deliveries or drivers, and investigating delayed or missed shipments
- supervising off-loading of incoming materials, including off hour deliveries
- reconciling the product and amount received with the product and amount ordered and the product and amount listed on the invoice and shipping documents, taking into account any sampling performed prior to receipt
- investigating shipping documents with suspicious alterations
- inspecting incoming materials, including ingredients, compressed gas, packaging, labels, product returns, and materials for research and development, for signs of tampering, contamination or damage (for example, abnormal powders, liquids, stains, or odors, evidence of resealing, compromised tamper-evident packaging) or "counterfeiting" (for example, inappropriate or mismatched product identity, labeling, product lot coding or specifications, absence of tamper-evident packaging when the label contains a tamper-evident notice), when appropriate
- evaluating the utility of testing incoming ingredients, compressed gas, packaging, labels, product returns, and materials for research and development for detecting tampering or other malicious, criminal, or terrorist action
- rejecting suspect food
- alerting appropriate law enforcement and public health authorities about evidence of tampering, "counterfeiting" or other malicious, criminal, or terrorist action

- Storage

    - having a system for receiving, storing, and handling distressed, damaged, returned, and rework products that minimizes their potential for being compromised or to compromise the security of other products (for example, destroying products that are unfit for human or animal consumption, products with illegible codes, products of questionable origin, and products returned by consumers to retail stores)
    - keeping track of incoming materials and materials in use, including ingredients, compressed gas, packaging, labels, salvage products, rework products, and product returns
    - investigating missing or extra stock or other irregularities outside a normal range of variability and reporting unresolved problems to appropriate law enforcement and public health authorities, when appropriate
    - storing product labels in a secure location and destroying outdated or discarded product labels
    - minimizing reuse of containers, shipping packages, cartons, etc., where practical
- Security of water and utilities
    - limiting, to the extent practical, access to controls for airflow, water, electricity, and refrigeration
    - securing non-municipal water wells, hydrants, storage, and handling facilities
    - ensuring that water systems and trucks are equipped with backflow prevention
    - chlorinating water systems and monitoring chlorination equipment, where practical, and especially for non-municipal water systems
    - testing non-municipal sources for potability regularly, as well as randomly, and being alert to changes in the profile of the results
    - staying attentive to the potential for media alerts about public water provider problems, when applicable
    - identifying alternate sources of potable water for use during emergency situations where normal water systems have been compromised (for example, trucking from an approved source, treating on-site or maintaining on-site storage)
- Finished products
    - ensuring that public storage warehousing and shipping operations (vehicles and vessels) practice appropriate security measures (for example, auditing, where practical, for compliance with food security measures that are contained in contracts or letters of guarantee)
    - performing random inspection of storage facilities, vehicles, and vessels

    - evaluating the utility of finished product testing for detecting tampering or other malicious, criminal, or terrorist actions
    - requesting locked and/or sealed vehicles/containers/railcars and providing the seal number to the consignee
    - requesting that the transporter have the capability to verify the location of the load at any time
    - establishing scheduled pickups, and not accepting unexplained, unscheduled pickups
    - keeping track of finished products
    - investigating missing or extra stock or other irregularities outside a normal range of variation and alerting appropriate law enforcement and public health authorities about unresolved problems, when appropriate
    - advising sales staff to be on the lookout for counterfeit products and to alert management if any problems are detected
- Mail/packages
    - implementing procedures to ensure the security of incoming mail and packages (for example, locating the mailroom away from food processing and storage areas, securing mailroom, visual or x-ray mail/package screening, following U.S. Postal Service guidance)
- Access to computer systems
    - restricting access to computer process control systems and critical data systems to those with appropriate clearance (for example, using passwords, firewalls)
    - eliminating computer access when a staff member is no longer associated with the establishment
    - establishing a system of traceability of computer transactions
    - reviewing the adequacy of virus protection systems and procedures for backing up critical computer based data systems
    - validating the computer security system

Emergency Point of Contact

U.S. Food and Drug Administration
5600 Fishers Lane
Rockville, MD 20857
301-443-1240

If a food establishment operator suspects that any of his/her products that are regulated by the FDA have been subject to tampering, "counterfeiting," or other malicious, criminal, or terrorist action, FDA recommends that he/she notify the FDA 24-hour emergency number at 301-443-1240 or call their local FDA District Office. FDA District Office telephone numbers are listed at: http://www.fda.gov/ora/inspect_ref/iom/iomoradir.html. FDA recommends that the operator also notify appropriate law enforcement and public health authorities.

*Reference to these documents is provided for informational purposes only. These documents are not incorporated by reference into this guidance and should not be considered to be FDA guidance.

Guidance Documents | Food Safety and Terrorism
Foods Home | FDA Home | HHS Home | Search/Subject Index | Disclaimers & Privacy Policy | Accessibility/Help

Appendix I

GUIDANCE FOR INDUSTRY

IMPORTERS AND FILERS: FOOD SECURITY PREVENTIVE MEASURES GUIDANCE*

This guidance represents the Agency's current thinking on the kinds of measures that food importers and filers may take to minimize the risk that food under their control will be subject to tampering or other malicious, criminal, or terrorist actions. It does not create or confer any rights for or on any person and does not operate to bind FDA or the public.

Purpose and Scope

This guidance is designed as an aid to operators of food importing establishments, storage warehouses, and filers. It identifies the kinds of preventive measures that they may take to minimize the risk that food under their control will be subject to tampering or other malicious, criminal, or terrorist actions. Operators of food importing establishments are encouraged to review their current procedures and controls in light of the potential for tampering or other malicious, criminal, or terrorist actions and make appropriate improvements.

This guidance is designed to focus operator's attention sequentially on each segment of the food delivery system that is within their control, to minimize the risk of tampering or other malicious, criminal, or terrorist action at each segment. To be successful, implementing enhanced preventive

* From the U.S. Department of Health and Human Services, U.S. Food and Drug Administration, Center for Food Safety and Applied Nutrition, March 21, 2003.

measures requires the commitment of management and staff. Accordingly, FDA recommends that both management and staff participate in the development and review of such measures.

Limitations

Not all of the guidance contained in this document may be appropriate or practical for every food importing establishment, particularly small facilities. FDA recommends that operators review the guidance in each section that relates to a component of their operation, and assess which preventive measures are suitable. Example approaches are provided for many of the preventive measures listed in this document. These examples should not be regarded as minimum standards. Nor should the examples provided be considered an inclusive list of all potential approaches to achieving the goal of the preventive measure. FDA recommends that operators consider the goal of the preventive measure, assess whether the goal is relevant to their operation, and, if it is, design an approach that is both efficient and effective to accomplish the goal under their conditions of operation.

Structure

This guidance is divided into five sections that relate to individual components of food importing operations and practices: Management; Human Element — Staff; Human Element — Public; Facility; and Operations.

Related Guidance

FDA has published a companion guidance document on food security, entitled, "Guidance for Food Producers, Processors, and Transporters: Food security preventive measures guidance." This document is available at: http://www.access.gpo.gov/su_docs/aces/aces140.html.

Additional Resources*

A process called Operational Risk Management (ORM) may help prioritize the preventive measures that are most likely to have the greatest impact on reducing the risk of tampering or other malicious, criminal, or terrorist actions against food. Information on ORM is available in the Federal Aviation Administration (FAA) System Safety Handbook, U.S. Department of Transportation, FAA, December 30, 2000, Chapter 15, Operational Risk Management. The handbook is available at: http://www.asy.faa.gov/Risk/SSHandbook/ Chap15_1200.PDF.

The U.S. Department of Transportation, Research and Special Programs Administration has published an advisory notice of voluntary measures to enhance the security of hazardous materials shipments. It is available at: http://frwebgate.access.gpo.gov/cgi-bin/getdoc.cgi?dbname=2002_register&docid=02-3636-filed.pdf. The notice provides guidance to shippers and carriers on personnel, facility and en route security issues.

The U.S. Postal Service has prepared guidance for identifying and handling suspicious mail. It is available at: http://www.usps.com/news/2001/press/mailsecurity/postcard.htm.

The Federal Anti-Tampering Act (18 USC 1365) makes it a federal crime to tamper with or taint a consumer product, or to attempt, threaten or conspire to tamper with or taint a consumer product, or make a false statement about having tampered with or tainted a consumer product. Conviction can lead to penalties of up to $100,000 in fines and up to life imprisonment. The Act is available at: http://www.fda.gov/opacom/laws/fedatact.htm.

The National Infrastructure Protection Center (NIPC) serves as the federal government's focal point for threat assessment, warning, investigation, and response for threats or attacks against U.S. critical infrastructure. The NIPC has identified the food system as one of the eight critical infrastructures, and has established a public-private partnership with the food industry, called the Food Industry Information and Analysis Center (Food Industry ISAC). The NIPC provides the Food Industry ISAC with access, information and analysis, enabling the food industry to report, identify, and reduce its vulnerabilities to malicious attacks, and to recover from such attacks as quickly as possible. In particular, the NIPC identifies credible threats and crafts specific warning messages to the food industry. Further information is available at http://www.nipc.gov/ and http://www.foodisac.org/.

Finally, FDA encourages trade associations to evaluate the preventive measures contained in this guidance document and adapt them to their specific products and operations and to supplement this guidance with additional preventive measures when appropriate. FDA welcomes dialogue on the content of sector specific guidance with appropriate trade associations.

Food Importing Operations

Management

FDA recommends that operators of food importing establishments consider:

- Preparing for the possibility of tampering or other malicious, criminal, or terrorist actions
  - assigning responsibility for security to knowledgeable individual(s)
  - conducting an initial assessment of food security procedures and operations, which we recommend be kept confidential

  - having a crisis management strategy to prepare for and respond to tampering and other malicious, criminal, or terrorist actions, both threats and actual events, including identifying, segregating and securing affected product
  - planning for emergency evacuation, including preventing security breaches during evacuation
  - becoming familiar with the emergency response system in the community
  - making management aware of 24-hour contact information for local, state, and federal police/fire/rescue/health/homeland security agencies
  - making staff aware of who in management they should alert about potential security problems (24-hour contacts)
  - maintaining any floor and food flow plan in a secure, off-site location
  - promoting food security awareness to encourage all staff to be alert to any signs of tampering or malicious, criminal, or terrorist actions or areas that may be vulnerable to such actions, and to report any findings to identified management (for example, providing training, instituting a system of rewards, building security into job performance standards)
  - having an internal communication system to inform and update staff about relevant security issues
  - having a strategy for communicating with the public (for example, identifying a media spokesperson, preparing generic press statements and background information, and coordinating press statements with appropriate authorities)
- Supervision
  - providing an appropriate level of supervision to all staff, including cleaning and maintenance staff, contract workers, data entry and computer support staff, and especially, new staff (for example, supervisor on duty, daily visits by supervisor, two staff on duty at all times, monitored video cameras, one way and two way windows)
  - conducting routine security checks of the premises and critical computer data systems (at a frequency appropriate to the operation) for signs of tampering or malicious, criminal, or terrorist actions, or areas that may be vulnerable to such actions
- Recall strategy
  - identifying the person responsible, and a backup person
  - providing for proper handling and disposition of recalled product
  - identifying customer contacts, addresses, and phone numbers

- Investigation of suspicious activity
  - investigating threats or information about signs of tampering or other malicious, criminal, or terrorist actions
  - alerting appropriate law enforcement and public health authorities about any threats of or suspected tampering or other malicious, criminal, or terrorist actions
- Evaluation program
  - evaluating the lessons learned from past tampering or other malicious, criminal, or terrorist actions and threats
  - reviewing and verifying, at least annually, the effectiveness of the security management program (for example, using knowledgeable in-house or third party staff to conduct tampering or other malicious, criminal, or terrorist action exercises and mock recalls and to challenge computer security systems), revising the program accordingly, and keeping this information confidential
  - performing random food security inspections of all appropriate areas of the facility (including receiving and storage, where applicable) using knowledgeable in-house or third party staff, and keeping this information confidential
  - verifying that security contractors are doing an appropriate job, when applicable

Human Element — Staff

Under Federal law, operators of food importing establishments are required to verify the employment eligibility of all new hires in accordance with the requirements of the Immigration and Nationality Act, by completing the INS Employment Eligibility Verification Form (INS Form I-9). Completion of Form I-9 for new hires is required by 8 USC 1324a and nondiscrimination provisions governing the verification process are set forth at 8 USC 1324b.

FDA recommends that operators of food importing establishments consider:

- Screening (pre-hiring, at hiring, post-hiring)
  - examining the background of all staff (including seasonal, temporary, contract, and volunteer staff, whether hired directly or through a recruitment firm) as appropriate to their position, considering candidates' access to sensitive areas of the facility and the degree to which they will be supervised and other relevant factors (for example, obtaining and verifying work references, addresses, and phone numbers, participating in one of the pilot programs managed by the Immigration and Naturalization Service and the

Social Security Administration [These programs provide electronic confirmation of employment eligibility for newly hired employees. For more information call the INS SAVE Program toll free at 1-888-464-4218, fax a request for information to (202) 514-9981, or write to US/INS, SAVE Program, 425 I Street, NW, ULLICO-4th Floor, Washington, D.C. 20536. These pilot programs may not be available in all states], having a criminal background check performed by local law enforcement or by a contract service provider [Remember to first consult any state or local laws that may apply to the performance of such checks])

- Note: screening procedures should be applied equally to all employees, regardless of race, national origin, religion, and citizenship or immigration status.
- Daily work assignments
    - knowing who is and who should be on premises, and where they should be located, for each shift
    - keeping assignment information updated
- Identification
    - establishing a system of positive identification and recognition that is appropriate to the nature of the workforce (for example, issuing uniforms, name tags, or photo identification badges, with individual control numbers, color coded by area of authorized access), when appropriate
    - collecting the uniforms, name tag, or identification badge when a staff member is no longer associated with the establishment
- Restricted access
    - identifying staff that require unlimited access to all areas of the facility
    - reassessing levels of access for all staff periodically
    - limiting access so staff enter only those areas or have access to only those segments of the operation necessary for their job functions and only during appropriate work hours, including access to data operating systems for purchasing, storing and distributing imported foods (for example, using key card or keyed or cipher locks for entry to sensitive areas, color coded uniforms [remember to consult any relevant federal, state or local fire or occupational safety codes before making any changes])
    - changing combinations, rekeying locks and/or collecting the retired key card when a staff member who is in possession of these is no longer associated with the establishment, and additionally as needed to maintain security
- Personal items

  - restricting the type of personal items allowed in non-public areas of the establishment
  - allowing in the establishment only those personal use medicines that are necessary for the health of staff and ensuring that these personal use medicines are properly labeled and stored away from food handling or storage areas
  - preventing staff from bringing personal items (for example, lunch containers, purses) into food preparation or storage areas
  - providing for regular inspection of contents of staff lockers (for example, providing metal mesh lockers, company issued locks), bags, packages, and vehicles when on company property (Remember to first consult and federal, state, or local laws that may related to such inspections)
- Training in food security procedures
  - incorporating food security awareness, including information on how to prevent, detect, and respond to tampering or other malicious, criminal, or terrorist actions or threats, into training programs for staff, including seasonal, temporary, contract, and volunteer staff providing periodic reminders of the importance of security procedures (for example, scheduled meetings, providing brochures, payroll stuffers)
  - providing periodic reminders of the importance of security procedures (for example, scheduled meetings, providing brochures, payroll stuffers)
  - encouraging staff support (for example, involving staff in food security planning and the food security awareness program, demonstrating the importance of security procedures to the staff)
- Unusual behavior
  - watching for unusual or suspicious behavior by staff (for example, staff who, without an identifiable purpose, stay unusually late after the end of their shift, arrive unusually early, access files/information/areas of the facility outside of the areas of their responsibility; remove documents from the facility; ask questions on sensitive subjects; bring cameras to work)
- Staff health
  - being alert for atypical staff health conditions that staff may voluntarily report and absences that could be an early indicator of tampering or other malicious, criminal, or terrorist actions (for example, an unusual number of staff who work in the same part of the facility reporting similar symptoms within a short time frame), and reporting such conditions to local health authorities

Human Element — Public

FDA recommends that operators of food importing establishments consider:

- Visitors (for example, contractors, supplier representatives, delivery drivers, customers, couriers, pest control representatives, third-party auditors, regulators, reporters, tours)
  - inspecting incoming and outgoing vehicles, packages and briefcases for suspicious, inappropriate or unusual items or activity, to the extent practical
  - restricting entry to the establishment (for example, checking visitors in and out at security or reception, requiring proof of identity, issuing visitors badges that are collected upon departure, accompanying visitors)
  - ensuring that there is a valid reason for the visit before providing access to the facility — beware of unsolicited visitors
  - verifying the identity of unknown visitors
  - restricting access to food handling and storage areas (for example, accompanying visitors, unless they are otherwise specifically authorized)
  - restricting access to locker rooms

Facility

FDA recommends that operators of food importing establishments consider:

- Physical security
  - protecting perimeter access with fencing or other deterrent, when appropriate
  - securing doors (including freight loading doors when not in use and not being monitored, and emergency exits), windows, roof openings/hatches, vent openings and trailer bodies, to the extent possible (for example, using locks, "jimmy plates," seals, alarms, intrusion detection sensors, guards, monitored video surveillance [remember to consult any relevant federal, state or local fire or occupational safety codes before making any changes])
  - using metal or metal-clad exterior doors to the extent possible when the facility is not in operation, except where visibility from public thoroughfares is an intended deterrent (remember to consult any relevant federal, state or local fire or occupational safety codes before making any changes)

- securing bulk unloading equipment (for example, augers, pipes, conveyor belts, and hoses) when not in use and inspecting the equipment before use
- minimizing the number of entrances to restricted areas (remember to consult any relevant federal, state or local fire or occupational safety codes before making any changes)
- accounting for all keys to establishment (for example, assigning responsibility for issuing, tracking and retrieving keys)
- monitoring the security of the premises using appropriate methods (for example, using security patrols [uniformed and/or plainclothed] and video surveillance)
- minimizing to the extent practical, places that can be used to temporarily hide intentional contaminants (for example, minimizing nooks and crannies, false ceilings)
- providing adequate interior and exterior lighting, including emergency lighting, where appropriate, to facilitate detection of suspicious or unusual activity
- implementing a system of controlling vehicles authorized to park on the premises (for example, using placards, decals, key cards, keyed or cipher locks, issuing passes for specific areas and times to visitors' vehicles)
- keeping parking areas separated from entrances to food storage and processing areas and utilities, where practical

- Storage and use of poisonous and toxic chemicals (for example, cleaning and sanitizing agents, pesticides)
  - limiting poisonous and toxic chemicals in the establishment to those that are required for the operation and maintenance of the facility and those that are being held for sale
  - storing poisonous and toxic chemicals as far away from food handling and storage areas as practical
  - limiting access to and securing storage areas for poisonous and toxic chemicals that are not being held for sale (for example, using keyed or cipher locks, keycards, seals, alarms, intrusion detection sensors, guards, monitored video surveillance [remember to consult any relevant state or local fire codes before making any changes])
  - ensuring that poisonous and toxic chemicals are properly labeled
  - using pesticides in accordance with the Federal Insecticide, Fungicide, and Rodenticide Act (for example, maintaining rodent bait that is in use in covered, tamper-resistant bait stations)
  - knowing what poisonous and toxic chemicals should be on the premises and keeping track of them

- investigating missing stock or other irregularities outside a normal range of variation and alerting appropriate law enforcement and public health authorities about unresolved problems, when appropriate

Operations

FDA recommends that operators of food importing establishments consider:

- Incoming products
  - using only known and appropriately licensed or permitted (where applicable) sources for all products
  - taking reasonable steps to encourage suppliers, distributors and transporters to practice appropriate food security measures (for example, auditing, where practical, for compliance with food security measures that are contained in purchase and shipping contracts or letters of credit or using a vendor approval program)
  - authenticating labeling, packaging configuration, tamper-evident packaging and product coding/expiration dating systems (where applicable) in advance of receipt of shipment, especially for new products
  - requesting locked and/or sealed vehicles/containers/railcars, and, if sealed, obtaining the seal number from the supplier, and verifying upon receipt, making arrangements to maintain the chain of custody when a seal is broken for inspection by a governmental agency or as a result of multiple deliveries
  - requesting that transporters have the capability to verify the location of the load at any time, when practical
  - establishing delivery schedules, not accepting unexplained, unscheduled deliveries or drivers, and investigating delayed or missed shipments
  - supervising off-loading of incoming materials, including off hour deliveries
  - reconciling the product and amount received with the product and amount ordered and the product and amount listed on the invoice and shipping documents, taking into account any sampling performed prior to receipt
  - investigating shipping documents with suspicious alterations
  - inspecting incoming products and product returns for signs of tampering, contamination or damage (for example, abnormal powders, liquids, stains, or odors, evidence of resealing, compromised tamper-evident packaging) or "counterfeiting" (inappropriate or mismatched product identity, labeling, product

lot coding or specifications, absence of tamper-evident packaging when the label contains a tamper-evident notice), when appropriate

- inspecting incoming products for authenticity, packaging/product integrity, and evidence of unauthorized relabeling/repackaging (for example, shipping cases and described contents not consistent with actual contents) and verifying batch/lot/container codes
- verifying conformance with FDA requirements for product safety, quality, effectiveness, and labeling (may require contact with and verification from the foreign manufacturer/processor)
- evaluating the utility of testing incoming products and product returns for detecting tampering or other malicious, criminal, or terrorist action
- developing and implementing procedures for inspecting shipping containers, vehicles
- investigating damage and loss and alerting appropriate authority of discrepancies
- rejecting suspect food
- alerting appropriate law enforcement and food public health authorities about evidence of tampering, "counterfeiting" or other malicious, criminal, or terrorist action

- Storage
    - having a system for receiving, storing and handling distressed, damaged, returned, and reworked products that minimizes their potential for being compromised or to compromise the security of other products (for example, destroying products that are unfit for human or animal consumption, products with illegible codes, products or questionable origin, and products returned by consumers to retail stores)
    - keeping track of incoming products, salvage products, and returned products
    - minimizing reuse of containers, shipping packages, cartons, etc., where practical
    - investigating missing or extra stock or other irregularities outside a normal range of variability and reporting unresolved problems to appropriate law enforcement and public health agencies, when appropriate
- Outgoing products
    - ensuring that public storage warehousing and shipping (vehicles and vessels) practice appropriate security measures (for example, auditing for compliance with food security measures that are contained in contracts or letters of guarantee)

    - performing random inspection of storage facilities, vehicles, and vessels
    - requesting locked and/or sealed vehicles/containers/railcars and providing the seal number to the consignee (remember to consult any relevant federal, state or local fire or occupational safety codes before making any changes)
    - establishing scheduled pickups and not accepting unexplained, unscheduled pickups
    - restricting access to distribution process to employees with appropriate clearance
    - requesting that the transporter have the capability to verify the location of the load at any time
    - advising sales staff to be on the lookout for counterfeit products during visits to customers and notify management if any problems are detected
    - investigating missing or extra stock or other irregularities outside a normal range of variation and alerting appropriate law enforcement and public health authorities about unresolved problems, when appropriate
- Security of water and utilities
    - limiting, to the extent practical, access to controls for airflow, water, electricity, and refrigeration securing non-municipal water wells, hydrants, storage, and handling facilities
    - ensuring that water systems and trucks are equipped with backflow prevention
    - chlorinating water systems and monitoring chlorination equipment, where practical, and especially for non-municipal water systems
    - testing non-municipal sources for potability regularly, as well as randomly, and being alert to changes in the profile of the results
    - staying attentive to the potential for media alerts about public water provider problems, when applicable
    - identifying alternate sources of potable water for use during emergency situations where normal water systems have been compromised (for example, bottled water, trucking from an approved source, treating on-site or maintaining on-site storage)
- Security of ventilation system (where applicable)
    - securing access to air intake points for the facility, to the extent possible (for example, using fences, sensors, guards, video surveillance)
    - examining air intake points for physical integrity routinely

- Mail/packages
  - implementing procedures to ensure the security of incoming mail and packages (for example, following U.S. Postal Service guidance, locating the mailroom away from food handling and storage areas, securing mailroom visual or x-ray mail/package screening)
- Access to computer systems
  - restricting access to critical computer data systems to those with appropriate clearance (for example, using passwords, firewalls)
  - eliminating computer access when a staff member is no longer associated with the establishment
  - establishing a system of traceability of computer transactions
  - reviewing the adequacy of virus protection systems and procedures for backing up critical computer based data systems
  - validating and periodically challenging the computer security system and procedures

Emergency Point of Contact

U.S. Food and Drug Administration
5600 Fishers Lane
Rockville, MD 20857

If a food import establishment operator suspects that any of his/her products that are regulated by the FDA have been subject to tampering, "counterfeiting," or other malicious, criminal, or terrorist action, FDA recommends that he/she notify the FDA 24-hour emergency number at 301-443-1240 or call their local FDA District Office. FDA District Office telephone numbers are listed at http://www.fda.gov/ora/inspect_ref/iom/iomoradir.html. FDA recommends that the operator also notify local law enforcement and public health agencies.

*Reference to these documents is provided for informational purposes only. These documents are not incorporated by reference into this guidance and should not be considered to be FDA guidance.

Appendix J

GUIDANCE FOR INDUSTRY

COSMETICS PROCESSORS AND TRANSPORTERS: COSMETICS SECURITY PREVENTIVE MEASURES GUIDANCE*

Draft Guidance

This draft guidance represents the Agency's current thinking on appropriate measures that cosmetics establishments may take to minimize the risk that cosmetics under their control will be subject to tampering or other malicious, criminal, or terrorist actions. It does not create or confer any rights for or on any person and does not operate to bind FDA or the public.

Purpose and Scope

This draft guidance is designed as an aid to operators of cosmetics establishments (for example, firms that process, store, repack, re-label, distribute, or transport cosmetics or cosmetics ingredients). This is a very diverse set of establishments, which includes both very large and very small entities.

This draft guidance identifies the kinds of preventive measures operators of cosmetics establishments may take to minimize the risk that cosmetics under their control will be subject to tampering or other malicious, criminal, or terrorist actions.

Operators of cosmetics establishments are encouraged to review their current procedures and controls in light of the potential for tampering or other malicious, criminal, or terrorist actions and make appropriate improve-

* From the U.S. Department of Health and Human Services, U.S. Food and Drug Administration, Center for Food Safety and Applied Nutrition, March 21, 2003.

ments. FDA recommends that the review include consideration of the role that unit and distribution packaging might have in a cosmetics security program. This guidance is designed to focus operator's attention sequentially on each segment of the cosmetic production system that is within their control, to minimize the risk of tampering or other malicious, criminal, or terrorist action at each segment. To be successful, implementing enhanced preventive measures requires the commitment of management and staff. Accordingly, FDA recommends that both management and staff participate in the development and review of such measures.

Limitations

Not all of the guidance contained in this document may be appropriate or practical for every cosmetics establishment, particularly smaller facilities and distributors. FDA recommends that operators review the guidance in each section that relates to a component of their operation, and assess which preventive measures are suitable. Example approaches are provided for many of the preventive measures listed in this document. These examples should not be regarded as minimum standards. Nor should the examples provided be considered an inclusive list of all potential approaches to achieving the goal of the preventive measure. FDA recommends that operators consider the goal of the preventive measure, assess whether the goal is relevant to their operation, and, if it is, design an approach that is both efficient and effective to accomplish the goal under their conditions of operation.

Structure

This draft guidance is divided into five sections that relate to individual components of a cosmetics establishment operation: management; human element — staff; human element — the public; facility; and operations.

Related Guidance

FDA has published two guidance documents on food security entitled, "Food producers, processors, and transporters: Food security preventive measures guidance," and "Importers and filers: Food security preventive measures guidance" to cover the farm-to-table spectrum of food production. The two documents are available at http://www.access.gpo.gov/su_docs/aces/aces140.html.

Additional Resources*

A process called Operational Risk Management (ORM) may help prioritize the preventive measures that are most likely to have the greatest impact on reducing the risk of tampering or other malicious criminal, or terrorist actions against cosmetics. Information on ORM is available in the Federal

Aviation Administration (FAA) System Safety Handbook, U.S. Department of Transportation, FAA, December 30, 2000, Chapter 15, Operational Risk Management. The handbook is available at: http://www.asy.faa.gov/Risk/SSHandbook/Chap15_1200.PDF.

The U.S. Department of Transportation, Research and Special Programs Administration published an advisory notice of voluntary measures to enhance the security of hazardous materials shipments. It is available at: http://frwebgate.access.gpo.gov/cgi-bin/getdoc.cgi?dbname=2002_register&docid=02-3636-filed.pdf. The notice provides guidance to shippers and carriers on personnel, facility and en route security issues.

The U.S. Postal Service has prepared guidance for identifying and handling suspicious mail. It is available at: http://www.usps.com/news/2001/press/mailsecurity/postcard.htm.

The Federal Anti-Tampering Act (18 USC 1365) makes it a federal crime to tamper with or taint a consumer product, or to attempt, threaten or conspire to tamper with or taint a consumer product, or make a false statement about having tampered with or tainted a consumer product. Conviction can lead to penalties of up to $100,000 in fines and up to life imprisonment. The Act is available at: http://www.fda.gov/opacom/laws/fedatact.htm.

The National Infrastructure Protection Center (NIPC) serves as the federal government's focal point for threat assessment, warning, investigation, and response for threats or attacks against U.S. critical infrastructure. The NIPC has identified the food system as one of the eight critical infrastructures, and has established a public-private partnership with the food industry, called the Food Industry Information and Analysis Center (Food Industry ISAC). The NIPC provides the Food Industry ISAC with access, information and analysis, enabling the food industry to report, identify, and reduce its vulnerabilities to malicious attacks, and to recover from such attacks as quickly as possible. In particular, the NIPC identifies credible threats and crafts specific warning messages to the food industry. Further information is available at http://www.nipc.gov/ and http://www.foodisac.org/.

FDA encourages cosmetics trade associations to evaluate the preventive measures contained in this guidance document and adapt them to their specific products and operations and to supplement this guidance with additional preventive measures when appropriate. FDA welcomes dialogue on the content of sector specific guidance with appropriate trade associations.

Cosmetics Establishment Operations

Management

FDA recommends that cosmetics establishment operators consider:

- Preparing for the possibility of tampering or other malicious, criminal, or terrorist actions
    - assigning responsibility for security to knowledgeable individual(s)
    - conducting an initial assessment of cosmetics security procedures and operations, which we recommend be kept confidential
    - having a security management strategy to prepare for and respond to tampering or other malicious, criminal, or terrorist actions, both threats and actual events, including identifying, segregating, and securing affected product
    - planning for emergency evacuation, including preventing security breaches during evacuation
    - maintaining any floor or flow plan in a secure, off-site location
    - becoming familiar with the emergency response system in the community
    - making management aware of 24-hour contact information for local, state, and federal police/fire/rescue/health/homeland security agencies
    - making staff aware of who in management they should alert about potential security problems (24-hour contacts)
    - promoting cosmetics security awareness to encourage all staff to be alert to any signs of tampering or other malicious, criminal, or terrorist actions, or areas that may be vulnerable to such actions, and reporting any findings to identified management (for example, providing training, instituting a system of rewards, building security into job performance standards)
    - having an internal communication system to inform and update staff about relevant security issues
    - having a strategy for communicating with the public (for example, identifying a media spokesperson, preparing generic press statements and background information, and coordinating press statements with appropriate authorities)
- Supervision
    - providing an appropriate level of supervision to all staff, including cleaning and maintenance staff, contract workers, data entry and computer support staff, and especially, new staff
    - conducting routine security checks of the premises, including utilities and critical computer systems (at a frequency appropriate to the operation), for signs of tampering or other malicious, criminal, or terrorist actions, or areas that may be vulnerable to such actions
- Recall strategy
    - identifying the person responsible, and a backup person
    - providing for proper handling and disposition of recalled product

  - identifying customer contacts, addresses and phone numbers
- Investigation of suspicious activity
  - investigating threats or information about signs of tampering or other malicious, criminal, or terrorist actions
  - alerting appropriate law enforcement and public health authorities about any threats of or suspected tampering or other malicious, criminal, or terrorist actions
- Evaluation program
  - evaluating the lessons learned from past tampering or other malicious, criminal, or terrorist actions and threats
  - reviewing and verifying, at least annually, the effectiveness of the security management program (for example, using knowledgeable in-house or third party staff to conduct tampering or other malicious, criminal, or terrorist action exercises and mock recalls and to challenge computer security systems), revising the program accordingly, and keeping this information confidential
  - performing random cosmetics security inspections of all appropriate areas of the facility (including receiving and warehousing areas) using knowledgeable in-house or third party staff and keeping this information confidential
  - verifying that security contractors are doing an appropriate job, when applicable

Human Element — Staff

Under Federal law, cosmetics establishment operators are required to verify the employment eligibility of all new hires in accordance with the requirements of the Immigration and Nationality Act, by completing the INS Employment Eligibility Verification Form (INS Form I-9). Completion of Form I-9 for new hires is required by 8 USC 1324a and nondiscrimination provisions governing the verification process are set forth at 1324b.

FDA recommends that cosmetics establishment operators consider:

- Screening (pre-hiring, at hiring, post-hiring)
  - examine the background of all staff (including seasonal, temporary, contract, and volunteer staff, whether hired directly or through a recruitment firm) as appropriate to their position, considering candidates' access to sensitive areas of the facility and the degree to which they will be supervised and other relevant factors (for example, obtaining and verifying work references, addresses, and phone numbers, participating in one of the pilot programs managed by the Immigration and Naturalization Service and the Social Security Administration [These programs provide electron-

ic confirmation of employment eligibility for newly hired employees. For more information call the INS SAVE Program toll free at 1-888-464-4218, fax a request for information to (202) 514-9981, or write to US/INS, SAVE Program, 425 I Street, NW, ULLICO-4th Floor, Washington, D.C. 20536. These pilot programs may not be available in all states], having a criminal background check performed by local law enforcement or by a contract service provider [Remember to first consult any state or local laws that may apply to the performance of such checks])

- Note: screening procedures should be applied equally to all staff, regardless of race, national origin, religion, and citizenship or immigration status.
- Daily work assignments
  - knowing who is and who should be on premises, and where they should be located, for each shift
  - keeping information updated
- Identification
  - establishing a system of positive identification and recognition (for example, issuing uniforms, name tags, or photo identification badges with individual control numbers, color coded by area of authorized access), when appropriate
  - collecting the uniforms, name tag or identification badge when a staff member is no longer associated with the establishment
- Restricted access
  - identifying staff that require unlimited access to all areas of the facility
  - reassessing levels of access for all staff periodically
  - limiting access so staff enter only those areas necessary for their job functions and only during appropriate work hours (for example, using key cards or keyed or cipher locks for entry to sensitive areas, color coded uniforms [remember to consult any relevant federal, state or local fire or occupational safety codes before making any changes])
  - changing combinations, rekeying locks and/or collecting the retired key card when a staff member is no longer associated with the establishment, and additionally as needed to maintain security
- Personal items
  - restricting the type of personal items allowed in establishment
  - allowing in the establishment only those personal use medicines that are necessary for the health of the staff and ensuring that these personal use medicines are properly labeled and stored away from cosmetics handling or storage

    - preventing staff from bringing personal items (for example, lunch containers, purses) into cosmetics manufacturing and storage areas
    - providing for regular inspection of contents of staff lockers (for example, providing metal mesh lockers, company issued locks), bags, packages, and vehicles when on company property (Remember to first consult any federal, state, or local laws that may relate to such inspections)
- Training in cosmetics security procedures
    - incorporating cosmetics security awareness, including information on how to prevent, detect, and respond to tampering or other malicious, criminal, or terrorist actions or threats, into training programs for staff, including seasonal, temporary, contract, and volunteer staff
    - providing periodic reminders of the importance of security procedures (for example, scheduling meetings, providing brochures or payroll stuffers)
    - encouraging staff support (for example, involving staff in cosmetics security planning and the cosmetics security awareness program, demonstrating the importance of security procedures to the staff)
- Unusual behavior
    - watching for unusual or suspicious behavior by staff (for example, staff who, without an identifiable purpose, stay unusually late after the end of their shift, arrive unusually early, access files/information/areas of the facility outside of the areas of their responsibility; remove documents from the facility; ask questions on sensitive subjects; bring cameras to work)
- Staff health
    - being alert for atypical staff health conditions that staff may voluntarily report and absences that could be an early indicator of tampering or other malicious, criminal, or terrorist actions (for example, an unusual number of staff who work in the same part of the facility reporting similar symptoms within a short time frame), and reporting such conditions to local health authorities

Human Element — The Public

FDA recommends that cosmetics establishment operators consider:

- Visitors (e.g., contractors, supplier representatives, delivery drivers, customers, couriers, pest control representatives, third-party auditors, regulators, reporters, tours)

- inspecting incoming and outgoing vehicles, packages and brief-cases for suspicious, inappropriate or unusual items or activity, to the extent practical
- restricting entry to the establishment (for example, checking visitors in and out at security or reception, requiring proof of identity, issuing visitors badges that are collected upon departure, accompanying visitors)
- ensuring that there is a valid reason for the visit before providing access to the facility — beware of unsolicited visitors
- verifying the identity of unknown visitors
- restricting access to cosmetics manufacturing and storage areas (for example, accompanying visitors, unless they are otherwise specifically authorized)
- restricting access to locker rooms

Facility

FDA recommends that cosmetics establishment operators consider:

- Physical security
    - protecting perimeter access with fencing or other deterrent, when appropriate
    - securing doors (including freight loading doors, when not in use and not being monitored, and emergency exits), windows, roof openings/hatches, vent openings, ventilation systems, utility rooms, loft areas, trailer bodies, tanker trucks, railcars, and bulk storage tanks for liquids, solids, and compressed gases, to the extent possible (for example, using locks, "jimmy plates," seals, alarms, intrusion detection sensors, guards, monitored video surveillance [remember to consult any relevant federal, state or local fire or occupational safety codes before making any changes])
    - using metal or metal-clad exterior doors to the extent possible, when the facility is not in operation, except where visibility from public thoroughfares is an intended deterrent (remember to consult any relevant federal, state or local fire or occupational safety codes before making any changes)
    - minimizing the number of entrances to restricted areas (remember to consult any relevant federal, state or local fire or occupational safety codes before making any changes)
    - securing bulk unloading equipment (for example, augers, pipes, conveyor belts, and hoses) when not in use and inspecting the equipment before use

    - accounting for all keys to establishment (for example, assigning responsibility for issuing, tracking, and retrieving keys)
    - monitoring the security of the premises using appropriate methods (for example, using security patrols [uniformed and/or plain-clothed], video surveillance)
    - minimizing, to the extent practical, places that can be used to temporarily hide intentional contaminants (for example, minimizing nooks and crannies, false ceilings)
    - providing adequate interior and exterior lighting, including emergency lighting, where appropriate, to facilitate detection of suspicious or unusual activities
    - implementing a system of controlling vehicles authorized to park on the premises (for example, using placards, decals, key cards, keyed or cipher locks, issuing passes for specific areas and times to visitors' vehicles)
    - keeping parking areas separated from cosmetics manufacturing and storage areas and utilities, where practical
- Laboratory safety
    - restricting access to the laboratory (for example, using key cards or keyed or cipher locks [remember to consult any relevant federal, state or local fire or occupational safety codes before making any changes])
    - restricting laboratory materials to the laboratory, except as needed for sampling or other appropriate activities
    - restricting access (e.g., using locks, seals, alarms, key cards, keyed or cipher locks) to sensitive materials (e.g., reagents and bacterial and toxin positive controls) to sensitive materials (for example, reagents and bacterial, drug and toxin positive controls)
    - assigning responsibility for integrity of positive controls to a qualified individual
    - knowing what reagents and positive controls should be on the premises and keeping track of them
    - investigating missing reagents or positive controls or other irregularities outside a normal range of variability immediately, and alerting appropriate law enforcement and public health authorities about unresolved problems, when appropriate
    - disposing of unneeded reagents and positive controls in a manner that minimizes the risk that they can be used as a contaminant
- Storage and use of poisonous and toxic chemicals (for example, cleaning and sanitizing agents, pesticides)
    - limiting poisonous and toxic chemicals in the establishment to those that are required for the operation and maintenance of the facility

- storing poisonous and toxic chemicals as far away from cosmetics manufacturing and storage areas as practical
- limiting access to and securing storage areas for poisonous and toxic chemicals (for example, using keyed or cipher locks, key cards, seals, alarms, intrusion detection sensors, guards, monitored video surveillance [remember to consult any relevant federal, state or local fire or occupational safety codes before making any changes])
- ensuring that poisonous and toxic chemicals are properly labeled
- using pesticides in accordance with the Federal Insecticide, Fungicide, and Rodenticide Act (for example, maintaining rodent bait that is in use in covered, tamper-resistant bait stations)
- knowing what poisonous and toxic chemicals should be on the premises and keeping track of them
- investigating missing stock or other irregularities outside a normal range of variation and alerting appropriate law enforcement and public health agencies about unresolved problems, when appropriate

Operations

FDA recommends that cosmetics establishment operators consider:

- Incoming materials and contract operations
    - using only known, appropriately licensed or permitted (where applicable) contract manufacturing and packaging operators and sources for all incoming materials, including ingredients, compressed gas, packaging, labels and materials for research and development
    - taking steps to ensure that suppliers, contract operators and transporters practice appropriate cosmetics security measures (for example, auditing, where practical, for compliance with cosmetics security measures that are contained in purchase and shipping contracts or letters of credit, or using a vendor approval program)
    - authenticating labeling and packaging configuration and product coding/expiration dating systems (where applicable) for incoming materials in advance of receipt of shipment, especially for new products
    - requesting locked and/or sealed vehicles/containers/railcars, and, if sealed, obtaining the seal number from the supplier and verifying upon receipt, making arrangements to maintain the chain of custody when a seal is broken for inspection by a governmental agency or as a result of multiple deliveries

  - requesting that the transporter have the capability to verify the location of the load at any time, when practical
  - establishing delivery schedules, not accepting unexplained, unscheduled deliveries or drivers, and investigating delayed or missed shipments
  - supervising off-loading of incoming materials, including off hour deliveries
  - reconciling the product and amount received with the product and amount ordered and the product and amount listed on the invoice and shipping documents, taking into account any sampling performed prior to receipt
  - investigating shipping documents with suspicious alterations
  - inspecting incoming materials, including ingredients, compressed gas, packaging, labels, product returns and materials for research and development for signs of tampering, contamination or damage (for example, abnormal powders, liquids, stains, or odors, evidence of resealing, compromised tamper-evident packaging) or "counterfeiting" (for example, inappropriate or mismatched product identity, labeling, product lot coding or specifications, absence of tamper-evident packaging when the label contains a tamper-evident notice), when appropriate
  - evaluating the utility of testing incoming ingredients, compressed gas, packaging, labels, product returns and materials for research and development for detecting tampering or other malicious, criminal, or terrorist action
  - rejecting suspect cosmetics or cosmetics ingredients
  - alerting appropriate law enforcement and public health authorities about evidence of tampering, "counterfeiting" or other malicious, criminal, or terrorist action
- Storage
  - having a system for receiving, storing and handling distressed, damaged, returned, and rework products that minimizes their potential for being compromised or to compromise the security of other products (for example, destroying products that are unfit for use, products with illegible codes, products of questionable origin, and products returned by consumers to retail stores)
  - keeping track of incoming materials and materials in use, including ingredients, compressed gas, packaging, labels, salvage products, rework products, and product returns
  - investigating missing or extra stock or other irregularities outside a normal range of variability and reporting unresolved problems

  to local appropriate enforcement and public health authorities, when appropriate
  - storing product labels in a secure location and destroying outdated or discarded product labels
  - minimizing reuse of containers, shipping packages, cartons, etc., where practical
- Security of water and utilities
  - securing, to the extent practical, access to controls for airflow, water, electricity, and refrigeration
  - securing non-municipal water wells, hydrants, storage and handling facilities
  - ensuring that water systems and trucks are equipped with backflow prevention
  - chlorinating water systems and monitoring chlorination equipment, where practical, and especially for non-municipal water systems
  - testing non-municipal sources for potability regularly, as well as randomly, and being alert to changes in the profile of the results
  - staying attentive to the potential for media alerts about public water provider problems, when applicable
  - identifying alternate sources of potable water for use during emergency situations where normal water systems have been compromised (for example, trucking from an approved source, treating on-site or maintaining on-site storage)
- Finished products
  - ensuring that contract warehousing and shipping operations (vehicles and vessels) practice appropriate security measures (for example, auditing, where practical, for compliance with cosmetics security measures that are contained in contracts or letters of guarantee)
  - performing random inspection of storage facilities, vehicles, and vessels
  - evaluating the utility of finished product testing for detecting tampering or other malicious, criminal, or terrorist action
  - requesting locked and/or sealed vehicles/containers/railcars and providing the seal number to the requesting that the transporter have the capability to verify the location of the load at any time
  - establishing scheduled pickups, and not accepting unexplained, unscheduled pickups
  - keeping track of finished products
  - investigating missing or extra stock or other irregularities outside a normal range of variation and alerting appropriate law enforcement and public health authorities about unresolved problems, when appropriate

  - advising sales staff to be on the lookout for counterfeit products and to alert management if any problems are detected
- Mail/packages
  - implementing procedures to ensure the security of incoming mail and packages (for example, locating the mailroom away from cosmetics manufacturing and storage areas, securing mailroom, visual or x-ray mail/package screening, following U.S. Postal Service guidance)
- Access to computer systems
  - restricting access to computer process control systems and critical data systems to those with appropriate clearance (for example, using passwords, firewalls)
  - eliminating computer access when a staff member is no longer associated with the establishment
  - establishing a system of traceability of computer transactions
  - reviewing the adequacy of virus protection and procedures for backing up critical computer based data systems
  - validating the computer security system

Emergency Point of Contact

U.S. Food and Drug Administration
5600 Fishers Lane
Rockville, MD 20857

If a cosmetics establishment operator suspects that any of his/her products that are regulated by the FDA have been subject to tampering, "counterfeiting," or other malicious, criminal, or terrorist action, FDA recommends that he/she notify the FDA 24-hour emergency number at 301-443-1240 or call their local FDA District Office. FDA District Office telephone numbers are listed at http://www.fda.gov/ora/inspect_ref/iom/iomoradir.html. FDA recommends that the operator also notify local law enforcement and public health agencies.

Appendix K

United States
Department of
Agriculture
USDA
Economic
Research
Service
Agricultural
Economic
Report
Number 830
Traceability in the U.S. Food Supply: Economic Theory and Industry Studies
Elise Golan, Barry Krissoff, Fred Kuchler, Linda Calvin,
Kenneth Nelson, and Gregory Price

Visit Our Website To Learn More!

Want to learn more about traceability? Visit our website at **www.ers.usda.gov.**

You can also find additional information, both paper and electronic, about ERS publications, databases, and other products at our website.

National Agricultural Library Cataloging Record:

Traceability in the U.S. food supply: economic theory and industry studies.
(Agricultural economic report ; no. 830)
1. Food supply--United States.
2. Total quality control--United States.
3. Food law and legislation--United States.
4. Food industry and trade--United States--
Quality control.
I. Golan, Elise H. II. United States. Dept. of Agriculture.
Economic Research Service.
III. Title.
HD9005

Traceability in the U.S. Food Supply:
Economic Theory and Industry Studies

Elise Golan, Barry Krissoff, Fred Kuchler,
Linda Calvin, Kenneth Nelson, and Gregory Price

I. Introduction and Methodology

Traceability systems are recordkeeping systems designed to track the flow of product or product attributes through the production process or supply chain. Recently, policymakers have begun weighing the usefulness of making such systems mandatory so as to address issues ranging from food safety and bioterrorism to consumers' right to know. For example, policymakers in many countries have proposed or adopted mandatory systems to track animal feed to control the risk of mad cow disease and to improve meat safety. Other proposals involve mandatory tracking of food transportation systems to reduce the risk of tampering. Numerous proposals involve mandating traceability to help provide consumers with information on a variety of food attributes including country of origin, animal welfare, and genetic engineering.

Food producers, manufacturers, and retailers have many of the same concerns as government policymakers and in fact already keep traceability records for a wide range of foods and food attributes. The questions before policymakers are, does the private sector provide enough traceability to meet social objectives? If not, what policy tools are best targeted to increasing the supply of traceability?

The objective of this study is to provide a framework to answer those questions. To do that, we first needed an accurate description of the extent and type of traceability maintained by private firms, that is, the traceability baseline. We could not begin to assess the adequacy of private sector traceability systems without a clear understanding of how typical it is for firms to establish these systems, why they establish them, and how they function. We began our investigation by reviewing market studies, interviewing government officials, and talking with industry associations. Next, we conducted telephone interviews with a wide range of food industry representatives, including grain and food processors, fast-food retailers, safety auditors, and food distributors. We conducted several site visits in each of the three major food sectors: fresh produce, grains, and livestock. During these visits, we interviewed owners, plant supervisors, and/or quality control managers in fruit and vegetable packing and processing plants, beef slaughter plants, grain elevators, mills and food manufacturing plants, and food distribution centers.

In each interview, we asked about the company's traceability system, including its bookkeeping records, lot or batch sizing, computer use, and tracking technologies. We asked about the cost of the traceability system and about how long it had been in use. We received a high level of voluntary cooperation from these firms, sometimes getting a tour of their facilities. However, our discussions were informal and we generally did not review firms' records to confirm the information provided. Our discussions were often broad based about the firm's recordkeeping systems and we did not systematically collect specific data about a firm's traceability system.

A number of our site interviews were with firms that are eligible to submit bids for U.S. procurement programs. We received access to these firms by accompanying USDA auditors on their inspections to ensure that the firms were complying with procurement regulations and guidelines. We asked the firms' managers whether they thought the firms' traceability systems were typical for their industries. While most indicated that their systems were characteristic for their industry, some pointed out their innovative and state-of-the-art approaches to traceability. Our site-visit sample, thus, may be skewed to firms that are at least average or better in their use of good manufacturing practices, although we are confident that our conclusions hold for the majority of firms in each sector.

Our investigations led us to conclude that 1) traceability is an objective-specific concept; 2) the private sector in the United States has developed a significant capacity to trace; and 3) industry/product characteristics lead to systematic variation in traceability systems. We found that efficient traceability systems vary across industries and over time as firms balance costs and benefits to determine the efficient breadth, depth, and precision of their traceability systems. We examine the evidence leading to these conclusions in the second section of the report, where we look at the factors that influence the costs and benefits of traceability. The three chapters in Section III provide further elaboration of these conclusions by describing in detail the supply chain and traceability systems characterizing the fresh produce, grains and oilseeds, and cattle/beef sectors.

While private sector traceability systems are extensive, gaps may nevertheless exist. Some gaps are the result of an efficient balancing of traceability costs and benefits. Others, however, are the result of market failures and may warrant government intervention. To examine the possibility that market failure has resulted in gaps in the supply of traceability, we qualitatively analyzed and compared social and private costs and benefits of traceability. We found that asymmetric information problems have the potential to dampen firms' supply of traceability for food safety and for product differentiation. Section IV contains our analysis of market failure in the provision of food traceability and our investigation into the types of government policy tools that may correct market failure and encourage the development of private traceability systems. We also consider the characteristics of a government-mandated traceability system that would most efficiently mesh with private systems. The appendix to this section lists selected mandatory traceability laws in the United States. In section V, we provide some concluding thoughts.

II. Efficient Traceability Systems Vary

The ISO 9000:2000 guidelines define traceability as the "ability to trace the history, application or location of that which is under consideration" (ISO, 2000).[1] The ISO guidelines further specify that traceability may refer to the origin of the materials and parts, the processing history, and the distribution and location of the product after delivery.

This definition of traceability is quite broad. It does not specify a standard measurement for "that which is under consideration" (a grain of wheat or a truckload), a standard location size (field, farm, or county), a list of processes that must be identified (pesticide applications or animal welfare), where the information is recorded (paper or electronic record, box, container or product itself), or a bookkeeping technology (pen and paper or computer). It does not specify that a hamburger be traceable to the cow or that the wheat in a loaf of bread be traceable to the field. It does not specify which type of system is necessary for identity preservation of tofu-quality soybeans, for quality control of cereal grains, or for guaranteeing correct payments to farmers for different grades of apples.

Complete Traceability is Impossible

The definition of traceability is necessarily broad because traceability is a tool for achieving a number of different objectives. No single approach is adequate for every objective. Even a hypothetical system for tracking beef, in which consumers scan their packet of beef at the check-out counter and receive information on the date and location of the animal's birth, lineage, vaccination records, acreage of pasturage, and use of mammalian protein supplements, is incomplete. It does not provide traceability with respect to pest control in the barn (a potential food safety issue), use of genetically engineered feed, or animal welfare attributes like pasturage hours and playtime. There are hundreds of inputs and processes in the production of beef. A system for tracking each and every input and process with a degree of precision adequate for every objective would be virtually impossible.

The characteristics of good traceability systems vary and cannot be defined without reference to the system's objectives. Different objectives help drive differences in the *breadth*, *depth*, and *precision* of traceability systems.

[1] ISO is a worldwide federation of national standards bodies which promotes the development of standardization and international standards for a wide range of products. ISO 9000 guidelines are quality management system standards.

Breadth describes the amount of information the traceability system records. There is a lot to know about the food we eat, and a recordkeeping system cataloging all of a food's attributes would be enormous, unnecessary, and expensive. Take for example, a cup of coffee. The beans could come from any number of countries; be grown with numerous pesticides or just a few; grown on huge corporate organic farms or small family-run conventional farms; harvested by children or by machines; stored in hygienic or pest-infested facilities; decaffeinated using a chemical solvent or hot water. A traceability system for one attribute does not require collecting information on other attributes.

The *depth* of a traceability system is how far back or forward the system tracks. In many cases, the depth of a system is largely determined by its breadth: once the firm or regulator has decided which attributes are worth tracking, the depth of the system is essentially determined. For example, a traceability system for decaffeinated coffee would only need to extend back to the processing stage (figure 1). A traceability system for fair trade coffee would only need to extend to information on price and terms of trade between coffee growers and processors. A traceability system for fair wage would only need to extend to harvest; for shade grown, to cultivation; and for non-genetically engineered (GE), to the bean or seed. In other cases, the depth of the system is determined by quality or safety control points along the supply chain. In these cases, traceability systems may only need to extend back to the last control point, that is the point where quality or safety was established or verified. For example, a firm's traceability system for pathogen control may only need to extend to the last "kill" step—where product was treated, cooked, or irradiated.

Precision reflects the degree of assurance with which the tracing system can pinpoint a particular food product's movement or characteristics. Precision is determined by the unit of analysis used in the system and the acceptable error rate. The unit of analysis, whether container, truck, crate, day of production, shift, or any other unit, is the tracking unit for the traceability system. Systems that have large tracking units, such as an entire feedlot or grain silo, will have poor precision in isolating safety or quality problems. Systems with smaller units, such as individual cows, will have greater precision. Likewise, systems with low acceptable error rates, such as low tolerances for GE kernels in a shipment of conventional corn, are more precise than systems with high acceptable

Figure 1
The depth of a traceability system depends on the attributes of interest

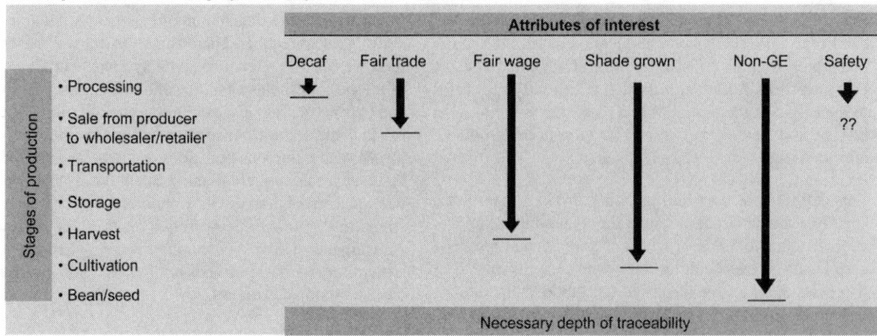

error rates. In some cases, the objectives of the system will dictate a precise system while for other objectives a less precise system will suffice.

The breadth, depth, and precision of private traceability systems will vary depending on the objectives of the systems and the corresponding benefits and costs to the firm. Though at first glance this variability may appear to indicate deficiencies in the private supply of traceability, it is actually an indication of efficiency. Firms collect information on an attribute and track its flow through the supply chain only if the net benefits (benefits minus costs) of doing so are positive. Likewise, they invest in precision only if the benefits outweigh the costs. Because firms balance the costs and benefits of traceability, they tend to efficiently allocate resources to building and maintaining these systems.

Firms Consider a Wide Range of Costs and Benefits

Traceability systems that yield positive net benefits to the firm are a worthwhile investment; those yielding negative net benefits are not worthwhile to the firm. Below, we examine the range of benefits and costs that firms consider when determining the efficient breadth, depth, and precision for their traceability systems.

Benefits of Traceability

Firms have three primary objectives in developing, implementing, and maintaining traceability systems: to improve supply management; to facilitate traceback for food safety and quality; and to differentiate and market foods with subtle or undetectable quality attributes. The benefits associated with these objectives range from lower-cost distribution systems, reduced recall expenses, and expanded sales of products with attributes that are difficult to discern. In every case, the benefits of traceability translate into larger net revenues for the firm. Firms establish traceability systems to achieve one or more traceability objectives—and to reap the benefits. These benefits are driving the widespread development of traceability systems across the food supply chain.

Objective/Benefits I: Traceability for Supply Management During 2000, American companies spent $1.6 trillion on supply-related activities, including the movement, storage, and control of products across the supply chain (State of Logistics Report, 2001). The ability to reduce these costs often marks the difference between successful and failed firms. In the food industry, where margins are thin, supply management is an increasingly important area of competition.

An indispensable element of any supply management strategy is the collection of information on each product from production to delivery or point of sale. The idea is "to have an information trail that follows the product's physical trail" (Simchi-Levi, 2003, pg. 267). Information trails, or in other words, traceability systems, provide the basis for good supply management. A business's traceability system is key to finding the most efficient ways to produce, assemble, warehouse, and distribute products. The benefits of traceability systems for supply management are greater the higher the value of coordination along the supply chain.

Electronic systems for tracking inventory, purchases, production, and sales have become an integral part of doing business in the United States. A few big retailers such as Wal-Mart and Target have even created proprietary supply-chain information systems that they require their suppliers to adopt. In addition to private systems, U.S. firms may also use industry-standard coding systems, such as UPC codes (see box, "From UPC to RSS: Tracking Technologies Drive Down the Costs of Precision"). These systems are not confined to packaged products. The food industry has developed a number of complex coding systems to track the flow of raw agricultural inputs to the products on grocery store shelves. These systems are helping to create a supply management system stretching from the farm to the retailer.

Evidence that American companies are embracing new sophisticated tracking systems can also be found in macro-economic statistics. The success of traceability systems in helping to control inventory costs is reflected in national inventory-to-sales ratio statistics. Over short time periods, inventories may rise or fall, but a consistent pattern in which inventories fall relative to a firm's total sales indicates that the firm is getting better at keeping track of its inputs and outputs and it is taking advantage of that knowledge. Figure 2, showing the ratio of private inventories to final sales of domestic business, displays a declining time trend, falling by half since the end of WWII (U.S. Department of Commerce, 2003).

The same trend can be observed in many sectors of the domestic food industry. Figure 3 shows the ratio of end-of-year inventories to total value of shipments for proxies for the dairy, grain, and sugar industries (Bartlesman, Becker, and Gray, 2000). In every case, the inventory-to-sales ratio fell, with the largest decline in the cereal sector, where the ratio fell from over 8 percent to approximately 3 percent. The downward trend in inventories in major components of the food industry reflects growing efficiencies in supply management, including traceability systems.

Across the economy, firms are adopting systems to more efficiently manage resources. In many cases, new tracking/information systems are at the heart of these efforts. The depth, breadth, and precision of these systems vary across industries and firms, mirroring the distribution requirements of the enterprise.

Objective/Benefits 2: Traceability for Food Safety and Quality Control Product-tracing systems are essential for food safety and quality control. Traceability systems help firms isolate the source and extent of safety or quality-control problems. The more precise the tracing system, the faster a producer can identify and resolve food safety or quality problems. Firms have an incentive to invest in traceability systems because they help minimize the production and distribution of unsafe or poor quality products, which in turn minimizes the potential for bad publicity, liability, and recalls.

Traceability systems can help track product distribution and target recall activities, thereby limiting the extent of damage and liability. Most, if not all, voluntary recalls listed on USDA's Food Safety and Inspection Service website refer consumers to coded information on products' packaging to identify the recalled items. The advent of grocery store or club cards to track sales enhances the potential for targeted recall information. Grocery stores could use their sales data to identify and then warn buyers of recalled products. Some have

Figure 2

Ratio of private inventories to final sales of domestic business--seasonally adjusted, 1946(4)-2003(1)

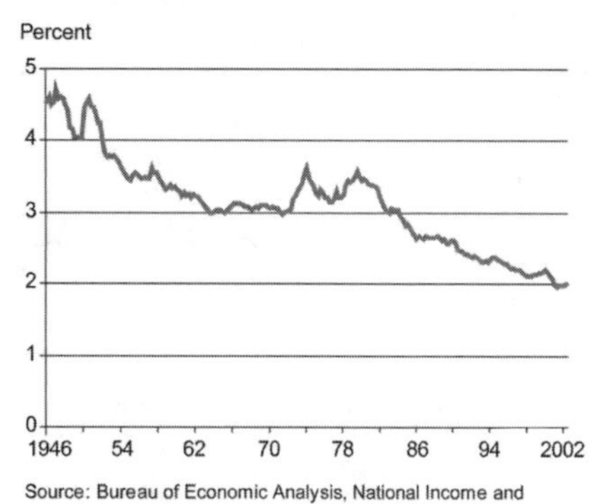

Source: Bureau of Economic Analysis, National Income and Product Accounts.

Figure 3

Ratio of inventories to total value of shipments for selected food industries, 1958-1996

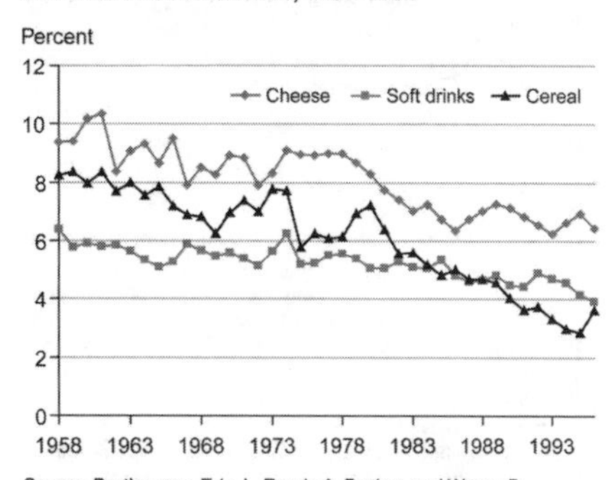

Source: Bartlesman, Eric J., Randy A. Becker, and Wayne B. Gray. "NBER-CES Manufacturing Industry Database." www.nber.org/nberces/nbprod96.htm. Inventories are measured at end of year.

From UPC to RSS: Tracking Technologies Drive Down the Cost of Precision

Ever since a 10-pack of Wrigley's Juicyfruit gum was scanned at the checkout counter in 1974, bar codes have become ubiquitous in the U.S. grocery stores. Almost everything we buy has packaging with printed bar codes. In the food industry, the vast majority of packaged products bear bar codes, as do a growing number of bulk foods, like bagged apples and oranges.

The Uniform Code Council (UCC), a non-profit private company that establishes and promotes multi-industry standards for product identification, created bar codes in response to the needs of food manufacturers and retailers who were interested in speeding the checkout process at the grocery store and for improving inventory management (Uniform Code Council, 2003). Bar codes contain a series of numbers reflecting type of product and manufacturer (the UPC 12-digit code), and a series of numbers assigned by the manufacturer to nonstandard production or distribution details. Each product, including those with different size packaging, contains a unique UPC code. When a package is scanned under a laser beam at the checkout counter, the store's central computer reads the UPC number, records the sale, and marks the change in inventory. Recently, UCC has developed an extension of UPC codes to 14 digits called the Global Trade Item Numbers (GTIN) system, which contains expanded information about companies, products, and product attributes worldwide (Global Trade Item Numbers Implementation Guide, 2003).

The success of the original UPC system has combined with technological advances and e-marketing to spur the development of integrated systems that code, track, and manage wholesale and retail transactions within the United States and in the global community. In some cases, buyers manage these systems to monitor supply flow. In other cases, firms establish systems to link suppliers and buyers. For example, EAN.UCC, which is a subsidiary of the UCC and EAN International, a European commercial standard setting organization, has developed an open integrated system to standardize and automate information systems across a supply chain that includes GTINs, along with an industry standard set of 62 product attributes (EAN.UCC, 2003). With an integrated system, the process of entering information into retailers' systems is automated so when new information is logged into the system by the producer, it's added in real time to all systems across a network. With such systems, anyone along the chain can track inputs, production, and inventory by an array of characteristics.

New technologies are spurring the development of even more precise systems. One example of an upcoming technology is the expansion of bar codes to reduced space symbology (RSS) (Rowe, 2001). Currently, stickers with 4-digit price look-up codes on fresh produce identify the product and assist the retailer in inventory management. With RSS, 14-digit GTIN bar codes could be attached to individual items. An apple, a box, and a pallet could all be linked by the same product and grower-shipper codes, with an additional numeric indicating "item," "box," or pallet." Similarly a package of ground beef could be linked to a packinghouse. Other bar code application identifiers and numbers could be used as well, including price, weight, sell date, and lot. Having an electronic lot number on a package of ground beef would facilitate a traceback in case of a quality or safety concern. Moreover, customers who purchase specific foods using frequent shopper cards can be quickly identified even if they have discarded the food package. Thus, tracing forward or backward to facilitate supply chain management or quality and safety control would be more easily and swiftly accomplished

Bar codes have a few disadvantages (Brain, 2003). In order to keep up with inventories, companies must scan the bar code under a laser beam. A more proactive technology would allow a reader to scan a smart label—a computer chip embedded in each product's package, box, or pallet—whether the item remains on the shelf (in the front part of the shelf or hidden in the back) or is sold. For even more efficiency in retail store management, the store could have a "smart setup." In these stores, a consumer could carry out their shopping and exit the store without going to a checkout counter. Instead, a radio frequency identification (RFID) reader embedded in an exit door could read the smart tags simultaneously for each food package. Even detailed attributes could be read such as a 1-quart container of non-fat organic milk with a sale date of January 14, 2004. Inventories on each product with all its unique attributes even including "must sell by date" could be efficiently traced and managed at manufacturing locations, warehouses, distribution centers and grocery stores.

Furthermore, computers at the grocery store and its suppliers' facilities would know automatically which items had been purchased and needed to be replenished. The computers would also be able to automatically notify the consumer's bank of charges and debit the consumer's account.

While this scenario sounds like it may be far in the future, RFID technology is not new and currently is used to track livestock and container cargo on trucks and ships. With RFID tags, ranchers can determine the location and movements of cattle and more quickly round up any particular heifer or steer. With RFID tags, a distributor can determine precisely the location of a cargo ship or truck and the condition of produce in a controlled-atmosphere container. In July 2003, Wal-Mart issued a mandate to its top suppliers requiring the use of RFID tags on pallets and cases by the end of 2004 (Dunn, 2003). As the cost of RFID technology falls, it is possible that, several years from now, we may see RFID tags on many individual food items.

UCC and EAN International are facilitating the use of RFID technology with the establishment of standardized Electronic Product Codes (EPC) and an EPC network (EPCglobal, 2003). Unlike other electronic networks that are proprietary, these will be open to any firm. Already Wal-Mart is requiring its top suppliers to be EPC-compliant. With the use of electronics and widely accepted standards, the number of attributes that can be traced for each food product is nearly limitless.

already done that. For example, during the recent mad cow beef recall, one supermarket chain used its preferred customer cards to identify and warn shoppers who had bought the suspect meat (Anderson, 2004). Likewise, credit card information could be used to track purchases of contaminated foods. In fact, the Food and Drug Administration (FDA) has used credit card information in its traceback investigations.

The benefits of precise traceability for food safety and quality control are greater the higher the likelihood and cost of safety or quality failures. Where the likelihood and cost of failure are high, manufacturers have large financial incentives to reduce the size of the standard recall lot and to adopt a more precise traceability system. The likelihood of failure differs among food industries because some foods are more perishable or more susceptible to contamination than others. The costs of safety or quality breaches also vary among firms because the value of products and the value of firms' reputations vary. For high-value products, recall costs per item are higher than for low-value products. For firms with valuable reputations, the costs of recall or safety breaches are higher than for firms with little name-brand equity. The costs of safety or quality failures may also be larger in industries where government or consumer-group oversight is more stringent, meaning that the likelihood of detection in the case of a food safety problem is greater.

The benefits of traceability are also likely to be high if other options for safety control are few. If a firm can eliminate safety problems with a simple kill step or through inexpensive testing, then the marginal benefits of a traceability system for monitoring safety are likely to be small. For example, if a firm could use a chemical dip on incoming produce that completely eliminated the risk of pathogen contamination, there would be little value in a traceability system to identify producers of product with high levels of pathogen contamination. Likewise, if safety or quality problems are unlikely to arise in a specific stretch of the production or supply chain, there is little value in establishing traceability systems for that stretch.

Another benefit of traceability systems is that they may help firms establish the extent of their liability in cases of food safety failure and potentially shift liability to others in the supply chain. If a firm can produce documentation to establish that safety failure did not occur in its plant, then it may be able to protect itself from liability or other negative consequences. Traceability systems in themselves do not determine liability, but because they provide information about the production process, including safety procedures, they have a role in providing evidence of negligence or improper production practices.

Despite the important safety role they play, traceability systems are, however, only one element of a firm's overall safety/quality control system and are designed to complement and reinforce the other elements of the safety/quality system. In themselves, traceability systems do not produce safer or high-quality products—or determine liability. Traceability systems provide information about whether control points in the production or supply chain are operating correctly or not. The breadth, depth, and precision of traceability systems for safety and quality necessarily reflect the control points in the overall safety/quality system and vary systematically across industries and over time depending on safety and quality technologies and innovations.

Objective/Benefits 3: Traceability To Differentiate and Market Foods with Credence Attributes The U.S. food industry is a powerhouse producer of homogenous bulk commodities such as wheat, corn, soybeans, and meats. Increasingly, the industry has also begun producing goods and services tailored to the tastes and preferences of various segments of the consumer population. In the competition over micromarkets, producers try to differentiate one product from otherwise similar products in ways that matter to customers.

Food producers differentiate products over a wide variety of quality attributes including taste, texture, nutritional content, cultivation techniques, and origin. Consumers can easily detect some attributes—green ketchup is hard to miss. However, other innovations involve credence attributes, characteristics that consumers cannot discern even after consuming the product (Darby and Karni, 1973). Consumers cannot, for example, taste or otherwise distinguish between oil made from GE corn and oil made from conventional corn.

Credence attributes can be content or process:

> ***Content attributes*** affect the physical properties of a product, although they can be difficult for consumers to perceive. For example, consumers are unable to determine the amount of isoflavones in a glass of soymilk or the amount of calcium in a glass of enriched orange juice by drinking these beverages.
>
> ***Process attributes*** do not affect final product content but refer to characteristics of the production process. Process attributes include country-of-origin, free-range, dolphin-safe, shade-grown, earth-friendly, and fair trade. In general, neither consumers nor special-

ized testing equipment can detect process attributes.

Traceability is an indispensable part of any market for process credence attributes—or content attributes that are difficult or costly to measure. The only way to verify the existence of these attributes is through a bookkeeping record that establishes their creation and preservation. For example, tuna caught with dolphin-safe nets can be distinguished from tuna caught using other methods only through the bookkeeping system that ties the dolphin-safe tuna to the observer on the boat from which the tuna was caught. No test conducted on a can of tuna could detect whether the tuna was caught using dolphin-safe technologies. Without traceability as evidence of value, no viable market could exist for dolphin-safe tuna, fair-trade coffee, non-GE corn oil, or any other process credence attribute.

The benefits of traceability (and third-party verification) for credence attributes are greater the more valuable the attribute is to processors or final consumers. Attributes tend to be more valuable the more marketable they are, the higher the expected premiums, and the larger the potential market. Firms will only find it worthwhile to establish traceability to market attributes with the potential to generate additional revenue—and the larger the potential revenue, the greater the benefits of traceability.

Costs of Traceability

Traceability costs include the costs of recordkeeping and product differentiation. Recordkeeping costs are those incurred in the collection and maintenance of information on product attributes as they move through production and distribution channels. In some cases, the recordkeeping system necessary for traceability is very similar to that already maintained by the firm for accounting or other purposes. For example, in the United States, most firms keep records of their receipts and bills. For these firms, one-up, one-down traceability for a standard set of attributes would require little if any change in the firm's accounting system. In other instances, new traceability objectives may require expensive additions to existing recordkeeping systems.

Product differentiation costs are those incurred in keeping products or sets of product attributes separate from one another for tracking purposes. Product differentiation for tracking is primarily achieved by breaking product flow into lots or any other discrete unit defined over a set of common processes or content attributes (see box, "What's a Lot?"). When traceability requirements accommodate production-based lot sizes such as the amount of production from one shift or the product from one field, traceability differentiation costs are minimal. Likewise, when new traceability objectives accommodate differentiation systems that are already in place for other traceability objectives, the costs of the new traceability systems will be relatively small.

When traceability differentiation requires firms to adopt different or additional criteria for product differentiation, firms could incur large costs—at least in the short run. Such a situation may arise when firms instigate traceability for new credence attributes. For example, the desire to distinguish GE corn from conventional corn has prompted a number of growers and processors to establish new systems to identify and keep the two types of corn separate.

The longrun cost of separating products with different attributes depends on a number of factors, including underlying production technologies and the level of demand. In some cases, a change or addition to existing production lines is the low-cost solution to meet demand. For example, a packer-shipper may determine that installing scanner equipment on conveyer belts to separate fruit by color or size is the most efficient technology. In other cases, firms may choose to differentiate production by establishing separate product lines within the same plant or by sequencing production and thoroughly cleaning production facilities between differentiated product batches. A packer-shipper could run lines at separate times for conventional and organic produce or build separate lines for each attribute. Firms facing large demand may dedicate a whole plant or distribution channel to the production or distribution of one specific product line. Average costs increase when the separation of product lines creates unused capacity, such as underutilized trucks and storage facilities, or requires stopping, cleaning, and restarting production lines. If demand for the differentiated products is sufficient, however, the firm may realize economies of scope and increased net profits.

The level of precision also affects the type and cost of product differentiation. Systems requiring a high degree of accuracy also tend to require stringent systems for separating crops or products. There are two primary approaches for separating attributes:

- A *segregation* system separates one crop or batch of food ingredients from others. Though segregation implies that specific crops and products are kept apart, segregation systems do not typically entail a high level of precision. In the United States, white corn is chan-

What's A Lot?

Product differentiation for tracking is achieved by breaking product flow into lots, or any other discrete unit defined over a set of common process or content attributes. Lots are the smallest quantity for which firms keep records. Firms may choose among an infinite array of unit sizes, shapes, or time, defining their own lot size by the quantity of product that fits in a container, that a forklift can move on a pallet, or that fills a truck. A lot may be an individual animal or group of animals, or production from an entire day or shift. Firms that choose a large lot size for tracking purposes, such as a feedlot or grain silo, will have more difficulty isolating safety or quality problems than a firm that chooses a smaller size. A smaller lot size, such as an individual cow or container, will allow greater precision.

In choosing lot size, firms typically consider a number of factors, including accounting procedures, production technologies, and transportation. As these factors vary within and among industries, lot size varies from plant to plant. There is no standard traceability unit. Furthermore, a firm is likely to have a different lot size for incoming and outgoing products. Firms add value in their production and marketing practices by commingling, transforming, and processing products. Clearly the incoming products for a meat processor and slaughterhouse (for example, group of pigs) differ from the outgoing product (boxes of primal cuts and consumer-ready products). The size and shape of a lot is therefore likely to change at each processing juncture. Some firms may find it efficient to maintain depth of traceability by linking incoming and outgoing lots, while others may not.

Consider two examples. An apple packer-shipper may use accounting procedures to choose the incoming lot size. The shipper may receive apples from a number of growers and must pay each grower based on the type, size, and grade of the product. Since these attributes are known only after the apples have been sorted, each grower's apples need to be kept separate in the packing line. These accounting procedures thus influence the lot size of product entering the packinghouse. As apples are sorted, packed, and shipped, a packinghouse may choose to make a lot the number of boxes that can be loaded onto a truck. One or several growers' apples could be loaded together—it is most cost-effective to fully pack a truck. There may be food safety and quality concerns that motivate a shipper to keep a lot size no larger than a truckload. In the case of a food safety problem—for example, a piece of metal or glass found in an apple—the shipper may want to limit the size of a recall and limit the number of affected growers.

The lot for a farrow-to-finish operation, a farm where pigs are born, raised, and prepared for slaughter, might be a batch or group of pigs. When the batch is moved from one stage of production to another, the all-in, all-out production system allows for cleaning the facilities between batches. This method meets the farmer's objective of preventing disease from spreading from one batch of pigs to another (Hayes and Meyer, 2003). Commingling batches of pigs raises the potential for disease. The slaughterhouse may process several batches of pigs in a shift or day, packing the outgoing product—various cuts of pork—in boxes. Each box may specify the name and address of the packer, the lot number, and place and time of production, allowing the firm to track similar products. This reflects the packer's objective of efficiently managing large volumes of meat and concern for food safety. If the packer or Federal or State authorities discover that there is contaminated pork, they can identify product by lot number and inform retailers and/or consumers.

neled through the bulk commodity infrastructure, but it is segregated from other types of corn.

- An *identity preservation* (IP) system identifies the source and/or nature of the crop or batch of food ingredients. IP systems are stricter than segregation systems and often require containerization or other physical barriers to guarantee that certain traits or qualities are maintained throughout the food supply chain. Tofu-quality soybeans are put into containers to preserve their identity. Produce treated to meet phytosanitary requirements of foreign countries is segregated by box to preserve its identity.

The distinction between "IP" and "segregation" is often blurred and a "strict segregation" system may be more precise than a loose IP system. Regardless of the exact terminology, precise systems requiring that products be strictly separated will likely be more expensive than others because such systems are usually more expensive to develop and maintain than loose systems.

The level of precision of the traceability system may also influence recordkeeping costs. Recordkeeping expenses tend to rise with smaller lot sizes. Five tons of production broken into 5 one-ton lots require less paperwork than the same quantity broken into 1,000 ten-pound lots. In addition, the bookkeeping records required to maintain a highly accurate traceability system tend to require more detail and expense than those for less exacting systems. For example, a traceability system for stringent pathogen control will require more sampling, testing, and verification paperwork than a system designed for less stringent control.

Both recordkeeping and differentiation expenses tend to rise with the complexity of the production and distribution systems. Products that undergo a large number of transformations on their way to market generate a lot of new information and are typically more difficult to track than products with little processing. Food products vary considerably with respect to the number of handlers and manufacturers and the degree of commingling and processing. Lettuce picked in the field and sold directly to retailers is relatively easy to track. Tracking a chicken potpie is more challenging. The process of transforming the wheat to wheat flour, the chicken and the vegetables to bite size pieces, and combining all the raw ingredients into a pie generates a trail of numerous different lots that themselves are composed of commingled lots.

Products that are bought and sold numerous times also tend to generate higher bookkeeping and differentiation expenses than those that remain within the same company. Any time product is passed from one firm to another, new paperwork is generated as firms link receipts with product and reconcile or adjust lot numbers and sizes. New coding and software technologies are helping to drive down the costs of linking supply-management records across the food chain and of coordinating the flow of product along the chain. In many sectors of the food supply chain, new information technologies are helping push down the cost of recordkeeping and stimulating investment in traceability systems. As mentioned before, electronic systems for tracking inventory, purchases, production, and sales are becoming an integral part of doing business in the United States.

Vertical integration and contracting are other methods for reducing the costs of tracing and supply management. Vertically integrated firms and firms that contract along the supply chain for specific attributes are often better able to coordinate production, transportation, processing, and marketing. They are able to respond to consumer preferences for select quality attributes and provide consistency of product. Vertically integrated firms can also adopt the same recordkeeping system across the chain to streamline product coordination. Thus, these food suppliers can attain value and limit the cost of traceability systems.

Benefits and Costs Vary Across Industries and Time

The development of traceability systems throughout the food supply system reflects a dynamic balancing of benefits and costs. Though many firms operate traceability systems for supply management, quality control, and product differentiation, these objectives have played different roles in driving the development of traceability systems in different sectors of the food supply system. In some sectors, food scares have been the primary motivation pushing firms to establish traceability systems; in others, the growth in demand for high-value attributes has pushed firms to differentiate and track attributes; in yet other sectors, supply management has been the key driving force in the creation of traceability systems. Different types and levels of costs, reflecting differences in industry organization, production processes, and distribution and accounting systems affect traceability adoption.

The dynamic interplay of objectives, benefits, and costs has spurred different rates of investment in breadth, depth, and precision of traceability across sectors—and continues to do so. Table 1 summarizes key factors affecting the benefits and costs of traceability systems. These factors vary across industries and across time, reflecting market dynamics, technological advances, and changes in consumer preferences. Changes in the factors influence traceability benefits and costs, thereby influencing the private sector's tracking capabilities.

Table 1—Major factors affecting the costs and benefits of traceability

Factors affecting benefits	Factors affecting costs
■ The higher the value of coordination along the supply chain, the larger the benefits of traceability for supply-side management	■ The wider the breadth of traceability, the more information to record and the higher the costs of traceability
■ The larger the market, the larger the benefits of traceability for supply side management, safety and quality control, and credence attribute marketing	■ The greater the depth and the number of transactions, the higher the costs of traceability
■ The higher the value of the food product, the larger the benefits of traceability for safety and quality control	■ The greater the precision, the smaller and more exacting the tracking units, the higher the costs of traceability
■ The higher the likelihood of safety or quality failures, the larger the benefits of reducing the extent of failure with traceability systems for safety and quality control	■ The greater the degree of product transformation, the more complex the traceability system, the higher the costs of traceability
■ The higher the penalty for safety or quality failures, where penalties include loss of market, legal expenses, or government-mandated fines, the greater the benefits of reducing the extent of safety or quality failures with traceability	■ The larger the number of new segregation or identity preservation activities, the higher the costs of traceability
■ The higher the expected premiums, the larger the benefits of traceability for credence attribute marketing	■ The larger the number of new accounting systems and procedures, the more expensive the start-up costs of traceability
	■ The greater the technological difficulties of tracking, the higher the cost of traceability

III. Industry Studies:

Private-Sector Traceability Systems Balance Private Costs and Benefits

In this section, we examine the development of traceability systems in three food sectors in the United States: fresh produce, grains and oilseeds, and cattle/beef. We describe the breadth, depth, and precision of each sector's traceability system and examine the influence that varying costs and benefits have had in the development of traceability in these sectors. We find that traceability systems are rapidly developing as traceability benefits increase in value and as technology drives down the cost of creating and managing information. We also find that the dynamic balancing of benefits and costs has led to wide variation in the development of traceability systems in the three food sectors.

Fresh Produce

The development of traceability systems in the fresh produce industry has been greatly influenced by the characteristics of the product. Perishability of and quality variation in fresh fruit and vegetables necessitate the boxing and identification of quality attributes early in the supply chain, either in the field or packinghouse. This has facilitated tracing capability for a number of objectives, including marketing, food safety, supply management, and differentiation of new quality attributes.

The history of traceability in the produce industry dates back to the early part of the 20th century. The development of refrigerated railcars in the late 1800s allowed produce from the West and other distant areas to be shipped to the major eastern population centers. As a result, local spot market produce sales with face-to-face transactions where both buyer and seller could verify the quality at the same time became less common. Instead, transactions over long distances became the norm (Dimitri, 2001). Problems began to arise due to the high perishability and fragility of most produce: produce quality could change substantially in transit. When produce deteriorated, it was not clear where the responsibility lay—the grower, shipper, transportation firm, intermediaries, or buyer. When delivered quality was less than expected, buyers demanded price adjustments. These long-distance transactions also introduced more intermediaries into the marketing chain. Buyers and sellers needed a system to verify quality at various points in the marketing chain and establish their legal rights in the case of a disputed transaction.

In response to these problems, produce growers urged Congress to provide legislation to regulate marketing practices for their industry, and in 1930 Congress passed the Perishable Agricultural Commodities Act (PACA). One part of the Act focused on recordkeeping requirements in produce transactions for shippers selling on behalf of growers—the most common marketing arrangement for fresh produce. The recordkeeping system provides growers with a paper trail to ensure they receive the proper price for their produce. A shipper must assign a lot number, or other positive identification, to all loads received so as to segregate and track produce from different growers from receipt of the product until the first sale. PACA regulations for shipper recordkeeping establish the first link in the fresh produce traceability system at the shipper level.

More recently, the impetus for further developing traceability systems for produce has come from the industry's concerns about food safety. In the event of a foodborne illness outbreak, damage can be limited if the contaminated product can be identified quickly, allowing other noncontaminated product to be marketed. In the mid-1990s a series of well-publicized outbreaks, traced back to microbial contamination of produce, raised public awareness of potential problems. In response, FDA developed voluntary guidelines for good agricultural practices (GAPs) for reducing the potential for microbial contamination of produce. One part of the guidelines focuses on improving traceability. Some retailers now want their produce growers to comply with GAPs and to provide third-party audits to verify compliance. Some farmers voluntarily provide these audits already. Third-party audits reduce the asymmetric information inherent in a transaction where food safety attributes are not obvious. But this new concern requires more traceback information than required by PACA. In a food safety crisis, retailers and the food service industry are concerned about identifying the shipper of a contaminated product but shippers and growers require more precision to uncover the source of the contamination problem and resolve it. Some have begun to track information on exactly where a product comes from—down to a part of a field in some cases.

The costs of establishing and maintaining traceability systems are generally lower for perishable produce than for other commodities because of the way produce is packaged. Most fresh produce is sold in small well-marked containers (generally boxes), as opposed to bulk sales, because much of it is easily damaged and must be protected during shipment. Containers are so small that they generally contain produce from only one grower. Compare this to the nut or dried bean industry, where the products can be stored in silos without damage until they are packed. In these industries, which are not covered by PACA since they are not considered perishable, product from more than one supplier may be mixed together in a silo.

Because produce is packed in boxes, the industry can easily segregate products with different characteristics of concern to buyers. Segregating various types of products has always been important to the produce industry. Unlike grains or meat, fresh produce is a consumer-ready product. Size and appearance matter. For some commodities, variety is also important. For example, a large Washington apple shipper today could be selling over 3,000 distinct apple products that vary by variety, grade, size, packaging, and other characteristics. Segregation is a necessity. The variation in products is also increasing. In 1987, the typical U.S. supermarket carried 173 produce items. By 2001, the number had grown to 350. The well-established ability to segregate and trace fruit and vegetables has allowed the produce industry to adjust relatively easily to new products with different characteristics such as organic or no-pesticide-residue items.

Tracing Produce Through the Marketing Chain

Figure 4 presents a diagram of the marketing chain for produce. In 2002, U.S. growers produced fruit and vegetables (both fresh and processed) worth $24.5 billion (see table 2). In general, growers can market their produce through shippers, sell it directly to consumers at farmers' markets and roadside stands, or sell it to processors. Shippers may sell directly to retailers and the food

Figure 4
Tracing fresh produce through the food marketing system

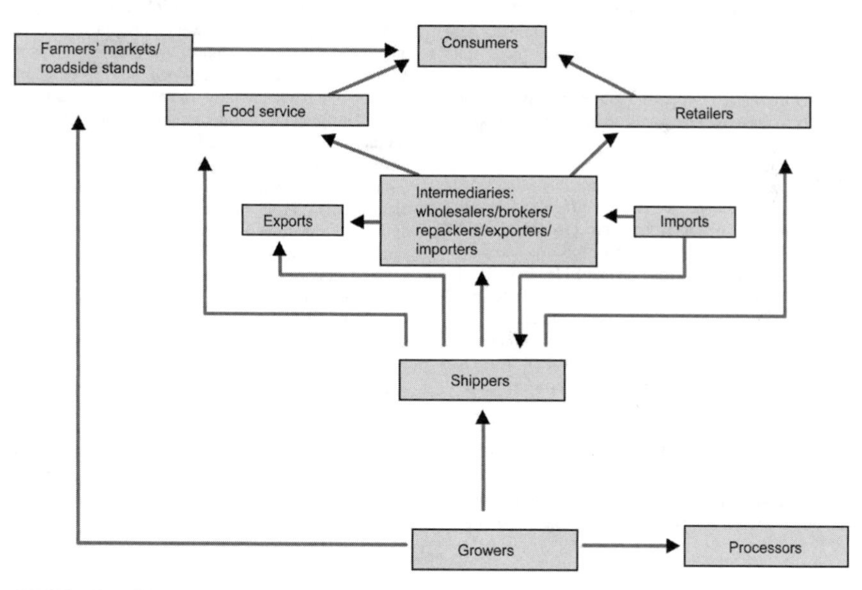

Table 2—U.S. fruit and vegetable industry, 2002

		Production		Imports	Exports
		$ billion	*million tons*	*$ billion*	*$ billion*
Vegetables[1]		13.7	21.4	3.3	1.8
	Fresh		3.0		
	Processed		18.4		
Fruit[2]		10.8	35.6	3.6	2.1
	Fresh		10.9		
	Processed		24.7		

[1] Vegetable trade numbers include fresh and frozen vegetables.
[2] Fruit production numbers contain information for 2002 for the noncitrus industry and the 2001/2002 season for the citrus industry. Fruit imports include fresh and frozen but exports include just fresh.

Sources: Noncitrus Fruits and Nuts, NASS; Citrus Fruits, NASS; Vegetables, NASS; Potatoes, NASS; and FATUS, ERS.

service industry (restaurants, hospitals, military institutions, schools, etc.) or to a range of market intermediaries who in turn sell to retailers and the food service industry. In 1997, 48 percent of fresh produce consumed in the United States was purchased at retail and 50 percent at food service establishments (Kaufman et al., 2000).

Direct sales to consumers are small, accounting for only about 2 percent of final fresh produce consumption in 1997. On the other hand, processing is an important part of the produce industry. In 2002, 86 percent of vegetables and 69 percent of fruit produced in the United States, by weight, went to processing. Trade is also important for the fresh produce industry. In 2002, fresh imports totaled 28 percent of the value of fruit and vegetable production, and the export share was 16 percent. Shippers and market intermediaries both import produce directly from foreign suppliers. Shippers may also sell directly to the export market or to intermediaries who then sell to that market.

This chapter focuses on fresh produce that is marketed by shippers. Fresh produce is more difficult to trace than a processed fruit or vegetable. A processed product, like a can of tomatoes in a consumer's cupboard, carries a wealth of traceback information embedded in its label and its product code printed on the bottom of the can (see box, "Traceability for Processed Fruit and Vegetables"). A fresh tomato on a consumer's countertop may display no identifying information at all. This chapter discusses how the produce industry provides traceability in a challenging environment.

The Grower to Shipper Link—Including Exports and Imports

The traceability chain begins with the grower to shipper link. Growers and shippers generally make marketing agreements before production begins. Growers want to be sure that someone is committed to selling their produce on their behalf. The shipper may want growers to follow specific practices since any problems traced back to the grower would damage the shipper's reputation too. The shipper markets the grower's produce and returns the proceeds to the grower after deducting the agreed-upon fees.

Typically, shippers market for growers and are covered by PACA regulations requiring produce to be identified by lot and accounted for until the first sale. PACA does not require a lot number to be marked on a box although many shippers do so. Also, PACA does not specify the size of a lot, it just requires that it be adequate to provide correct payment to growers. Lots can vary depending on the needs of the shipper and grower. At one end of the spectrum, a lot could be one grower's entire production of a particular crop over the length of a season. But identifying lots by smaller production units can be an important business tool. For example, a grower with several apple orchards may want each to be a separate lot to be able to compare yields with different production practices. From a food safety perspective, it is also important to narrow down where a contaminated product comes from and limit potential losses. If all contaminated product comes from a lot representing one orchard, a grower may be able to continue marketing from the others. On the other hand, there are diminishing benefits to precision. No one traces apples back to a particular tree. So far, there is no reason to do so. The costs would be high, and the benefits, compared to just being able to trace back to an orchard block (or part of one), would appear to be negligible, if not zero. Most things that would affect apples would generally affect more than one tree. So if an apple from a particular block had a problem, the entire block would be treated to be sure the problem was resolved.

Traceability for Processed Fruit and Vegetables

There is a critical difference in traceability between fresh and processed fruit and vegetables. Each processed item a consumer buys is generally individually identified unlike fresh produce. For example, when the consumer gets a fresh tomato home, he may not know where it came from, but a canned tomato product will be labeled and almost always has a product code. Processed products often have consumer-recognized brand names that are also helpful in a traceback situation even if the can or other container is no longer available.

Produce for processing is usually contracted for in advance with very specific requirements for varieties, production practices, and harvest time. The processor may harvest the product and take it directly to the processing facility. Like the fresh shipper, the processor records information about the grower and field for all arrivals. Each load is processed and the time noted by the processor. PACA rules apply for fresh-cut produce like bagged salads and frozen produce, but canned fruit and vegetables are generally exempt.

For canned tomatoes, for example, the recordkeeping challenge is to link the fresh tomatoes coming in to the canned tomatoes going out. Product codes are an important component of traceability. In the canned-tomato case, a product code would generally be inkjet printed or embossed on the bottom of the can. Then a firm would be able to say that the finished product with a certain range of product codes corresponds to fresh tomatoes processed at a certain time that came from a particular grower.

FDA provides guidelines for product codes that would aid a firm with a potential recall situation, but there are no requirements to use product codes. However, the benefits of product codes are so great that most firms use some kind of product code. A typical product code might contain information such as firm, plant, line, date, and time. If there is a recall, FDA needs to know which product is a problem. If firms cannot identify particular product codes that are contaminated, FDA would have no alternative but to recall all the firm's products. Firms want to keep any potential recall as small as possible, which requires more precise identification information. A firm would have to balance the costs of more precise information with the cost of a potential recall to determine the appropriate amount of information.

PACA also does not specify the form of the recordkeeping or accounting system. Some systems are quite sophisticated and others less so, depending on the firm's capabilities and needs. A large company may have a state-of-the-art computer system. In some cases, retailers require their suppliers to use specific computer software to aid invoicing or electronic ordering and other procurement activities. A smaller company may have a less complex system. Some firms do not need much information. If they sell produce for just a few growers or sell to a limited number of buyers, a simple system may be adequate.

PACA establishes the depth of traceability in the fresh produce industry—generally produce can be traced back to the individual grower. But there are some exceptions. For example, growers may agree to pool their produce and receive an average price for the pooled product. In this case, traceback would be less precise, going back to a small group of growers rather than the actual grower. If produce is sold and then repacked by another shipper or market intermediary, PACA laws would not apply, and the origin of the produce could be lost if careful records were not kept. However, in a traceback situation, a repacker could identify the sources of the different items packed on a particular day and narrow the search to several growers.

Shippers who do not sell on behalf of growers are not covered by PACA requirements to identify produce by lot. These include vertically integrated grower/shippers who market only their own production. However, most grower/shippers market for at least a few other growers. Produce purchased by shippers instead of marketed for growers would also not be covered. Both of these groups are probably quite small. But the general business benefits of a traceability system are so great that most firms likely maintain a level of traceability even if not required to do so.

At harvest time, growers send their produce for the fresh market to shippers. Some fruit and vegetables are harvested and transported to a central packinghouse or shed for cleaning, grading, and boxing. Apples, citrus, stone fruit, tomatoes, and potatoes are examples of crops that are shed-packed. When a grower brings in a load of fruit or vegetables to a central packinghouse, the packing line is cleared of all other loads. The grower's whole load then goes through the packing line all at once or, in the case of storable products, like apples or potatoes, the produce may first go into storage until packing at a later date. Information about how much is graded into different qualities and sizes, including culls, is recorded for each lot. The shipper may also collect other data on the lot such as specific field or orchard, pickers, harvest date, etc. This information facilitates payment to the grower, operations management, and, if necessary, traceback. The shipper packs and labels the cartons, usually with an ink jet printer.

Other fruit and vegetables are packed in the field. For example, lettuce, berries, broccoli, and melons are typically harvested, wrapped, and boxed in the field. The shipper uses stickers on each box, or on each pallet of boxes, that generally identify the grower, packing crew, and date. Handheld ink jet printers are available for use in field packing but are used infrequently because they are expensive.

Containers are printed with various types of information relevant to different people along the marketing chain. Pallets of boxes may also be labeled. Because fresh produce is not transformed before it gets to the consumer (unlike grains and livestock), it is easy to add stickers, tags, and other special labels to the produce to appeal directly to consumers. Each of these methods of identifying produce is discussed below. The exact type of information provided will depend on various laws that apply and the needs of the shipper and buyer.

Information on Boxes PACA does not require any information on boxes, just that everything printed on the box be true. In practice, boxes provide a wealth of information, some required by law and some voluntary. Typically, States require that certain information be included on a box. For example, California State law requires each produce box to identify the commodity and variety, responsible party (entity, town, and State), and quantity (weight, count, or size).

Although not required, most shippers voluntarily mark boxes with lot numbers. It is easier to look up records by lot number than to have to search through other records to identify a particular grower's product. FDA would like to see growers also add lot information to invoices to help speed up traceback in a food safety outbreak (FDA, 1998). If a shipper is selling only for himself and a neighbor and can keep the boxes separate, there would be no need for lot numbers on boxes. Recordkeeping alone could indicate whose boxes were sold to which buyer.

In addition to ensuring proper payment, the traceability system that identifies boxes by lot can also be important for general business operations because not all produce is of equal quality. For example, if someone liked a particular purchase and wanted more from the same grower, a shipper would need to know whose product, identified by lot number, was sent. Alternatively, if a product does not hold up well and a buyer complains, a shipper wants to know which grower's product was involved. The shipper may dock the price for that load, decide to not ship for that grower again, or ship only to nearby markets. Similarly, if produce is exported but fails phytosanitary inspections because of the presence of pests, an exporter might request no more loads from the lot with problems.

Labeling on boxes is important for marketing. Produce growers and shippers are always looking for ways to distinguish their product and raise its price above that of an undifferentiated commodity. Currently there are several characteristics that consumers are particularly concerned about. If organic produce is to be marketed as such, it must be marked to verify that production practices meet USDA's organic standards. Similarly, produce with no pesticide residues can be marked with a third-party certifying seal to verify its status.

Marketing orders, which allow producers to collectively regulate certain marketing activities for an industry, may also require additional label information. A marketing order may require that shippers market only produce of a certain quality or size. Quality standards can bolster a product's reputation, which benefits all growers in the order. Restriction of supply can also raise the price for all producers. This type of program can involve additional mandatory markings on boxes to ensure that the marketing order can regulate the program by identifying producers who are not complying and undermining the integrity of the program.

In the case of California peaches, the marketing order requires positive lot identification (PLI) which means that each box of peaches is inspected by a USDA inspector to verify that the quality meets the marketing order specifications. The size of the lot is specified in the marketing order and is not necessarily the same lot used by shippers to comply with PACA. Some marketing orders require additional information. The California peach marketing order also requires that each box be marked with the packinghouse number and date. The additional information allows the shipper to identify whose product was packed at that location and time.

Marketing orders can also be used to provide more precision in traceback. In addition to individual grower efforts to improve traceback capabilities, grower organizations have become more concerned about the reputation of their crops for food safety. Several grower organizations have developed systems to strengthen traceability, which encourages grower responsibility and reduces the free-rider problem in developing a positive industry reputation—a public good. In the case of an outbreak, a grower organization that encourages traceback can prove to the public that their product is not responsible for the problem. Or, when the industry is responsible for the outbreak, the problem grower or growers can be identified and damage can be limited to that group. The California cantaloupe industry has developed a more precise trace-back system to deal with potential food safety issues (see box, "Cantaloupe Industry's Response to Food Safety Problems").

Another set of mandatory labels relates to products exported to other countries. Produce that is grown or treated for export may be required to bear a mark from USDA's Animal and Plant Health Inspection Service verifying that the product meets certain phytosanitary provisions. Foreign countries requiring the phytosanitary provisions would not accept a box without the correct markings. In the case of Washington apples, only those that have passed a cold treatment process may be exported to Mexico, and boxes must be marked with the number of the registered treatment facility.

Shippers may also import produce directly to market with their domestic production as a means to extend their marketing season or provide more variety in product offerings. Almost all imported produce, like domestic produce, is marketed on behalf of the foreign growers so the transactions are also covered by PACA. Typically, the produce is packed and labeled in the foreign country to comply with U.S. labeling requirements, but it may be repacked in the United States as well. The only additional labeling requirement for a box of imported fresh produce is that it show a country-of-origin label. For produce in consumer-ready containers, such as raspberries in plastic boxes, grapes in bags, and shrink-wrapped greenhouse cucumbers, each container must be labeled with the country of origin.

Information on Pallet Tags After initial packing, boxes are formed into pallets, and a pallet tag with a barcode is sometimes attached. The number of shippers using pallet tags is increasing. Pallet tags are for internal accounting and logistics; they are not required by law. The tags reflect shipper needs. A typical pallet tag might indicate the date packed, packing shift, grower, lot number, variety, grade, style of pack, and size. Pallet tags allow staff moving pallets in cold storage with forklifts to easily find the exact product they are looking for without having to read the small print on the boxes. Scannable pallet tags are also used to verify that orders contain the correct products. Pallet tags are also useful in narrowing the scope of a quality or food safety problem beyond just the lot. If the only problem products in a lot were on a pallet shipped to one distribution center, the focus of the investigation would concentrate on contamination sometime after the pallet left the shipper. If the only problem pallets from the lot were packed during a particular shift at the packinghouse, some kind of postharvest contamination might be suspected. While there are voluntary Universal Code Council standards for pallet tags, very few U.S. produce firms use them. Most barcodes are internal systems that can be read only by the shipper. Pallet tags are discussed again below.

Cantaloupe Industry's Response to Food Safety Problems

Beginning in 2000, the California Cantaloupe Advisory Board (a marketing order for California cantaloupe grown north of Bakersfield) began requiring additional traceback information on cantaloupe boxes as part of the State marketing order (this program was voluntary in 1999). This was not a very difficult process. California cantaloupe is field-packed and the Board had already contracted with the California Department of Food and Agriculture to inspect cantaloupe during harvest for quality control and apply an inspection sticker to every box (growers pay the Board a per-box fee for this service). Cantaloupe from this area cannot be sold without the sticker identifying the county and shipper.

The new program requires information on the packing date, field, and packing crew which allows a grower to trace a problem back to a particular part of a field. This would allow a grower to determine if contamination perhaps originated with a sanitation problem with a particular packing crew or was more widespread and perhaps originated with irrigation water. Some growers had already been providing this additional information on a voluntary basis. Adding this additional traceback information to the box was neither particularly costly nor complicated. It did take some administrative changes, however. To be able to require traceback, the members of the Board had to propose a change to the marketing order and vote on it. The original marketing order covered grades and quality standards.

The new marketing order specifically approves "such grade and quality standards of cantaloupes as necessary, including the marking or certification of cantaloupes or their shipping containers to expedite and implement industry practices related to food safety" (California Department of Food and Agriculture, 2003). If a foodborne illness outbreak were to occur, this program would allow the industry to immediately confirm or deny that the problem is due to California cantaloupe and help growers pinpoint the source of the problem. This may be the only grower-organized program for produce in the United States that requires such detailed traceback information on each box. To date, the system has not been necessary for a food safety outbreak.

Information on Individual Produce Items By the time fresh produce reaches retailer shelves, many products have lost their identity. A bin of loose potatoes is completely anonymous unless displayed in its shipping box. But some products do retain at least some of their identity—potatoes packed in bags, bagged salads, berries in plastic consumer-ready containers, and items, such as bananas, that are marked with stickers emblazoned with their brands. The trends toward more fresh-cut produce, consumer-ready packaging, and branded products ensure a continued increase in the information available to consumers when selecting fresh produce. In 1997, 19 percent of retail produce sales were branded products, compared with only 7 percent in 1987 (Kaufman et al., 2000).

Retailers often request that fruit and vegetables sold loose (as opposed to those in consumer-ready packages like a bag of carrots) carry stickers with the product's price-look-up (PLU) code. Stickers work relatively well for some products such as large tomatoes and apples. Stickers do not adhere as well to other products with a rough texture such as cantaloupe. Some products, such as chili peppers, are too small to use stickers although they could be packaged instead of being sold loose and then a sticker could be applied. The primary motivation for PLU codes on loose produce is to ensure that the retail cashier rings up the right price code for each item and charges the right price—identification of the item, not traceability *per se*. Shippers charge for this service. Some shippers use stickers with their company, brand name, or additional product attributes, as well as the PLU code. This can also convey useful information to the savvy consumer. For example, greenhouse tomatoes sold loose in bins usually have stickers with the firm name applied. Consumers may prefer one firm's tomatoes to another's. Such information could prove useful in a food safety traceback situation, if consumers paid attention to it.

Since the only product information for produce sold loose that actually reaches the consumer is the PLU sticker, there is some interest in trying to put more traceability information on it. Retailers want scannable PLU stickers to reduce labor costs and cash register keying errors. With reduced space symbology (RSS), additional information such as a shipper code, and perhaps even lot, could be incorporated into a barcode. There are, however, constraints to sticker size and the amount of information that can be included. The newest stickers also require newer scanning machines; that requirement could delay retail adoption of RSS.

The Shipper to Retailer or Food Service Establishment Link—Direct Sales and Intermediate Sales

Shippers sell produce to a wide range of final commercial customers—retailers and food service establishments—and market intermediaries. If a shipper sells directly to a retailer or a food service buyer (an increasing trend in the industry), traceability can be straightforward since PACA requires recordkeeping to the first buyer. Recent research shows that shippers' share of sales made directly to retailers and mass merchandisers increased between 1994 and 1999. For example, 63 percent of total grape sales and 54 percent of orange sales were direct sales in 1999, up from 60 percent and 48 percent, respectively, in 1994 (Calvin et al., 2001).

While the shipper has a wealth of information about the product, only a limited amount of information is forwarded to commercial buyers in accounting records. Information on the box and pallet is generally not entered into the buyer's database since it is not in a standardized machine-readable form. The commercial buyer creates a new tracking system. The link between the shipper and buyer databases is the purchase order number for each transaction. If the buyer calls up about an order and has the purchase order number, the shipper can access all his records about the product including lot and pallet numbers.

As a commercial buyer receives each load, information is entered into the firm's data system that tracks the entry and eventual disposition of the product. For example, a large retailer might have a central warehouse that receives produce from shippers and then distributes produce in smaller volume to its local stores. The more sophisticated distribution centers add new internal pallet tags specific to the retailer's tracing system. For example, it would link to information on the purchase order number, the date of receipt for use in rotation of the stock, and information on storage location in the warehouse. The pallets received from the shipper may be broken down and then reformed into mixed pallets (a pallet of different products and/or different suppliers) to be shipped off to a local retail store or food service firm. The outgoing pallets also need pallet tags. These outgoing tags do not, however, link individual boxes back to their purchase order number, so the commercial buyer does not necessarily know which suppliers' product went where. In a traceback, commercial buyers would look through their records to see what they had in stock in the warehouse during the relevant time period, identify the purchase order numbers associated with that product, and contact the shippers. If there is only one supplier, there is no problem. If there are two or more, traceback becomes more difficult.

In foodborne illness cases where there is more than one supplier, multiple outbreaks may provide additional information to identify the source of contamination. Consider a hypothetical example of a traceback where multiple outbreaks would help to pinpoint the most likely source of contaminated product (see figure 5). Looking at just the Food Service Outlet or just Retailer 2, both of which received produce from multiple sources, would provide insufficient information to allow FDA to determine the source of contamination. Looking at both together, however, shows that Shipper 4 is likely to be the source of the problem. If FDA had information only on Retailer 1, which received produce from just one shipper, that information alone would be sufficient to identify the probable source.

Better traceback requires a system that maps out the exact path a box of produce follows through the distribution center. If there were a standardized machine-readable data system, the shipper's pallet tag could be read as the pallet entered the system and linked to the buyer's pallet tag to carry data such as shipper pallet and lot number. Similarly, as boxes left the buyer in new mixed pallets, the lot information on the box could be tracked to record exactly where that box went. If such a system were in place, a food safety problem in a particular store could be uniquely linked back to the distribution center and the original shipper's pallet, lot, and purchase order.

There is growing recognition in the industry of the potential efficiency gains from developing a traceability system that is standardized across individuals up and down the marketing chain. The U.S. Produce Marketing Association and the Canadian Produce Marketing Association are collaborating to develop a strategy for adopting the UCC.EAN standardized barcode system. Bolstering the shipper-commercial buyer link involves standardized machine-readable information on pallet tags and boxes (*The Packer*, 2003).

In a more complicated transaction, produce may also pass through other hands, including one or more intermediaries such as brokers, wholesalers, repackers, terminal markets, or exporters before reaching the final point of consumer sale. These indirect sales can sometimes pose traceability challenges. Nearly all firms in the produce marketing chain require a PACA license which imposes recordkeeping requirements, but each layer of transaction adds another chance for human error, and a different tracking system may be used at each stage in the marketing chain. Traceability depends on the recordkeeping standards of the market intermediaries. Many of these intermediaries are large companies with sophisticated traceability systems that track incoming and outgoing shipments in the same way that large retailers do. Some are smaller firms and may have less comprehensive systems. As produce passes through many hands, the information on the box becomes potentially more important for identifying its source.

Figure 5
Hypothetical traceback scenario with multiple outbreaks

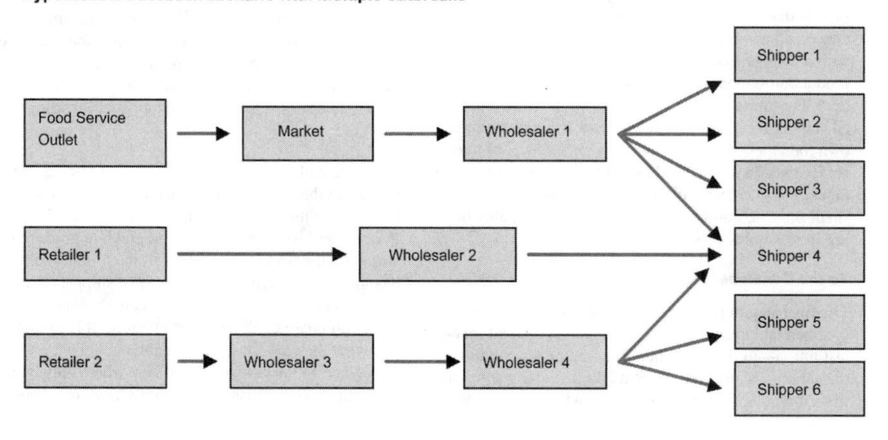

A standardized traceability system up and down the marketing chain would make traceback for sales with intermediate buyers much easier. If the last commercial buyer could identify the lot, pallet, and shipper immediately, FDA could avoid the delay of having investigators wade through information on several transactions to determine the original shipper.

Certain types of markets can also pose problems for tracing. Terminal wholesale markets, for example, serve a number of types of buyers: large retailers or food service firms that need to make an emergency purchase to fill in a sudden hole in their supplies, small firms that rely on the terminal market for their main purchases, and sidewalk food stands. The last category, although probably a very small share of sales, can pose a particular problem for tracing sales because they are often cash transactions that are not necessarily well documented.

Likewise, certain types of market intermediaries (repacking operations for example) can present traceability difficulties. Frequently, tomatoes are sold and shipped from their production regions to repackers or wholesalers who ripen, resort, and repackage for uniform color and then sell to local retailers and food service buyers. On any day, repackers may use tomatoes from several different sources to create a new box of tomatoes. In a traceback situation, a repacker might be unable to identify the exact grower but could at least identify a small group of growers whose tomatoes could have been in the box.

Shippers also sell fresh produce to the food service industry, either directly to the food service firms or their specialized warehouses, or via wholesalers and other intermediaries. Big fast food companies are particularly concerned about food safety and will often deal directly with a shipper to ensure the product meets their exact production standards, which could be specified in a contract. For tomatoes, fast food firms might use an integrated shipper/repacker (one that is repacking only with its own tomatoes), which maintains a higher level of traceability than an unaffiliated repacker. But the food service industry also consists of many small restaurants with small produce purchases. These firms are probably buying from wholesalers or other intermediaries.

To the Consumer

The final step in produce traceability is from the last commercial buyer—generally the retailer or food service institution—to the consumer. This can be a weak link in traceability. Many consumers might be uneasy about the idea of retailers' keeping records of what they buy. But this information is important for traceback, particularly for a food safety problem.

Many observable quality issues can be resolved if a consumer returns produce in poor condition to the retailer. For example, if a consumer brings in a package of bagged lettuce that has spoiled before its sell-by date, traceback would also be a routine process since all the information is printed on the bag. Even a head of lettuce may have a plastic sleeve with the shipper name or a twist tie with a firm name to identify its origin. Traceback for a food safety problem is more problematic. Food with microbial contamination generally looks fine. Even testing cannot always pick up contamination problems because microbial contamination is often sporadic and present at low levels. By the time someone becomes ill and consults a physician, and health authorities identify the contaminated product and the place and date of purchase (or consumption in the case of food service institutions), the perishable produce is usually long gone. Even when the produce comes in consumer-ready packages, such as a bag of apples marked with the shipper's name, the packaging is also usually discarded. For a branded processed product, consumers may know that they always buy a particular brand, but for a fresh product, most people have no idea who provided it. In cases where the box or other container is no longer available, traceback relies on good recordkeeping by all the firms in the marketing chain.

If the Centers for Disease Control and Prevention or State/local health departments can identify the contaminated product and the place and date of purchase, commercial buyers can usually identify the shipper. In the best case scenario, where a firm was using only one supplier of the problem product on that date, the retailer or food service firm could call up the shipper, who would have all the information about the product. But in practice there can be a lot of uncertainty about whose product was sold.

One potential solution to this problem of tracing from the retailer to the consumer and back is the RSS sticker with barcode identifying the shipper as well as the PLU code. If a retailer knew only the day the problem produce item was sold, the firm could look at all the product sold that day and perhaps reduce the number of shippers that could potentially be involved. If a consumer used a consumer purchase card, a retailer might be able to look up just what the sick consumer bought and know the shipper to contact. In the case of club stores, where only members can make purchases, traceability is more complete.

Conclusions

Traceability has been a critical component of the produce industry for many years. Historically, the perishability of produce and the potential for deterioration during cross-country shipment demanded better recordkeeping to ensure correct payment to growers. Because produce must be packed in relatively small boxes to minimize damage, implementation of traceability has also been relatively low cost. The industry is in a much better position to adapt to new concerns than industries where bulk sales have been the norm and segregation and traceability would involve new costs.

Currently, there are two systems of information involved in produce. First, there are physical labels on boxes and sometimes on pallets. For general business purposes, it is important to be able to identify the product in the boxes. There are various State laws requiring box information, and marketing orders also often require additional box information. Pallet tags are completely voluntary. Second, a paper or electronic trail allows traceback between different links in the marketing chain, though each link may use a different traceability system. U.S. and Canadian produce organizations are looking at ways to promote a universal traceability system between links in the chain. They recommend that shipper name, pallet tag number (if available), and lot number be part of the paperwork at each link. This would effectively combine information on boxes and the paper or electronic trail. Such a system would require developing a standardized system of barcodes or other machine-readable information, as well as shipper and buyer investment in machines to apply and read codes. One of the challenges to developing a compelling technical solution that all market participants would use voluntarily is to ensure that all segments of the industry can afford the costs of a new system.

Grain and Oilseeds

Virtually all grains and oilseeds produced in the United States are traceable from production to consumption. For the most part, however, quality and safety variation in grain and oilseeds has not warranted the cost of precise traceability systems. Systems to track product to elevators, the point at which quality and safety are monitored, have been largely sufficient for the efficient operation of grain and oilseed markets. Growing demand for specialty crops, including products not genetically engineered, has spurred the development of more precise traceability systems, although the elevator still operates as an important quality-control point.

The history of the grain supply chain in the United States chronicles the growth of an infrastructure built to manage large flows of product differentiated on a limited number of variety or class attributes and then blended or processed to meet quality and safety standards. In most cases, the blending and homogenization of product begins as soon as farmers deliver their crop to the local elevator and continues until the crop is transformed into animal feed or into the loaf of bread, cereal, or other grain product on grocery-store shelves. In most cases, grain and oilseeds are mixed and transformed all along the chain, so that safety and quality characteristics are redefined at each step. As a result, processors need information on the characteristics of the product as delivered only from the last stage of processing. The high level of processing necessary to produce consumer-ready grain products eliminates most safety and quality problems stemming from mishandling or contamination early in the supply chain and often eliminates the need to establish traceback to the farm for safety or quality reasons.

More recently, consumer and processor demand for specialty grains, including products not genetically engineered, has introduced the need to differentiate product over a new set of quality characteristics. In a few cases, these new quality demands are accompanied by demands for traceability systems to track product back to the farm. For the most part, just as it has many times before, the grain and oilseed infrastructure is adjusting to accommodate new quality variations and ensure the delivery of homogeneous product meeting new quality and safety standards.

From the Farm to the Elevator

With the exception of a small amount of on-farm feed use (mainly corn), most grains and oilseeds are marketed through a supply chain that includes country elevators, sub-terminal elevators, processors, river elevators, export port elevators, and retailers (fig. 6). This supply chain handles a wide range of bulk commodities distinguished by variety or class, such as No. 2 yellow dent corn and hard red winter, hard red spring, soft red winter, white, and durum wheat. Large-scale marketing affords efficiencies in terms of lower per-unit handling costs.

Conventional Crops

When farmers harvest standardized crops, they usually store the grains and oilseeds in large storage units (or bins) on their farms. Crops of a certain type—for example, wheat—are typically commingled, even though producers may have grown several different varieties. These may differ in terms of yield, maturity, resistance to adverse weather conditions (e.g., drought), and other factors, but often do not have quality attributes valued by buyers and are not sold at a premium.

Producers sell their crops to local (country) elevators. In 1997, there were 9,378 wholesale handlers (particularly country and export elevators) of grains and oilseeds operating in the United States (U.S. Dept. Commerce, Bureau of the Census, 2000). When farmers deliver their crops to local elevators, they are given receipts that indicate the commodity sold, its weight, price received, time of purchase, and any premiums or discounts for quality factors such as extra moisture, damage, pests, or dockage (easily removable foreign material). Country elevators keep this information, thus establishing a record-keeping link from the product in an elevator at a point in time to the farmers who supplied the product. An elevator operator knows the farmers who delivered grain and oilseeds at that location and the geographic area from which they came.

This rather imprecise system of traceability from the elevator to the farm is sufficient because quality variations that may exist at the farm level are mostly eliminated at the elevator level. The elevator serves as a key quality control point for the grain supply chain. Elevators clean each shipment to remove the foreign material and lower quality kernels or beans. If the moisture level is too high, the shipment may be dried before being placed in the silo. Elevators also sort deliveries by variety and quality, such as protein level. Different quality, variety, or classes of crop are either segregated at the silo or bin level

Figure 6
The grain marketing system

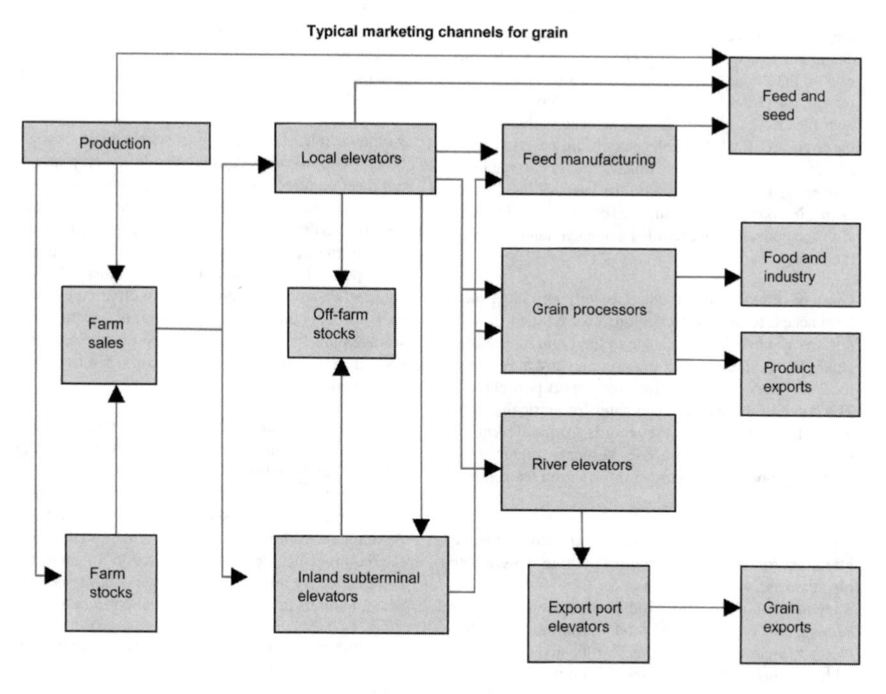

Source: Adapted from *The Organization and Performance of the U.S. Food System* by Bruce W. Marion, 1986

depending on the size of the elevator and anticipated volumes of production. Elevators then blend shipments to achieve a homogeneous quality. Once blended, only the new grading information is relevant—there is no need to track back to the farm to control for quality problems. Strict segregation by farm would thwart the ability of elevators to mix shipments for homogeneous product and would not be necessary for safety or quality assurance.

Country elevators strive to market crops of homogeneous quality to millers, feed manufacturers, and oilseed crushing facilities. Millers and crushers, in turn, sell processed grains (such as corn grits), flours, and oil to food processors. Crushers also sell soybean meal to feed manufacturers. Country elevators send grain and oilseeds to inland sub-terminal and/or river elevators, which collect crops from different regions. River elevators then ship crops to port elevators that load grain and oilseeds onto vessels for export to foreign countries.

Precision in traceback to the farm declines the further one goes down the production chain. As grain is funneled from a wider geographic area, it is more difficult to pinpoint from where and from whom the commodities came. For example, grain held at port elevators may have originated from a number of country elevators serving a large number of farmers across a wide geographic area. Traceability at the port elevator level typically extends only back to the country or sub-terminal elevator.

Recordkeeping systems for conventional grains and oilseeds can therefore be best characterized as "one step forward, one step backward." That is, handlers know from whom they bought grain and to whom it was sold. This one-step-forward, one-step-backward system means that a given handler is acquainted only with the entities that it deals with directly. Retrieving information from further up or down the marketing chain forces the handler to rely on the recordkeeping ability of others in the chain. For example, if a river elevator needed information on the farmers who produced the soybeans stored in its silo, the river elevator would need to look up in its own files the identity of the local elevators that supplied the soybeans. Each local elevator would have to check its accounting information on which farmers had made deliveries. Thus, traceability to the farm, or handful of farms, in conventional grain marketing is possible only with the collection of records from each handler along the supply chain.

Grain or oilseed handlers that are vertically integrated have access to more information. That is, such firms operate at more than one stage in the grain marketing chain. For example, a large grain company may own local elevators as well as river and export port elevators. The depth of information is greater for vertically integrated firms simply because records from different stages are maintained in-house. Vertically integrated firms can more easily retrieve information from their operating units.

Whether vertically integrated or not, elevators serve an important role as a quality-control point in the grain supply chain and as the linchpin in the traceability system. They monitor and control product quality and safety and keep records on the flow of product from farms to the elevator. Since the bulk system fulfills buyers' demands with cleaning and blending, there is no need for information to be collected throughout the supply chain: information from the next immediate step in the supply chain is sufficient.

Specialty Crops

While most grains and oilseeds in the United States are produced and marketed in bulk, there are growing markets for more specialized commodities. Some examples include high-value crops (e.g., high-oil corn), organic foods, and non-genetically engineered crops (Dimitri and Richman; 2000; Lin, Chambers, and Harwood, 2000). Traceability systems are becoming more extensive in these markets, reflecting customers' demands to verify the presence of the specialty attribute, particularly when it is a credence attribute. These traceability systems document the efforts of each segment in the supply chain to segregate the high-value specialty product from conventional or other specialty products.

Segregation and traceability documentation for specialty attributes may begin as early as the seed. At this point, documentation verifies the existence of specific crop traits and purity levels. In general, seed is tested and lots are tracked using identification numbers. If necessary, specific information about parent genes is obtained from the seed developers.

At the farm level, farmers must segregate crops to ensure that cross-pollination does not result in a crop that does not meet required specifications. For example, producers of non-genetically engineered crops, particularly corn, may be required to keep genetically engineered varieties away from other fields by a minimum distance to prevent cross-pollination. In addition, farmers must either dedicate certain storage, harvesting, and other equipment to each specialty crop or thoroughly clean equipment and storage units between different crop types. Some farmers specialize in particular specialty crops thereby avoiding commingling problems.

To verify that adequate precautions have been taken at the farm level to assure the quality of the specialty grain, farmers may be asked to provide elevators with third-party certification. For example, for organic crops, third-party certifiers accredited by the U.S. Department of Agriculture work with individual farmers to determine the requirements for organic production for each crop and then verify that these requirements have been fulfilled. Farmers provide this certification to buyers.

For some crops, farmers may be asked to submit their shipments for testing. For example, the oil content of corn and the protein level in wheat are routinely tested. Tests may be performed by the elevator or by independent third-party verifiers. Elevators usually keep records of test results, including the identity of the farms that sold the commodities to them. For some specialty crops, buyers may simply require farmers to "certify" that the crops are as specified. This was the case early in the development of differentiated markets for non-genetically engineered crops.

As the repository of documentation certifying attributes or the point of attribute testing, elevators play an important quality-control function in the specialty crop supply chain. In many cases, testing results and certifications are not sent further up the supply chain because elevators essentially certify the quality and homogeneity of their products. As with the conventional supply chain, elevators blend shipments to achieve a homogeneous quality and meet sanitation and quality standards. Once blended, only the new attribute information is relevant;

there is no need to track back to the farm to control for quality problems.

At the elevator level, segregation of specialty crops is achieved with dedicated elevators (those specializing in one type of specialty crop, such as organic, waxy corn, non-genetically engineered crops, and food-grade soybeans), multiple bins, or by thoroughly cleaning bins and equipment after each crop has passed through. If identity preservation is required, shipments may be containerized in order to minimize handling and the number of points at which quality could be compromised.

A key constraint in the ability of the bulk-system infrastructure to supply specialty grains is the ability of elevators to adjust their product flow in response to consumer demand. Large grain companies with a large infrastructure at their disposal, including country and export elevators as well as railcars and barges, may have more flexibility in managing flows and creating segregated systems. Likewise, smaller producers with access to a number of small elevators may be able to efficiently manage specialty flows. However, as the number of specialty attributes grows, investments in elevator infrastructure may be required, raising the costs of segregation.

Segregation and documentation for specialty crops continue from the elevator to the final producer or consumer. Trucks, railcars, and barges must all be thoroughly cleaned between specialty crops or be dedicated to a particular specialty crop, as must sub-terminal, river, and export port elevators. All along the line, either testing or process certification guarantees that quality attributes are maintained. As with conventional crops, such verification is usually of the "one-step-forward, one-step-back" variety. Each player in the specialty chain is usually required to retain information on product identity, volume, lot numbers, test results, and suppliers/customers to ensure quality and allow for traceback if problems arise in the marketing chain. How far back a given elevator can trace a shipment depends on the extent to which the firm is vertically integrated. As with conventional grain production, vertical integration in handling—whereby a firm owns operations in more than one level of the marketing chain (e.g., country and export elevators)—eases traceback, since information can be retrieved from internal suppliers and/or buyers. If elevators are not vertically integrated, they must rely on other handlers to retain much of the information.

A number of third-party certifiers offer services to verify that specialty quality attributes have been adequately safeguarded throughout the supply chain. In the case of organic products, farmers, handlers, processors, and retailers are certified by third-party firms that must be accredited by the U.S. Department of Agriculture. Wholesalers and retailers must prove that the organic product came from certified sources satisfying the organic labeling and handling requirements. As a result, organic products can be traced throughout the supply chain.

Generally, the cost of establishing and verifying supply chains for specialty grains makes them more expensive to produce than conventional grains. As a result, farmers, elevators, and handlers may be reluctant to construct these chains and produce these grains without some guarantee that they will receive adequate compensation. A large segment of the specialty crop market is therefore built on contracts. Contracts not only allow buyers to specify the attributes they desire, they also provide sellers with assurances that their costs will be covered through price premiums or long-term sales. Premiums must cover the additional physical costs associated with segregation and traceability, and also customer service and coordination activities.

Elevators typically contract with producers to grow certain varieties, such as high-oil corn or food-grade soybeans, with the delivery volumes and times being predetermined. The contracts may specify that producers follow certain production and handling practices that are consistent with the traced products. Contracts are also drawn up between the elevator and the buyer. Contracts provide a type of paper trail by which commodities can be traced.

Manufacturers may require information on a host of characteristics, such as color, variety, grind, etc. For example, a cereal manufacturer that uses a specific class and grade of wheat to produce the desired flake curl may require special coding. Larger food processors may also require that suppliers use codes that signify that the ingredients are specifically for the food manufacturer. All these steps are taken to ensure high and consistent quality over time—and to facilitate efficient ingredient management. For efficient output management, firms may also track final products. This information allows companies to understand which products are popular and where they are selling well. This information helps companies produce the right mix of products and the best distribution.

In general, traceability systems for specialty crops are more precise than for conventional ones. The paperwork generated with contracting and the existence of relatively few producers and handlers who deal with specialty crops make it easier to track shipments; a railcar filled with a certain commodity can be traced back to a small set of handlers and producers. However, in most cases, one could not likely associate an individual kernel or bean with a particular producer, since even specialty crops are commingled by elevators. There are a few cases for which one can trace shipments back to individual farmers. For example, food-grade soybeans are containerized on-farm and shipped directly to Japan.

Conclusion

Regardless of whether they involve specialty or conventional grains, vertically integrated firms or independent operators, most traceability systems for grains do not extend back beyond the country elevator. For most manufacturers and consumers, this depth of traceback is sufficient to ensure quality and safety, even for specialty quality attributes. As long as elevators continue to ensure the safety and quality of the shipments they receive from farmers, manufacturers will likely not demand farm-level traceability.

If elevators fail to monitor the safety of the system, manufacturers and consumers may demand better control and maybe even farm-level traceability. The StarLink incident in 2000 highlights the economic consequences of inadequate quality control at the farm and elevator level. StarLink is a genetically engineered corn variety that was approved for animal feed and industrial uses but not for human consumption (Lin, Price, and Allen, 2003). In 2000, a portion of the StarLink crop was commingled with other corn varieties, contaminating millions of bushels stored on farms and in elevators. Moreover, as a precaution, food manufacturers took hundreds of food products off the market along with nearly 100 products served at restaurants. Disruptions occurred in domestic marketing and exports to foreign countries in the initial stage of the incident as commingled corn was rerouted to approved uses and contaminated food was removed from shelves. Had StarLink been properly segregated at the elevator, this incident would probably have been at most a minor issue.

In the wake of the StarLink incident, many consumer groups called for complete traceability for StarLink and other genetically engineered crops. Better quality control at elevators may actually be a more cost-effective means of ensuring the quality of the Nation's grain and oilseed supply. However, with the growth in the variety and type of credence quality characteristics, the ability of elevators to continue to serve as the system's quality-control monitors hinges on advances in testing technologies and improvements in verification services.

Cattle and Beef

The cattle/beef sector has a long history of identifying and tracking animals to establish rights of ownership and to control the spread of animal diseases. Producers in the meat sector have also developed traceability systems to improve product flow and to limit quality and safety failures. Recent developments are motivating firms to bridge animal and meat traceability systems and establish systems for tracking meat from the farm to the retailer. Though technological innovations are helping to reduce the costs of such systems, institutional and philosophical barriers are slowing their adoption.

A number of recent events, including the emergence of bovine spongiform encephalopathy (BSE, commonly known as mad cow disease) and the country-of-origin labeling provisions included in the 2002 U.S. Farm Bill, have focused attention on traceability in the cattle/beef sector. Policymakers, producers, and consumers are reassessing the value of systems to track animals and meat from the farm to the consumer. These events, however, are not the first to motivate livestock owners and meat processors and retailers to establish traceability systems for livestock and meat. Ownership disputes, animal health concerns, and meat foodborne illness outbreaks have all motivated the development of systems to identify the ownership and health status of animals and the safety attributes of meat and meat products.

The result of these historical motivations has been to create two largely distinct sets of traceability systems in the livestock/meat sector: one set for live animals and another for meat. The current challenge for the cattle/beef sector is to link these systems and develop a system for identifying farm-level attributes in finished meat products—in other words, to trace meat back to the farm.

Traceability for Live Animals

Livestock owners have three primary motives for establishing traceability systems for live animals. First and foremost, owners want to protect their property from theft or loss by clearly identifying which animals belong to them. Whenever animals are commingled, as is common in the open ranges of the United States, owners may be motivated to use identifying marks to distinguish their cattle from those belonging to others.

A second primary motive driving livestock owners to establish traceability systems for live animals is to control the spread of animal diseases. Efficient control or eradication of disease depends on the ability of owners to identify and track healthy and unhealthy animals. This information is vital in calculating contagion and in designing effective vaccination, segregation, and indemnity programs.

A third motive for establishing traceability systems for cattle lies in the fact that many valuable animal attributes are not evident to the naked eye—or even to specialized testing equipment. Credence attributes such as up-to-date vaccinations, proper medical care, animal welfare provisions, or feeding regimens may increase the value of an animal. Farmers who can prove, through traceability documentation, that their animals possess such valuable attributes are more likely to be able to negotiate higher prices for their animals.

These three motives have influenced the development of traceability systems in the livestock sector in the United States. Livestock owners have established animal traceability systems to meet one or many of these objectives—and have expanded or contracted systems to reflect dynamics in animal management, disease outbreaks, and consumer preferences for credence attributes.

Traceability at the Cow-Calf and Stocker Level

Most of the beef that Americans consume originates from cattle born and raised on one of the country's 800,000 cow-calf farms (fig. 7), with lesser amounts coming from U.S. dairies (culled dairy cows) and from Mexico and Canada. While the American West is traditionally recognized as "cattle country," the majority of the beef cattle in this country are in fact raised in the center of the country between the Mississippi River and the 100th meridian. And, contrary to general perceptions, the majority of cows are raised by small and mid-sized operators. In 2002, the 5,390 large cow-calf operations, those with more than 500 head, accounted for only 14 percent of the beef cows in this country. The 630,000 smallest operations, those with fewer than 50 head, accounted for 29 percent of cows. In 2002, the average herd of beef cows in the United States totaled only about 41 animals (USDA/NASS, 2003).

Cow-calf operations require large amounts of pasture and range land to feed the cows and growing calves. The cows and calves may graze on land owned or leased by the cow-calf operator or, for a fee, on Federal lands.

Figure 7
Cattle/meat marketing system

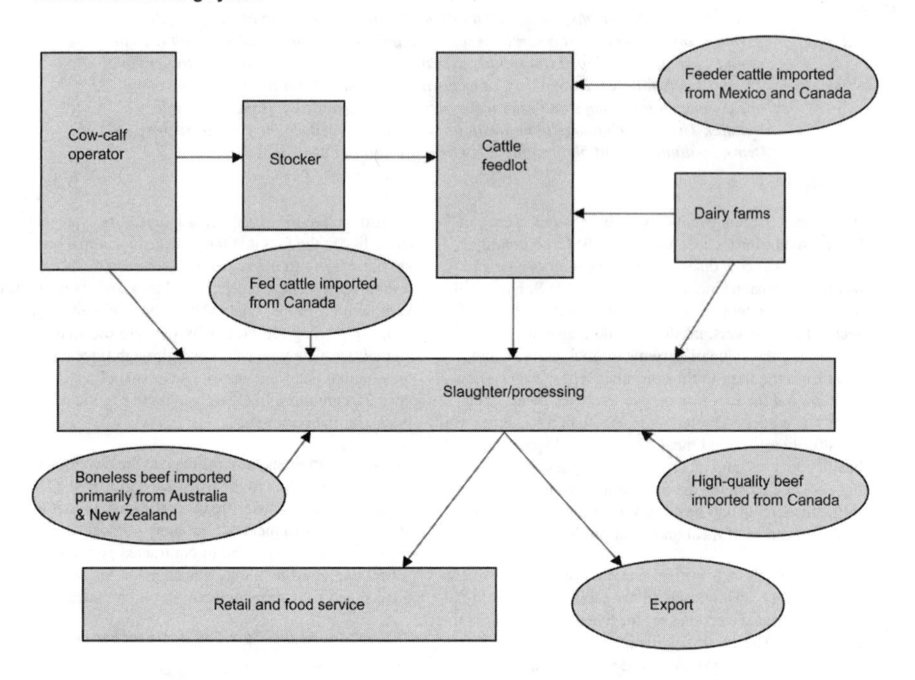

Grazing lands may be adjacent and not separated by fencing, meaning that animals belonging to different people may get mixed. Many farmers find it worthwhile to brand or otherwise identify their cattle to avoid ownership disputes.

The traditional method of identification for cattle is branding, whether hot branding, freeze branding, hide branding, or horn branding. As early as the Roman Empire, competitors employed branding irons to burn their names onto horses used in chariot races (Blancou, 2002). In the 7th century, the Chinese used branding irons to identify horses used by the postal service. Branding is also the traditional method of animal identification used in the United States. Most Western States still have branding laws that require brands to be registered and to be inspected when animals are moved or sold.

Other methods of animal identification include tattooing, retina scanning (Optibranding™), iris imaging, and, currently the most common method, tagging. Tags may have simple printed numbers, imbedded microchips, or machine-readable codes, such as radio frequency identification (RFID). Ear tags cost in the neighborhood of $1 or $2 apiece. RFID technology is more costly, with instruments for reading RFID tags costing several hundred dollars apiece, though prices have been rapidly falling.

Increasingly, tags include more information than just animal ownership. Coded information on tags may provide information on vaccination records, health history, breeding characteristics, and other process attributes. This information is either encoded directly on the tag or kept in separate records that are linked to the animal via codes on the tag. Larger cow-calf operations are much more likely to use individual or group calf identification systems than smaller operations because it is more difficult to remember characteristics of individual cattle when there are many animals. Information on individual animal characteristics is also valuable in cases where calves are sold to other cow-calf operators—a common occurrence as calves are moved to operations with available forage. New owners may demand information on vaccination records and other animal characteristics.

APHIS/USDA estimates that in 1997, 65 percent of calves were individually identified on large cow-calf operations (USDA/APHIS, 2000a,b). Overall estimates suggest that about 49 percent of all cow-calf operators use some form of individual identification with an estimated 52 percent of calves and about 65 percent of beef cows individually identified. More operations use some form of group identification, so that about 74 percent of cows are group identified at the cow-calf level.

Identification systems not only facilitate transactions between sellers and buyers, they also help safeguard the health of the livestock sector as a whole. Animal identification and tracking systems help ensure that unhealthy animals are not allowed to contaminate healthy herds. Nearly all States require a Certificate of Veterinary Inspection (CVI) for livestock entering the State. The CVI for interstate commerce is an official document, issued and signed by a licensed, accredited, and deputized veterinarian. The CVI provides documentation that an animal or a group of animals was apparently healthy and showed no signs of contagious or communicable diseases on the date the inspection took place.

Animal identification is also an important element of Federal programs for animal disease control and eradication. For example the program targeted at eradicating brucellosis, a costly and contagious disease that can affect ruminant animals and also humans (USDA/APHIS, Dec. 2003), hinges on "Market Cattle Identification (MCI)." With MCI, numbered tags called backtags are placed on the shoulders of marketed breeding animals from beef, dairy, and bison herds. MCI, along with testing procedures, provides a means of determining the brucellosis status of animals marketed from a large area and eliminates the need to round up cattle in all herds for routine testing. In the case of test-positive animals, ownership can be more easily identified and herds that may be affected can be efficiently isolated and tested. For cattle and bison in heavily infected areas or replacement animals added to such herds, officials recommend vaccination. At the time of vaccination, a tattoo is applied in the ear; identifying the animal as an "official vaccinate." The tattoo identifies the year in which vaccination took place.

The brucellosis eradication program has had dramatic results. In 1956, testing identified 124,000 affected herds in the United States. By 1992, this number had dropped to 700 herds, and as of June 30, 2000, there were only 6 known affected herds remaining in the entire United States (USDA/APHIS, Dec. 2003).

The success of the Federal animal-disease eradication programs has not only dramatically reduced the number of diseased livestock but also reduced the motivation for animal identification for these diseases. These programs demonstrate the ability of the industry to establish traceability systems for disease control—and the ability of the industry to jettison such systems when the benefits no longer outweigh the costs.

Traceability at the Feedlot

At 6 to 18 months old and weighing 500 to 900 pounds, calves are moved to a cattle feeding operation. Cattle feeding operations, or feedlots, are enterprises largely unique to the United States and Canada. The extensive production of soymeal and corn in the United States provides an inexpensive source of animal feed and an economic rationale for feedlots. Animals are fed until they reach slaughter weights in the 1,200-1,300 lb. range—for most cattle this corresponds to 90 to 180 days in the feedlot depending on their initial weight.

Feedlots are of two major types: farmer feedlots and commercial feedlots, with the latter gaining greatly in dominance over the last three decades. The approximately 93,000 small farmer feedlots (under 1,000 head capacity) are typically one part of a grain-farm operation and may feed home-raised or purchased calves with home-raised feed. The average small farmer feedlot had an average inventory of only about 25 head in 2002 (USDA, NASS Dec. 2003).

Most commercial feedlots are located in the Western Cornbelt and Plains States of Texas, Kansas, Nebraska, Colorado, and Iowa. Commercial feedlots feed both cattle owned by the feedlot as well as other people's cattle for a fee (custom feeding). Custom-fed cattle can be owned by a cow-calf producer (called retained ownership) or by outside investors. Because of mixed ownership, identification of cattle on large commercial feedlots is more important than on farmer-owned feedlots, and consequently there is likely to be more branding or ear-tagging on commercial operations. Branding or ear-tagging also helps feedlot operators to more easily sort animals by vaccination records and breeding and other characteristics. Table 3 shows that over 98 percent of cattle on large commercial feedlots (8,000 head of cattle or more) have individual or group identifiers (large commercial lots account for 66 percent of cattle) while almost 80 percent of cattle on small commercial feedlots have such identifiers (USDA, APHIS, 2000).

Table 3—Percent of cattle identified in commercial feedlots, by size of operation[1]

	Small feedlots	Large feedlots
	Percent of cattle	
Tagged with a unique number such that each animal was individually identifiable (excluding tagging of sick animals)	29.6	31.1
Individually identified using a method other than tagging such that each animal was individually identifiable (excluding tagging of sick animals)	1.6	2.1
Identified with a group or owner identifier (pen tag, brand, hot tag, ear notch, etc.)	49.7	80.0
Not identified	21.9	1.6

[1]Small operations 1,000-7,999 head, large operations, 8,000 head or more.
Source: USDA, APHIS, 2000.

Traceability from Feedlot to Slaughter

Cattle ready for slaughter are trucked to slaughter plants. Most fed cattle are sold in direct transactions between the cattle owner (or agent) and the packing company. A typical transaction for cattle sold on a liveweight basis involves the feedlot's placing cattle on a "showlist" and packer-buyers' viewing and placing bids on cattle with a final spot price arrived at by negotiation. Many other cattle are sold on a carcass basis (payment delayed until animal is slaughtered and carcass weighed), increasingly under a contract or agreement specifying the source for a base published price and an agreed-upon schedule or "grid" of premiums and discounts based on actual carcass characteristics. Of course, in these cases, the carcass basis can be determined only after the packer slaughters the animal. The base price and adjustments produce the final "formula" price adjusted for quality.

When valuable animal characteristics are unobservable at the point of sale, traceability records linking a particular animal to records on health and other characteristics help establish the premium grid and facilitate efficient market transactions. At sale from feedlot to slaughter plant—and at every point of sale in the chain—traceability documentation enables producers to sell their cattle at a price that more accurately reflects quality. Traceability documentation is the only way to verify the existence of credence attributes such as animal "playtime" and non-genetically engineered feed.

Though traceability documentation is a valuable tool for farmers who wish to appropriate the benefits of investments in animal health or quality, it may also entail some unwelcome side effects. Traceability documentation may force farmers to "appropriate" the costs of failures in animal health or quality. The possibility that traceability could be used to place liability for unhealthy or low-quality animals on farmers makes many in the livestock sector uncomfortable. Many producers adhere to an ethic that a seller should not knowingly sell diseased or defective feeder or breeder livestock without disclosing such to the buyer, but that after an honest sale, if any problems arise with the animals' health or fitness, including death, the seller is not liable. The buyer assumes all risks associated with long-term animal health.

Livestock producers have accordingly long enjoyed some legal protection from liability for factors over which they have little or no control after the sale. Livestock has traditionally been exempt from commercial implied-warranty laws partly because farmers were considered not to be "merchants." As farms became more commercialized, and buyers more litigious, this protection has become less secure; in response many States passed specific exemptions for livestock. Some version of the statutory exclusion of implied warranties has now been adopted in almost half of the States, in particular those States where the livestock industry is of major economic importance. The Kansas statute is typical of the modification (McEowen, 1996, p. 7):

> Kan. Stat. Ann. § 84-2-316(3)(d):
>
> *[W]ith respect to the sale of livestock, other than the sale of livestock for immediate slaughter, there shall be no implied warranties, except that the provisions of this paragraph shall not apply in any case where the seller knowingly sells livestock which is diseased.*

Traceability at Slaughter

There are over 3,000 small and large firms slaughtering cattle in the United States. Most cattle slaughtered are fed steers and heifers, typically slaughtered by one of the four large major packers located in the feeding States that dominate the industry and account for about 82 percent of steer and heifer slaughter and 69 percent of all cattle. Culled cow and bull slaughter tends to occur in smaller firms, less concentrated geographically and less likely to be vertically integrated (USDA/GIPSA, 2001). In addition to domestic cattle, U.S. plants slaughter imported cattle, mainly from Canada,[2] although calves are also imported from Mexico and fed on pasture and in feedlots to slaughter weights.

FSIS (Food Safety and Inspection Service) regulations require that slaughter plants keep the head and certain organs of slaughtered animals, plus all identifying tags, until all parts of the animal pass inspection. Slaughter plants must be able to identify which head and organs belong with which carcass. In most plants this is done by keeping them physically synchronized on separate chain and conveyors. The identity of individual animals is frequently lost once inspection takes place. At this point, the health and safety of the animal has been "verified" and the focus shifts to the safety of the meat.

Traceability for Meat

Two primary motives have driven the development of traceability systems for meat and meat products: supply management and safety and quality control. Traceability systems enable slaughter plants and processors to more efficiently track the flow of product and coordinate production. Traceability systems also help plants minimize the extent of safety or quality failures, thereby minimizing damages.

A number of large foodborne illness outbreaks and heightened awareness of food safety issues have led many producers to adopt increasingly precise traceability systems. These systems reflect not just the fact that the benefits of traceability are rising, but also the fact that technological innovations are reducing the costs of traceability. These trends are expected to continue as retailers and importers demand safer food and as the science and technology of pathogen control improves, thereby spurring additional demand for traceability and additional incentives for innovation.

Meat Tracking from Slaughter/Processor to Retailer

Most large firms convert beef carcasses into primal and subprimal cuts or "boxed beef." Ground beef is processed from mixes of boneless beef imported primarily from Australia and New Zealand and trimmings from domestic animals to attain a desired fat content. Boxed beef and ground beef are shipped to retailers, food service firms, and exporters, sometimes through specialty processors, institutional processor/distributors, and meat wholesalers. Increasingly, most large firms also further cut and package "case-ready" retail cuts ready to drop into the display case in the grocery store.

Slaughter plants and processors have developed a number of sophisticated systems for tracking the flow of production and monitoring quality and safety. In accordance with ISO 9000 guidelines, most track inputs by batch or lot and then assign new batch or lot numbers to track product as it is transformed. To control foodborne pathogens such as *E. coli* O157:H7 and *Salmonella*, a number of processors have established very precise sampling, testing, and tracking protocols.

For example, one of the largest independent ground beef producers in the United States begins its traceability documentation with the trimmings entering the plant. Incoming combo bins (2,000 lbs.) of raw material are sampled at least every 100,000 pounds, which for most raw material suppliers is daily. All raw materials are routinely screened for Aerobic Plate Counts (APC), generic coliforms, generic *E. coli*, *Staphylococcus aureus*, *Salmonella*, and *Listeria monocytogenes*. If lots test higher than standards, the supplier is notified immediately and testing is intensified. Samples are next taken at the final grind head, where each batch of 3,000 pounds of ground beef is tested for *E. coli* O157:H7. Finally, samples of the finished product are taken from each process line every 15 minutes. Every hour, composites of the four samples are tested to detect *E. coli* O157:H7. These samples are also combined to make a "half-shift" composite, which is tested for an entire microbial profile. If the half-shift composites show spikes or high counts, more tests are run on the backup samples also collected every 15 minutes (Golan et al., 2004).

As a result of its testing protocol, traceability documentation is extensive for this producer. This documentation enables the producer to monitor the quality and safety of its inputs and to work with suppliers to improve the quality of inputs, or drop suppliers that cannot comply. The producer's documentation also serves to provide buyers with assurances about the quality and safety of the producer's products. As a result, this producer has been able to shift from being a commodity producer selling on a week-to-week basis to being a contract supplier to major hamburger restaurants. This shift has allowed this producer to improve its operational efficiency through better planning for capacity utilization, capital investment, spending plans, and other business activities.

[2] Or they did until the ban on importation of animals from Canada due to the discovery of mad cow disease in a single cow in western Canada, May 2003. The United States and Canada are negotiating to begin bringing cattle under 30 months of age into the United States for immediate slaughter, or to designated feedlots for slaughter at less than 30 months of age (www.usda.gov/news/releases/2003/10/0372.htm).

Though not every processor or slaughter plant maintains records as precise as in the above example, virtually all meat sold in the United States is traceable from retail back to the processor or slaughter plant. Regulations require that USDA inspection numbers for the processing plant remain on the labels of meat as they pass through the distribution systems along with other information, depending on ingredients in the meat product and marketing chain. Other firm and lot number information can be placed on labels to identify a particular processing batch from a package of meat. Most, if not all, voluntary recalls listed on USDA's Food Safety and Inspection Service website, refer consumers to coded information on products' packaging specifying the lot or batch of items included in the recall. Good product tracing systems help minimize the production and distribution of unsafe or poor-quality products, thereby minimizing the potential for bad publicity, liability, and recalls.

Linking Animal and Meat Traceability Systems

Traditionally, once carcasses have passed USDA inspection, slaughter plants have not maintained information on the identity or characteristics of each animal. Until very recently, there have not been market or human-health reasons to do so. Now, however, meat-quality pricing has begun to expand beyond characteristics that can be judged by examining the meat itself. Meat prices have begun to reflect credence attributes related to farm-level, live-animal characteristics, such as animal welfare, type of feed, and use of antibiotics and growth hormones. In addition, diseases such as mad cow have established a link between animal health and human health—and have motivated many consumers, including those represented by foreign governments, to demand traceability back to the farm and animal feeding records (see box, "Animal Identification").

In response to these new motivations, the livestock sector has begun to build traceability systems to bridge animal tracking systems with those for meat tracking. Several systems can and have been incorporated into slaughtering lines to link group or individual animals with their meat products. These include sequence-in-slaughter order, carcass tagging, trolley-tracking, and RFID devices. Some systems are capital intensive and favor larger firms that can capture economies of scale, while others are labor intensive and may actually confer an advantage to smaller operations. For example, carcass tagging may require a human to apply the tag(s) while trolleys can be tracked optically and electronically. Small low-speed operations may have an advantage in maintaining animal identification because they can more likely use physical separation and tagging. Regardless of which technology is cost effective, the success of the system depends on appropriate operating procedures and traceability recordkeeping to keep sequences and identification numbers synchronized.

Animal Identification

A national animal identification plan is being developed through a cooperative effort of USDA, State animal health officials, and livestock industry groups (see: http://www.usaip.info#.). Called the National Identification Development Team, its goal is to develop a national standardized program that can identify all premises and animals that had direct contact with a foreign animal disease within 48 hours of discovery. The plan is aimed at quickly identifying animals exposed to disease and the history of their movements in order to rapidly detect, contain, and eliminate disease threats (Wiemers, 2003). The first phase of the work requires establishing standardized premise identification numbers for all production operations, markets, assembly points, exhibitions and processing plants. The second phase calls for individual identification for cattle in commerce. Other food animal and livestock species in commerce would be required to be identified through individual or group/lot identification.

Scientific advances in animal identification will continue to reduce the cost and increase diffusion of animal-to-meat traceability. A variety of high-tech, rapid animal identification methods such as electronic implants, banding, or tagging have been developed and science is advancing to a point where DNA testing could be used to help identify and trace animal products. Unlike electronic tags and animal "passports," biological signatures would be virtually impossible to falsify and could follow the product after processing.

Though technological barriers to animal-meat traceability are rapidly dissolving, philosophical and in some cases, legal barriers remain or are being erected. As previously mentioned, livestock has traditionally been exempt from commercial implied-warranty laws. Many in the livestock sector worry that traceability systems linking meat to animals will break this tradition and shift at least some of the liability for foodborne illness back to cow-calf operators and feedlots. Some livestock organizations have even publicly called for limits on liability that may arise from animal identification. For example, the Kansas Livestock Association (2003), a nonprofit trade association representing nearly 6,000 livestock producers, has recommended:

WHEREAS, livestock producers and government officials are researching the feasibility of a national individual animal identification program, and

WHEREAS, such a program, on a voluntary or mandatory basis, could provide the livestock industry a tool to quickly trace animal disease sources and enhance a breeder's ability to identify genetics that meet consumer demands, and

WHEREAS, animal trace-back technology can increase the liability exposure for owners of animals whose food and by-products threaten or cause damages to consumers, and

WHEREAS, liability in these circumstances can often be classified as "strict liability," even though an animal owner may not be at fault for such damages.

THEREFORE, BE IT RESOLVED, the Kansas Livestock Association supports state and federal legislation to limit animal owners' liability exposure that may arise under a private or public animal identification program.

In part to overcome some of the gaps in tracing documentation and quality assurance that may arise in the system, a small but growing segment of the cattle/beef industry has entered into alliances, associations, cooperatives, or marketing groups in which groups of cattle raisers, cattle feeders, producers, slaughter plants, and processors share some combination of decisions, responsibilities, information, costs, and returns. In many cases, alliances set quality and/or safety standards and provide systems to verify that quality standards for credence attributes exist. These types of alliances or vertically integrated operations such as those found in the pork (see box, "Traceability in Hogs and Pork") and poultry sectors, use contracts and incentives to link stages of production. Links are created between entities under separate ownership to help coordinate the efforts of those entities. Alliances attempt to create a market identity with a goal of producing a product that consumers desire and for which they are willing to pay a premium and sharing that premium with upstream entities (Florida Cooperative Extension Service, 2002).

Many of the products marketed through alliances entail credence attributes that the alliance certifies to exist. In some cases, alliances or even individual producers choose to use third-party certifiers to help establish credible claims. One such certifier is USDA's Agricultural Marketing Service (AMS). AMS's service is a voluntary fee-based program that certifies claims on items such as breed, feeding practices, or other process claims. The AMS "USDA Process Verified" label provides buyers with assurances that the advertised credence attributes actually exist (USDA, AMS 2004).

Traceability in Hogs and Pork

There are traditionally three basic types of hog enterprises. The first is feeder-pig production in which the farmer specializes in farrowing operations that produce 10- to 40-pound pigs. Feeder pig producers sell or transfer pigs to others for finishing. At farrow-to-finish operations, all phases of slaughter hog production are carried out by the same operation, though not necessarily in the same physical location. Third is feeder-pig finishing, in which feeder pigs are obtained from others and fed to slaughter weights. In the last decade, hog production has become even more specialized with separate nursery and growing phases appearing between farrowing and finishing. Increasingly, hogs are raised on a batch basis—"all in all out" which facilitates cleaning facilities between batches.

In 1950, over 2 million U.S. farmers sold hogs and pigs with average sales of 31 head per farm per year. By 2002, the number of farms had fallen to around 75,000 operations. More than half of these operations had fewer than 100 head, but this small-size group had only 1 percent of the hogs. In contrast, the 2,300 operations with more than 5,000 head accounted for more than half of the hogs in 2002 (USDA, NASS, Dec. 2003). Much larger mega-farms have been evolving into more important players; the 200 or so mega-farms are highly integrated. Some have more than 30,000 sows under tightly contracted or integrated arrangements from breeding to slaughter or even retail. Identification by herd or batch is therefore much higher today than 50 years ago.

Many hog operations, both large and small, not just mega-farms, are integrated by ownership or contractually connected to slaughtering firms. Less than 20 percent of slaughter hogs were sold in the spot market in 2002. Another large group of slaughter hogs are sold on a formula basis, sometimes under a continuing agreement. Hogs produced by or under contract for slaughter firms require no market transaction between the finisher and the slaughtering firm. Thus, the road from hogs to pork is far more integrated than in the cattle/beef sector—as are traceability systems.

Conclusions

The livestock industry has successfully developed and maintained a host of traceability systems: some for live animals and some for meat. Ranchers, cow-calf operators, and feedlot operators have had at least three motives in developing live-animal traceability: to establish ownership; to control animal diseases and quality; and to facilitate quality-based pricing. Livestock owners have established animal traceability systems to meet one or more of these objectives—and have expanded or contracted systems to reflect dynamics in animal management, disease outbreaks, and consumer preferences for credence attributes.

Slaughter plants and processors have had two primary motives for establishing traceability for meat products: to manage their supply chains and assure quality control and food safety. Traceability systems enable slaughter plants and processors to more efficiently track the flow of product and to coordinate production. Traceability systems also help plants minimize the extent of safety or quality failures, thereby minimizing damages. A number of large foodborne illness outbreaks and heightened awareness of food safety issues have led many producers to adopt increasingly precise meat traceability systems—a trend that is expected to continue with ever-increasing demands for food safety.

The challenge facing the industry now is to coordinate and link many disparate animal and meat traceability systems and priorities and develop a standardized system for identifying farm-level, live-animal attributes in finished meat products. Two institutional barriers may hinder these efforts. First, because USDA determines and certifies an animal's health and its suitability for the human food chain, meat processors may not have as much of an incentive to retain information on the origin of each piece of meat as they would if they were solely responsible for ensuring animal health.

Second, livestock has traditionally been exempt from commercial implied-warranty laws and many institutional or legal barriers are being constructed to safeguard this tradition. Limiting the liability of the cow-calf operator or feedlot will dampen incentives to establish traceability from meat to animal. Traceability to the animal supplier is less valuable if the supplier cannot be held legally accountable for diseased animals.

In part to overcome some of the gaps in tracing documentation and quality assurance that may arise in the system because of limits to liability, a small but growing segment of the cattle/beef industry has turned to alliances, associations, cooperatives, or marketing groups to help establish and enforce quality and safety standards and facilitate linking animal-tracking systems and traceability of meat products. The U.S. Animal Identification Plan is another major effort in this direction.

IV. Market Failure in the Supply of Traceability:

Industry and Government Response

It is not surprising to find systematic variation in traceability systems across sectors of the food industry because the costs and benefits of traceability vary systematically. Each sector has confronted different motivations for and constraints to erecting traceability systems. Different food safety problems, supply management concerns, and demands for credence attributes have motivated different sectors of the food industry to build traceability systems that vary in breadth, depth, and precision. Differences in product characteristics and infrastructure have led to differences in traceability costs that have also influenced the breadth, depth, and precision of the different systems.

Variation in traceability systems tends to reflect an efficient balancing of private costs and benefits. Are there, however, cases where variation actually signals market failure? Does the private sector supply of traceability fail to satisfy important social objectives?

The economic literature on market failure suggests that insufficient traceability in food markets could arise as a result of asymmetric or missing information problems in markets for food or as a result of externality or public good aspects of traceability. We find that though these possibilities arise, they do not typify the three food sectors we investigated. In all three food sectors, the private sector has developed methods to address costly market failure problems. We do find, however, that public good aspects of traceability may result in a less than optimal supply of traceability for identifying contaminated food once it is has been bought and consumed. In the sections below, we examine areas of potential market failure and industry and government response.

Market Failure and Differentiated Markets for Quality and Safety

Though firms have an incentive to use traceability systems to help generate information on credence attributes of value, they do not have an incentive to generate information about credence attributes that are not of value or have a negative value. As a result, the market may produce too little information about negative attributes. This potential is mitigated through the process of competitive disclosure. For example, though a food product may not sport a "high fat" label, the fact that rival brands are labeled "low fat" may lead consumers to conclude that the unlabeled product is in fact high in fat. This competitive disclosure, which Ippolito and Mathios (1990) named the "unfolding" theory, results in explicit claims for all positive aspects of products and allows consumers to make appropriate inferences about foods without claims.

However, competitive unfolding tends not to work when an entire product category has an undesirable characteristic that cannot be changed appreciably or for which the costs of alteration are too high, or where disclosure of the attribute may have negative repercussions. One area where product differentiation may be lacking is food safety. Very few firms seek to differentiate their product for consumers with respect to food safety (Golan et al., 2004). This may reflect the fact that foodborne pathogens are a commonly shared problem that is difficult to control with precision (Roberts et al., 2001). Firms may want to avoid specific safety guarantees that could expose them to additional liability because there is always the possibility that even the most careful producer could experience a safety problem. As a result, even the best producers may refrain from marketing safety to final consumers or trying to differentiate themselves from less safe producers.

Firms may also shy away from differentiating themselves and their safety records through traceability or other mechanisms if there is value in some level of anonymity (Starbird and Amanor-Boadu, 2003). If traceability systems increase the probability that a firm will be identified as a source of food safety problems and exposed to liability and bad publicity, then the firm may have an incentive to remain anonymous even if it has a good safety record. The benefits of product differentiation may not outweigh the costs of being more easily linked to a food product in the case of safety problems. In these cases, the market solution results in less disclosure than desired by consumers or less traceability than is socially optimal.

The amount of traceability offered by private firms for product differentiation may also be less than socially optimal if the benefits to the firm of establishing traceability for credible product differentiation is dampened by the existence of partial disclosure and innuendo. In some cases, the possibility of deception may erode producers' incentives to establish traceability systems because widespread deception makes consumers doubt the veracity of claims made by all producers, even honest ones. For some honest producers, the benefits of overcoming this high degree of consumer doubt will not outweigh the costs. For example, prior to the introduction of national organic standards, the proliferation of organic standards and labels—some more "organic" than others—may have made it difficult and costly for true organic producers to differentiate their product. Since credence attributes are inherently difficult to verify, they may be especially susceptible to fraud and unfair competition.

Industry Efforts To Bolster Differentiation

In the three food sectors we investigated, producers seem to be responsive to consumer demand for product differentiation. When consumer demand was strong enough to cover the cost of product differentiation, producers responded with new products and new traceability systems to substantiate credence attribute claims. While producers have difficulties marketing safety attributes directly to consumers, producers routinely market safety at earlier stages in the supply chain. The rich variety of differentiated products for sale in the fresh fruit and vegetable, grain, and livestock and meat sectors of the food industry—and the size and diversity of the industry—argue against the conclusion that market failure is stifling product differentiation in any of these markets. And, where market failure may have begun to emerge with respect to credence attributes, individual firms and industry groups have developed systems for policing the veracity of credence claims and for creating markets for differentiated products. Third-party safety/quality auditors are at the heart of these efforts.

Third-party entities (neither the buyer nor the seller) provide objective validation of quality attributes and traceability systems. They reassure input buyers and final consumers that the product's attributes are as advertised. Third-party verification of credence attributes can be provided by a wide variety of entities, including consumer groups, producer associations, private third-party entities, and international organizations. For example, Food Alliance and Veri-Pure, private for-profit entities, provide independent verification of food products that are grown in accordance with the principles of sustainable agriculture. Third-party entities certify attributes as wide ranging as kosher, free-range, predator-friendly, no-hormone use, location of production, and "slow food." Governments can also provide voluntary third-party verification services. For example, to facilitate marketing, producers may voluntarily abide by commodity grading systems established and monitored by the government.

Third-party entities also offer services to validate safety procedures and bolster market differentiation with respect to food safety. A growing number of buyers, including many restaurants and some grocery stores, are beginning to require that their suppliers establish safety/quality traceability systems and to verify, often through third-party certification, that such systems function as necessary. A growing number of firms are beginning to try to differentiate the safety of their products and processes for input buyers.

Most, if not all, third-party food-safety/quality certifiers such as the Swiss-based Société Générale de Surveillance (SGS) and the American Institute of Baking (AIB) recognize traceability as the centerpiece of a firm's safety management system. For example, AIB's standard food safety audit specifies a number of traceability-specific activities including (American Institute of Baking, 2003):

- Records were maintained for all incoming materials indicating date of receipt, carrier, lot number, temperature, amounts, and product condition.
- A documented, regularly reviewed, recall program was on file for all products manufactured. All products were coded, and lot or batch number records were maintained. Distribution records were maintained to identify the initial distribution and to facilitate segregation and recall of specific lots.
- All raw materials were identified in the program and work in progress, re-work, and finished products were traceable at all stages of manufacture, storage, dispatch and, where appropriate, distribution to the customer.

Third-party standards and certifying agencies are employed across the food industry. In 2002, AIB audited 5,954 food facilities in the United States and was slated to audit 6,697 in 2003 (Wohler, 2003); SGS expected to perform over 1,000 U.S. food safety audits in 2003 (Guidry and Muliyil, 2003); and ISO management standards are implemented by more than 430,000 organizations in 158 countries (ISO website). Food sectors employing third-party verifiers cover the spectrum from spices and seasoning to fruit and vegetables to meat and seafood to bakery products and dough. The growth of third-party standards and certifying agencies is helping to push the whole food industry—not just those firms that employ third-party auditors—toward documented, verifiable traceability systems.

Third-party audits provide customers, buyers, and in some cases, governments with assurances that a firm's safety management systems, including its traceability systems, have met some objective standards for quality. These assurances have potential to translate into increased demand because they foster confidence in the safety of the firm's products on the part of downstream and final customers. These assurances are helping to reduce the potential for market failure and to bolster markets for safety and quality.

Government Efforts To Bolster Differentiation

Government may also try to stimulate the supply of information and product differentiation. Mandatory traceability has been suggested as one possible policy option for supplying consumers with more information about credence attributes, including such diverse attributes as country of origin and genetic composition. One difficulty with such proposals is that they often fail to differentiate between valuable quality attributes, those for which verification is needed, and other less valuable attributes. For example, a government policy requiring that producers of valuable organic foods provide verification that these foods are indeed organic could protect consumers from fraud and producers from unfair competition. No such verification would be necessary for conventionally produced foods. Consumers do not need proof that conventional foods are indeed conventional—there is no potential for fraud in this case, no danger that producers would try to cheat consumers by misidentifying organic as conventional. A mandatory traceability system for both organic and conventional foods is unnecessary to protect consumers from fraud or producers from unfair competition.

Likewise, government may have an incentive to require that producers of foods that are not genetically engineered verify that these foods are in fact not genetically engineered, if that attribute is of value to some consumers. However, no such verification would be necessary for the genetically engineered foods currently on the market, because this attribute is not of value to consumers (most genetically engineered products currently on the market have producer, not consumer attributes). A mandatory traceability system for both genetically engineered and non-genetically engineered foods is unnecessary to protect consumers from fraud or producers from unfair competition. Such a system would raise costs without generating compensating benefits. Mandatory traceability for product differentiation that is not targeted to specific attributes of value to consumers will be costly and unnecessary.

Another difficulty with mandatory traceability lies in the propensity for government programs to require uniformity. As our industry review illustrates, private firms operate a wide variety of complex, highly sophisticated traceability systems. A government-mandated system that required all firms to adopt the same template could be highly costly and inefficient. For example, mandatory traceability systems requiring a common or standard lot size could result in enormous, unnecessary costs to industry. One meat processor found that, by working with USDA to develop a sub-lot sampling system, it was able to reduce the amount of product that needed to be destroyed in cases of contamination and, as a result, substantially reduced its destruction costs. In another case, a fruit producer found that USDA safety requirements specifying a particular lot size led to the development of a complicated traceability system that did not mesh with the plant's production/transportation system.

A flexible government-mandated system would likely be more efficient and less burdensome than one that required that all firms revamp their traceability systems to conform to a standard template. In the United States, both AMS and FSA rely on industry-developed traceability and bookkeeping systems to monitor the domestic origin of food purchased for Federal procurement programs. Programs such as the U.S. national organic food standard depend on private certifiers to provide flexibility to the system. Organic food certifiers, approved by the U.S. Department of Agriculture, work with growers and handlers to develop individualized recordkeeping systems to assure traceability of food products grown, marketed, and distributed in accordance with national organic standards.

Market Failure and Traceability for Food Safety

Though failure by private markets to supply adequate traceability for product differentiation is a concern to regulators, an even bigger concern is failure by private markets to supply adequate traceability systems for basic food safety control and monitoring. In some cases, the amount of traceability supplied by firms may be less than the social optimum because the public health benefits of traceability for food safety are larger than the firm's benefits. A firm's food safety traceability benefits include the reduction in the potential for lost markets, liability costs, and recalls, while the potential social benefits include a long list of avoided costs, including medical expenditures and productivity losses due to foodborne illness, costs of pain and suffering, and the costs of premature death.

Social benefits may also include the avoided costs to firms that produce safe products but lose sales because of safety problems in the industry. A firm's traceability system not only helps minimize potential damages for the individual firm, it also helps minimize damages to the whole industry and to upstream and downstream industries as well. For example, a series of widespread ground meat recalls has the potential to hurt the reputation and sales of the entire meat industry, including downstream industries such as fast food restaurants and upstream suppliers such as ranchers. The benefits to the industry of a traceability system pinpointing the source of the bad meat and minimizing recall (and bad publicity) could therefore be much larger than the benefits to the individual firm.

As mentioned in the section on differentiation, the amount of traceability supplied by firms may also be lower than the social optimum because firms may find value in some level of anonymity. If traceability systems increase the probability that a firm will be identified in the case of food safety problems and exposed to liability, then the firm may have an incentive to underinvest in traceability: the value of anonymity may reduce the firm's incentives to invest in traceability systems.

Private cost-benefit calculations may also differ from social calculations if the costs of erecting traceability systems are lower when industry groups or governments undertake these projects than when individual firms build them on their own. Or, once built, the marginal cost of including other firms or foods in the traceability system may be small or nothing. In these cases, the private benefits of such systems may not outweigh the private costs while the social benefits do outweigh the social costs. Public defense and libraries are classic examples of such a situation; traceability systems for detecting and tracing foodborne illness outbreaks to their source may be another.

Firms have an incentive to identify and isolate unsafe foods and to remove them from the supply chain as quickly as possible. Few firms, however, have an incentive to monitor the health of the Nation's consumers in order to speed the detection of unsafe product. Such a traceability system would be extremely expensive and would be poorly targeted to any individual firm's needs. The benefits to an individual firm of building a system to monitor all foodborne illness outbreaks just in case one is linked to the firm's product would certainly not outweigh the costs. However, the collective benefits to industry and to consumers may well outweigh the costs. Early detection and removal of contaminated foods can reduce the incidence of foodborne illness and save lives.

Industry Efforts To Increase Traceability for Food Safety

A host of new food safety concerns have pushed food industries to reevaluate their safety protocols, including their traceability systems. For the most part, industry has worked to strengthen safety systems in response to new threats, though the speed and success of industry response has varied. The fresh fruit and vegetable sector has probably been the most successful in adjusting traceability systems in response to new safety problems. This reflects the fact that firms in the sector have already established robust traceability systems and that the industry has experienced a series of foodborne illness outbreaks.

In the mid-1990s a series of well-publicized outbreaks, traced back to microbial contamination of produce, raised public awareness of potential problems. Recent outbreaks like the one traced to scallions served at a restaurant chain, continue to focus public attention on safety of fresh fruit and vegetables. Good-agricultural-practice audits, including traceability audits, are becoming a necessary part of doing business, as more and more buyers demand safety assurances. In addition, several grower organizations have developed systems to strengthen traceability. In the case of an outbreak, a grower organization that encourages traceback can prove to the public that their product is not responsible for the problem. Or, in the unfortunate case where the industry is responsible for the outbreak, the problem grower or growers can be identified and damage can be limited to that group.

The grain industry has yet to experience a well-publicized, pivotal safety problem. There have not been any major safety scares that would warrant the reevaluation of the industry's safety system, including its traceability systems. The highly processed nature of the product, and the large number of critical safety points along the production chain, largely eliminate safety problems that may arise early in the production process, thereby reducing the need for detailed traceability systems.

The beef sector may be experiencing the most difficulty of the three sectors in responding to new safety threats. These difficulties can be traced to uncertainties in the science of food safety and pathogen control in meat and institutional and philosophical barriers to traceability in the sector. Despite these difficulties, the industry has developed a number of approaches for strengthening food safety accountability and traceability. For example, the Beef Industry Food Safety Council (BIFSCo) has taken on the task of organizing representatives from all segments of the beef industry to develop industry-wide, science-based strategies to solve the problem of *E. coli* O157:H7 and other foodborne pathogens in beef. Industry groups are also cooperating to develop the national animal identification plan (see box, "Animal Identification," p. 32).

Buyers in the beef industry are also increasingly relying on contracting or associations to improve product traceability and safety. Fast food restaurants and other retailers have begun adopting the role of channel captains, monitoring the safety of products up and down the supply chain. By demanding safer products from their suppliers, these restaurants have successfully created markets for food safety. The success of these markets rests on the ability of these large buyers to enforce standards through testing and process audits—and to identify and reward suppliers who meet safety standards and punish those who do not. These large buyers have spurred the development of traceability systems throughout the industry.

Government Efforts To Increase Traceability for Food Safety

Mandatory traceability is one possible policy tool for increasing the food system's traceback capability. However, since the government's primary objective for food safety traceback is the swift identification and removal of unsafe foods, other policy tools may be more efficient than mandatory traceability. Policy aimed at ensuring that foods are quickly removed from the system, while allowing firms the flexibility to determine the manner, will likely be more efficient than mandatory traceability systems. For some firms, plant closure and total product recall may be the most efficient method for isolating production problems and removing contaminated food from the market. For other firms, detailed traceback, allowing the firm to pinpoint the production problem and minimize the extent of recall may be the most efficient solution. In either case, contaminated food is quickly removed from distribution channels and the social objective is achieved.

A performance standard, such as a standard for mock recall speed, is one possible policy tool for providing firms with incentives to establish efficient traceability systems. Mock recalls are a good tool for checking the ability of a system to quickly and accurately identify and remove contaminated product. In the United States, the two Federal agencies responsible for food safety, the U.S. Department of Agriculture (USDA) and the U.S. Food and Drug Administration (FDA), encourage firms to perform mock or simulated recalls to ensure that potentially contaminated foods can be tracked and removed from the system in an expedient manner. In addition, most, if not all, third-party safety/quality control certifiers require traceability documentation and mock recalls as part of their safety audits. Depending on the needs of the client, many also monitor and time mock recalls to evaluate the speed and precision with which facilities can identify potentially contaminated product. Société Générale de Surveillance monitors a 2-hour mock recall for many of its clients.

One area where industry has not had any incentive to create traceability systems is in tracking food once it has been sold and consumed. Firms have an incentive to identify and isolate unsafe foods and to remove them from the supply chain as quickly as possible. But, few firms have an incentive to monitor the health of the Nation's consumers in order to speed the detection of unsafe product. Such a traceability system would be extremely expensive and would be poorly targeted to any individual firm's needs. The benefits of building a system to monitor all foodborne illness outbreaks just in case one is linked to the firm's product would certainly not outweigh the costs. However, the collective benefits to industry and to consumers may outweigh the costs. Government-supplied foodborne illness sentinel systems could, therefore, play an important role in closing gaps in the food systems traceability system. By providing this public good, the government could increase the capability of the whole food supply chain to efficiently and quickly respond to food safety problems.

In the United States, the Federal Government and other public health entities have taken strides in building the infrastructure for tracking the incidence and sources of foodborne illness. The Foodborne Diseases Active Surveillance Network (FoodNet) combines active surveillance for foodborne diseases with related epidemiologic studies to help public health officials better respond to new and emerging foodborne diseases. FoodNet is a collaborative project of the Centers for Disease Control and Prevention (CDC), nine States, USDA, and the FDA. Another network, *PulseNet*, based at CDC, connects public health laboratories in 26 States, Los Angeles County, New York City, the FDA, and USDA to a system of standardized testing and information sharing.

With better surveillance of foodborne illness outbreaks, regulators can increase the likelihood that unsafe foods and unsafe producers will be more quickly identified. Better surveillance therefore reduces the risk of foodborne illness in two ways: by more quickly removing unsafe food from the food supply and by putting additional pressure on suppliers to produce safe foods. By increasing the likelihood that unsafe producers are identified, surveillance systems increase the likelihood that these producers will bear some of the costs of unsafe production, including recall, liability, and bad publicity. Increased surveillance therefore increases the potential costs of selling unsafe food, providing producers with increased incentive to invest in safety systems, including traceability systems.

V. Conclusions

In our investigation into the adequacy of the private-sector supply of traceability, we found that the private sector has a number of reasons to establish and maintain traceability systems and, as a result, the private sector has a substantial capacity to trace. This does not mean that the wheat in every slice of bread is traceable to the field or that the apples in every glass of apple juice are traceable to the tree. Firms evaluate their costs and benefits with respect to supply management, safety, and credence-attribute marketing to determine the efficient breadth, depth, and precision for their traceability systems. The net benefits of establishing and maintaining traceability systems are not necessarily positive for every attribute, for every step of the supply chain, or for the highest degree of precision.

Traceability systems are a tool to help firms manage the flow of inputs and product to improve efficiency, food safety and product quality, and product differentiation. However, traceability systems do not accomplish any of these objectives by themselves. Simply knowing where a product is in the supply chain does not improve supply management unless the traceability system is paired with a real-time delivery system or some other inventory-control system. Tracking food by lot in the production process does not improve safety unless the tracking system is linked to an effective safety control system. And of course, traceability systems do not create credence attributes, they simply verify their existence. Traceability systems are one element of a firm's supply side management system, safety system, and production strategy. Traceability systems are built to complement the other elements in each system.

The development of traceability systems throughout the food supply system reflects a dynamic balancing of benefits and costs. Though many firms operate traceability systems for supply management, quality control, and product differentiation, these objectives have played varying roles in driving the development of traceability systems in different sectors of the food supply system. In the fresh produce sector, quality control and food scare problems have been the primary motivation pushing firms to establish traceability systems. In the grain sector, supply management and growing demand for high-value attributes is pushing firms to differentiate and track production. In the beef sector, food scares and demand for high-value traceability systems have only recently begun to motivate firms to adopt traceability systems tracking production from animal to final meat product.

The varying costs of traceability systems, reflecting different product characteristics, industry organization, production processes, and distribution and accounting systems, have also influenced the development of traceability systems across the food supply. The development of traceability systems in the fresh produce industry has been greatly influenced by the characteristics of the product. Perishability of and quality variation in fresh fruits and vegetables necessitate that the product be boxed and its quality attributes identified early in the supply chain, either in the field or in the packinghouse. This practice has facilitated the establishment of traceability for a number of objectives including marketing, food safety, supply management, and differentiation of new quality attributes. In grains, safety and quality are largely controlled at the elevator level, greatly reducing the need for traceability throughout the sector. For beef, institutional and philosophical barriers have slowed the adoption of traceability systems for tracking animals from farm to table. In every sector, technological innovations are helping to reduce traceability costs and to spur the adoption of sophisticated systems.

Our investigation of the private supply of traceability in the United States has led us to conclude that for the most part, the food industry is successfully developing and maintaining traceability systems to meet changing objectives. In the three food sectors we investigated, producers seem to be responding to consumer demand for product differentiation. When final or input demand is strong enough to cover the cost of product differentiation, producers have responded with new products and new traceability systems to substantiate credence attribute claims, including food safety claims. To control for potential fraud or unfair competition, industry groups and individual firms are increasingly relying on the services of third-party auditors to verify the existence of credence attributes.

For the most part, industry has also worked to strengthen food safety systems in response to new threats, though the speed and success of the response has varied. The fresh fruit and vegetable sector has probably been the most successful in adjusting traceability systems in response to new safety problems, while the beef industry, with its history of limited liability, seems to have had the most difficulties. In all three food sectors, alliances, vertical integration, and contracts are facilitating traceability for safety and other quality attributes.

Our analysis suggests that government mandated and managed traceability is usually not the best-targeted policy response to potential market failures involving traceability. Even in those cases where traceability is necessary for the development of differentiated markets, mandatory traceability systems often miss the mark. Systems that include attributes that are not of value to consumers generate costs without any corresponding benefits. Only systems that focus on attributes of value to consumers actually facilitate market development. In addition, the widespread voluntary adoption of traceability may complicate the application of mandatory systems. Mandatory systems that prescribe one traceability template and fail to allow for variation across systems are likely to impose costs that are not justified by efficiency gains.

One area where the government may be able to increase the supply of a valuable public good is by augmenting tracking systems for contaminated food once it has been bought and consumed. By strengthening foodborne illness surveillance systems to speed the detection of foodborne illness outbreaks and the identification of the source of illness, the government could increase the capability of the whole food supply chain to efficiently and quickly respond to food safety problems. In addition, because they increase the likelihood that unsafe producers are identified, surveillance systems may provide producers with increased incentive to invest in safety systems, including traceability systems. In fact, any policy that increases the cost and probability of getting caught selling unsafe food provides producers with incentives to increase their traceback capabilities. These types of policies will encourage the development of more efficient systems for the swift removal of unsafe foods and for investment in safer food systems—which is the ultimate objective of food safety policy.

Appendix

Select Milestones in U.S Traceability Requirements for Foods

The U.S. Federal Government has a long, albeit limited, history of mandating programs that contain traceability requirements. Government regulations have a diverse set of objectives. Often, they take into consideration ensuring a level of food safety, preventing and limiting animal diseases, or facilitating market transactions. Some of these regulations entail establishing traceability systems for select attributes in particular food subsectors, while other regulations have broader objectives but, in effect, require firms to develop tracing capacity. Whether the intent of the regulation is to address food safety or animal disease concerns or other issues, Government-imposed demands for traceability usually require information about the sellers and buyers (name, address, phone, etc.) and product-related information. The demands on recordkeeping are usually one-up, one-back traceability. Less frequently required are traceability systems for quality credence attributes that have become more prevalent in the private sector, although there are exceptions, such as the national organic food standard.

Below we briefly highlight some important regulations that require traceability systems. We indicate the relevant legislation, the objectives of the regulations, the product coverage and the recordkeeping that is required. The list is not intended to be encyclopedic but, instead, illustrative of important and recent legislation that affects tracing by food suppliers.

Meat, Poultry, and Egg Inspection Acts

Key Legislation and Dates: Legislation was passed in 1906 for meats, 1957 for poultry, and 1970 for eggs. The Wholesome Meat and Poultry Acts of 1967 and 1968 substantially amended the initial legislation.

Objective: The Meat, Poultry, and Egg Inspection Acts have the primary goals of preventing adulterated or misbranded livestock and products from being sold as food and to ensure that meat and meat products are slaughtered and processed under sanitary conditions. The Food Safety and Inspection Service (FSIS), USDA, is responsible for ensuring that these products are safe and accurately labeled.

Coverage: Livestock, meat, poultry, and shell eggs and egg products.

Recordkeeping Required: The Acts call for complete and accurate recordkeeping and disclosure of all transactions in conducting commerce in livestock, meat, poultry, and eggs.

For example, packers, renderers, animal food manufacturers, or other businesses slaughtering, preparing, freezing, packaging, or labeling any carcasses must keep records of their transactions. Businesses only need to maintain one-up, one-back records.

For imported meat, poultry, and egg products, importers must satisfy requirements of two USDA agencies—FSIS and the Animal and Plant Health Inspection Service (APHIS)—and the U.S. Customs Service (USDA, FSIS, October 2003). Imported meat and poultry must be certified, not only by country but by individual establishment within a country. Certificates are issued by the government of the exporting country and are required to accompany imported meat, poultry, and egg products to identify products by country and plants of origin, destination, shipping marks, and amount. FSIS demands that the country of origin provide a health certificate indicating the product was inspected and passed by the country's inspection service and is eligible for export to the United States. To meet APHIS requirements, the product must not come from countries where certain animal diseases are present. USDA requirements are binding as the U.S. Customs Service demands that the importer post a bond, including the value of the product plus duties and fees, until FSIS notifies the Service of the results of its reinspection. Failure to meet U.S. requirements may lead to forfeiture plus penalties.

Perishable Agricultural Commodities Act

Key Legislation and Dates: Perishable Agricultural Commodities Act (PACA) was enacted in 1930.

Objective: PACA was enacted to promote fair trading practices in the fruit and vegetable industry. The objective of the recordkeeping is to help facilitate the marketing of fruit and vegetables, to verify claims, and to minimize any misrepresentation of the condition of the item, particularly when long distances separate the traders.

Coverage: Fruit and vegetables.

Recordkeeping Required: PACA calls for complete and accurate recordkeeping and disclosure for shippers, brokers, and other first handlers of produce selling on behalf of growers. PACA has extensive recordkeeping requirements on who buyers and sellers are, what quantities and kinds of produce is transacted, and when and how the transaction takes place. PACA regulations recognize that the varied fruit and vegetable industries will have different types of recordkeeping needs, and the regulations allow for this variance. Records need to be kept for 2 years from the closing date of the transaction.

National Shellfish Sanitation Program

Key Legislation and Dates: Federal Food, Drug, and Cosmetic Act, portions revised or new as amended by the Food and Drug Administration (FDA) Modernization Act and various State health regulations.

Shellfish must comply with the general requirements of the Federal Food, Drug, and Cosmetic Act and also with requirements of State health agencies cooperating in the National Shellfish Sanitation Program (NSSP) administered by the FDA in cooperation with the Interstate Shellfish Sanitation Conference (ISSC) (FDA, CFSAN, January 2003).

Objective: A key objective is to mitigate the adverse effects of a disease outbreak. Regional FDA specialists with expert knowledge about shellfish assist State officials with traceback. When notified rapidly about cases, they are able to sample harvest waters to discover possible sources of infection and to close waters when problems are identified.

Coverage: Shellfish.

Recordkeeping Required: Shellfish plants certified by the State Shellfish Sanitation Control Authority are required to place their certification number on each container or package of shellfish shipped. The number indicates that the shipper is under State inspection, and that it meets the applicable State requirements. It is central to tracing and identifying contaminated shipments. Shippers are also required to keep records showing the origin and disposition of all shellfish handled and to make these records available to the control authorities.

Organic Foods Production Act

Key Legislation and Dates: Organic Foods Production Act was enacted in 1990. Act was subsequently amended and rules went into effect October 2002.

Objective: The objective is to establish national standards governing the marketing of certain agricultural products as organically produced products, to assure consumers that organically produced products meet national production, handling, and labeling standards, and to facilitate commerce in fresh and processed food that are organically produced.

Coverage: Organic foods.

Recordkeeping Required: Organic food certifiers work with growers and handlers to develop an individualized recordkeeping system to assure traceability of food products grown, marketed, and distributed in accordance with national organic standards (USDA, AMS, October 2002). Records can be adapted to the particular business as long as they fully disclose all activities and transactions in sufficient detail to be readily understood, have an audit trail sufficient to prove that they are in compliance with the Act, and are maintained for at least 5 years. Many different types of records are acceptable. For example, documents supporting an organic system may include field, storage, breeding, animal purchase, and health records, sales invoices, general ledgers, and financial statements.

In order for the attribute "organic" to be preserved, growers and handlers must maintain traceability from receiving point to point of sale and ensure that only organic or approved materials are used throughout the supply chain. Thus, for a traceability system for organic products to be viable it must confer depth.

Food Assistance Programs

Key Legislation and Dates: The National School Lunch Act was enacted in 1946 after World War II.

Objective: To reduce malnutrition and improve poor eating habits, the U.S. Department of Agriculture provides food assistance to schools, Native American reservations, and needy families, the elderly, and the homeless through Federal Food Assistance Programs. In addition to financial subsidies for food purchases, the institutions receive entitlement and bonus commodities. The bonus commodities are procured to support the farm community in specific commodity markets that are experiencing weak market conditions.

Coverage: Flour, grains, oils and shortenings, dairy, red meat, fish, poultry, egg, fruit, vegetable, and peanut products.

Recordkeeping Required: To guarantee that foods are strictly American, producers who win U.S. Department of Agriculture contracts must provide documentation establishing the origin of each ingredient in a food product (USDA, AMS, 2003). The producer pays USDA inspectors to review the traceability documents and certify the origin of each food. Starting with the "code" or lot number on a processed product, inspectors use producer-supplied documentation to trace product origins all the way back to a grower's name and address.

Country of Origin Labeling

Key Legislation and Dates: The legislation amends the Agricultural Marketing Act of 1946 by incorporating country of origin labeling (COOL) in the Farm Security and Rural Investment Act of 2002 (Public Law 107-171). Specific guidelines for voluntary labeling were issued in 2002 and are currently in effect (USDA, AMS, October 11, 2002). Mandatory labeling rules were proposed in October 2003. The Farm Act states that mandatory COOL is to be promulgated no later than September 30, 2004. However, the 2004 Omnibus Appropriations Act delays until September 20, 2006, implementation for all covered commodities, except wild and farm raised fish, which must be labeled beginning September 30, 2004.

Objective: The objective is to provide consumers with more information regarding the country where covered commodities originate.

Coverage: The legislation affects the labeling of beef, pork, lamb, fish, shellfish, fresh fruit, vegetables, and peanuts. COOL is not required if these foods are ingredients in processed food items or are a combination of substantive food components. Examples include bacon, orange juice, peanut butter, bagged salad, seafood medley, and mixed nuts.

Food service establishments such as restaurants, food stands, and similar facilities including those within retail stores (delicatessens and salad bars, for example) are exempt from the requirements. Moreover, grocery stores that have an annual invoice value of less than $230,000 of fruits and vegetables are exempt from COOL requirements. Consequently, retail food outlets, like butcher shops and fish markets that do not sell fruit and vegetables, are not included under COOL requirements.

Recordkeeping Required: Retailers may use a label, stamp, mark, placard, or other clear and visible sign on the covered commodity, or on the package, display, holding unit, or bin containing the commodity at the final point of sale.

The Act and the proposed rules have stringent requirements on the depth of recordkeeping. First, the supplier responsible for initiating the country-of-origin declaration must establish and maintain records that substantiate the claim. If a firm already possesses records, then it is not necessary to create and maintain additional information. As a vertical supply chain, there must be a verifiable audit trail to ensure the integrity of the traceability system, that is, firms must assure the transfer of information of the country-of-origin claim. As a consequence, firms along the supply chain must maintain records to establish and identify the immediate previous source and the immediate subsequent recipient of the transaction. For an imported product, the traceability system must extend back to at least the port of entry into the United States. Firms have flexibility in the types of records that need to be maintained and systems that transfer information. Records need to be kept for 2 years.

The proposed rules provide flexibility in the type of recordkeeping. The Act states that the Secretary shall not use a mandatory identification system to verify country of origin. The U.S. Department of Agriculture provides examples of documents and records that may be useful to verify compliance with the Country of Origin Labeling provisions of the 2002 Farm Bill. (See http://www.ams.usda.gov/cool/records.htm.) These records vary depending on the business activities. As an example, a ship catching wild fish may keep records of site maps, and vessel, harvesting, and U.S. flagged vessel identification records. A distributor of wild fish may keep records of invoices, receiving and purchase records, sales receipts, inventories, labeling requirements, a segregation plan, and UPC codes.

Public Health Security and Bioterrorism Preparedness and Response Act

Key Legislation and Dates: The Public Health Security and Bioterrorism Preparedness and Response Act of 2002 provides new authority to the Federal Drug Administration.

Objective: The objective is to protect the Nation's food supply against the threat of serious adverse health consequences to human and animal health from intentional contamination.

Coverage: All foods are subject to the legislation except meat, poultry, and eggs (which are under U.S. Department of Agriculture's jurisdiction).

The Act requires both domestic and foreign facilities to register with the FDA no later than December 12, 2003 (FDA, CFSAN, 2002). Facilities subject to these provisions are those that manufacture, process, pack, transport, distribute, receive, hold or import food. The Act exempts farms, restaurants, other retail food establishments, nonprofit food establishments in which food is prepared for or served directly to the consumer; and fishing vessels from the requirement to register. Also, foreign facilities subject to the registration requirement are limited to those that manufacture, process, pack, or hold food, only if food from such facility is exported to the United States without further processing or packaging outside the United States.

Recordkeeping Required: The Act requires the creation and maintenance of records needed to determine the immediate previous sources and the immediate subsequent recipients of food (i.e., one-up, one-down). For imported food the rules also require prior notice of shipment and a description of the article including code identifiers, the name, address, telephone, fax, and email of the manufacturer, shipper, and the grower (if known), the country of origin, the country from which the article is shipped, and anticipated arrival information. Records are required to be retained for 2 years except for perishable products and animal foods (for example, pet foods) where 1 year of recordkeeping is allowed. Records may be stored offsite.

Food Safety and Inspection Service, USDA, has jurisdiction of meats, poultry, and eggs. FSIS has been issuing guidance to businesses engaged in production and distribution of these USDA-regulated foods. Among the guidance principles for slaughter and processing facilities, FSIS recommends validated procedures to ensure the traceback and traceforward of all raw materials and finished products.

References

American Institute of Baking. *Food Safety Audit Report*. Accessed September 14, 2003 < www.aibonline.org >.

Anderson, Peggy. (2004). "Supermarket Chain Uses Discount Cards to Inform Customers of Recalls." Associate Press Article, *Agweek*, Jan 26. pg 50 and *Grand Forks Herald*, Jan 24. pg 3.

Anton, T.E. (2002). "Beef Alliances: A Basic Economic Overview." EDIS document FE 337, Florida Cooperative Extension Service, Department of Food and Resource Economics, Institute of Food and Agricultural Sciences, University of Florida, Gainesville, FL. May, Accessed November 25, 2003 < http://edis.ifas.ufl.edu/BODY_FE337 >.

Bartlesman, Eric J., Randy A. Becker, and Wayne B. Gray (2000). "NBER-CES Manufacturing Industry Database." Accessed May 2, 2003 < http://www.nber.org/nberces/nbprod96.htm >.

Blancou. J. (2002). "A History of the Traceability of Animals and Animal Products." Accessed October 20, 2003 < http://www.oie.int/eng/publicat/rt/2002/BLANCOUA.PDF >.

Brain, Marshall. "How UPC Bar Codes Work." *How Stuff Works* Accessed November 24, 2003 < http://www.howstuffworks.com/upc.htm >.

Branaman, Brenda (1999). "The Perishable Agricultural Commodities Act (PACA)." Congressional Report for Congress. October 26, Accessed March 4, 2003 <http://www.ncseonline.org/NLE/CRSreports/Agriculture/ag-85.cfm >.

California Department of Food and Agriculture, Marketing Branch. "California Cantaloupe Program: Effective February 25, 1988, Incorporating Amendments through May 10, 2000." Accessed March 4, 2003 < www.cdfa.ca.gov/ >.

Calvin, Linda, and Roberta Cook (coordinators), Mark Denbaly, Carolyn Dimitri, Lewrene Glaser, Charles Handy, Mark Jekanowski, Phil Kaufman, Barry Krissoff, Gary Thompson, and Suzanne Thornsbury (2001). *U.S. Fresh Fruit and Vegetable Marketing: Emerging Trade Practices, Trends, and Issues*. U.S. Department of Agriculture, Economic Research Service AER-795, January.

Colorado State University (1998). "Multi-State Outbreak of *Salmonella Agona.*" *Safefood News* 2:3 (Spring) Accessed November 24, 2003 < http://www.colostate.edu/Orgs/safefood/NEWSLTR/v2n3s01.html >.

Darby, Michael R., and Edi Karni (1973). "Free Competition and the Optimal Amount of Fraud," *Journal of Law and Economics* 16(1): 67-88.

Dimitri, Carolyn (2001). "Contract Evolution and Institutional Innovation: Marketing Pacific-Grown Apples from 1890 to 1930." *The Journal of Economic History*, Vol. 62, No. 1 (March).

Dimitri, C. and N. J. Richman (2000). "Organic Foods: Niche Marketers Venture into the Mainstream." *Agricultural Outlook* (June-July): 11-14.

Dunn, Darrell (2003). "Wal-Mart RFID Rollout To Start In Texas." *InformationWeek* (Nov. 5) Accessed November 24, 2003 < http://www.informationweek.com/story/showArticle.jhtml?articleID=16000271 >.

EAN.UCC. "About EAN International and the Uniform Code Council, Inc." Accessed November 24, 2003 < http://www.ean-ucc.org/home.htm >.

EPCglobal. "Welcome to EPCglobal" Accessed November 24, 2003 < http://www.epcglobalinc.org/ >.

Food and Drug Administration, Center for Food Safety and Applied Nutrition (2002). "Public Health Security and Bioterrorism Preparedness and Response Act of 2002 (PL107-188)." Accessed February 5, 2003 < http://www.cfsan.fda.gov/~dms/sec-ltr.html >.

Food and Drug Administration, Center for Food Safety and Applied Nutrition (1998). "Guide to Minimize Microbial Food Safety Hazards for Fresh Fruits and Vegetables." (October 26) Accessed March 4, 2003 < http://www.cfsan.fda.gov/~dms/prodguid.html >.

Food and Drug Administration, Center for Food Safety and Applied Nutrition. "What Are FDA Requirements Regarding Shellfish Certification?" Accessed January 9, 2003 < http://www.cfsan.fda.gov/~dms/qa-ind6f.html >.

"Global Trade Item Numbers Implementation Guide." Accessed April 28, 2003 < http://www.uc-council.org/documents/pdf/GTIN_0205.pdf >.

Golan, Elise, Tanya Roberts, Elisabete Salay, Julie Caswell, Michael Ollinger, and Danna Moore (2004). *Food Safety Innovation in the United States: Evidence from the Meat Industry*, Forthcoming Economic Research Service, U.S. Department of Agriculture. AER-831

Guidry, Steve, and Victor Muliyil. Société Générale de Surveillance, Personal discussion with Elise Golan, May 28, 2003.

Hayes, Dermot, and Steve Meyer (2003). "Impact of Mandatory Country of Origin Labeling on U.S. Pork Exports." Manuscript. February.

International Organization for Standardization. http://www.iso.ch/iso/en/aboutiso/introduction/index.html#six

ISO (2000). *ISO 9000:2000. Quality Management Standards.*

Ippolito, Pauline M., and Alan D. Mathios (1990). "The Regulation of Science-Based Claims in Advertising," *Journal of Consumer Policy* 13: 413-445.

Kansas Livestock Association, Key KLA Policy, http://www.kla.org/policy.htm. Sept. 8 2003.

Kaufman, Phil, Charles Handy, Edward McLaughlin, Kristen Park, and Geoffrey Green (2000). *Understanding the Dynamics of Produce Markets: Consumption and Consolidation Grow.* Economic Research Service, U.S. Department of Agriculture, AIB-758, August.

Karst, Tom (2003). "Produce Groups Oppose FDA Import Regulation Plan," *The Packer Online* February 5.

Lin, W. W., W. Chambers, and J. Harwood (2000). "Biotechnology: Grain Handlers Look Ahead." *Agricultural Outlook* (April) pp. 29-34.

Lin, W., G. K. Price, and E. Allen (2003). "StarLink: Impacts on the U.S. Corn Market and World Trade." *Agribusiness: An International Journal* 19(4) pp.473-488

Lusk, Jayson, Randall Little, Allen Williams, John Anderson, and Blair McKinley (2003). "Utilizing Ultrasound Technology to Improve Livestock Marketing Decisions." *Review of Agricultural Economics*, 25(1): 203-217.

Marion, Bruce W. (1986). *The Organization and Performance of the U.S. Food System* Lexington, Mass.: Lexington Books.

McEowen, Roger A. (1996). "Rights and Liabilities Arising from the Sale of Defective Agricultural Goods." MF-2035, Kansas State University. Accessed November 25, 2003 < http://www.oznet.ksu.edu/library/agec2/mf2035.pdf >. (April).

Organization for Economic Cooperation and Development (2003). "Costs and Benefits of Food Safety Regulation," Working Paper on Agricultural Policies and Markets AGRA/CA/APM(2002)18/REV1 17 (February).

The Packer. (2003) "Tackle traceability standards soon, speakers urge." October 27: C4.

Price, G. K., F. Kuchler, and B. Krissoff (2004). "E.U. Traceability and the U.S. Soybean Sector." In *The Regulation of Agricultural Biotechnology*, R. Evenson and V. Santaniello, eds. CABI Publishing, Oxford.

Price, G. K., F. Kuchler, and B. Krissoff. (2003) *Crop Values*, Pr 2(03) (February).

Roberts, T., C. Narrod, S. Malcolm, and M. Modarres (2001). "An Interdisciplinary Approach to Developing a Probabilistic Risk Analysis Model: Applications to a Beef Slaughterhouse," *Interdisciplinary Food Safety Research*, ed. by N. Hooker & E. Murano, CRC Press: Boca Raton, FL, 21-23.

Rowe, Greg (2001). "Reduced Space Symbology (RSS): Overview" (February) Accessed May 12, 2003 < http://www.cpma.ca/pdf/2002/Reduced%20Space%20Symbology.pdf >.

Schroeder, T.C. and J.L. Graff (2000). "Estimated Value of Increased Pricing Accuracy for Fed Cattle." *Review of Agricultural Economics* 22(1): 89-101.

Simchi-Levi, David, Philip Kaminsky, and Edith Simchi-Levi 2003. *Designing and Managing the Supply Chain.* New York: McGraw-Hill Irwin.

Simms, B.T. (1962). "A Century of Progress in Livestock Health." presented at the 66th annual meeting of the U.S. Animal Health Association in Washington, DC, Oct. 29-Nov. 2.

"State of Logistics Report" (2001). Published by Rosalyn Wilson and Robert V. Delaney and cited in Simchi-Levi et al. *Designing and Managing the Supply Chain* New York: McGraw-Hill Irwin.

Starbird, S. Andrew, and Vincent Amanor-Boadu (2003). "The Value of Anonymity: What are They Hiding?" Paper presented at the American Agricultural Economics Association annual meeting in Montreal, July 2, 2003.

Uniform Code Council. "Global Trade Item Numbers^{TM} (GTIN^{TM}) Implementation Guide." Accessed November 24, 2003 < http://www1.uc-council.org/ean_ucc_system/pdf/GTIN_0205.pdf >.

U.S. Department of Agriculture, Agricultural Marketing Service. "USDA Process Verified Program." Accessed Jan. 24, 2004 < http://processverified.usda.gov/ >.

U.S. Department of Agriculture, Agricultural Marketing Service. "National School Lunch Program." Accessed April 7, 2003 < http://www.ams.usda.gov/nslpfact.htm >.

U.S. Department of Agriculture, Agricultural Marketing Service. "Notice of Request for Emergency Approval of a New Information Collection," *Federal Register* 67(225) November 21, 2002.

U.S. Department of Agriculture, Agricultural Marketing Service. "Establishment of Guidelines for the Interim Voluntary Country of Origin Labeling of Beef, Lamb, Pork, Fish, Perishable Agricultural Commodities and Peanuts Under the Authority of the Agricultural Marketing Act of 1946." *Federal Register* 67(198) October 11, 2002.

U.S. Department of Agriculture, Agricultural Marketing Service (2002). "The National Organic Program." Accessed September 3, 2003 < http://www.ams.usda.gov/nop/FactSheets/Backgrounder.html >. (October).

U.S. Department of Agriculture, Animal and Plant Health Inspection Service. "Facts About Brucellosis". Accessed December 3, 2003 < http://www.aphis.usda.gov/vs/nahps/brucellosis >.

U.S. Department of Agriculture, Animal and Plant Health Inspection Service. "Animal Identification." Accessed October 24, 2003 < http://www.aphis.usda.gov/vs/highlights/section3/section3-10.html# >.

U.S. Department of Agriculture, Animal and Plant Health Inspection Service (2000a). Beef Cow-Calf Management Practices, National Animal Health Monitoring System (data from survey during 1997, http://www.aphis.usda.gov/vs/ceah/cahm/Beef_Cow-Calf/chapa/chapdes1.pdf)

U.S. Department of Agriculture, Animal and Plant Health Inspection Service (2000b). *Identification in Beef Cow-Calf Herds*. Info Sheet (http://www.aphis.usda.gov/vs/ceah/cahm/Beef_Cow-Calf/bf97id.htm)

U.S. Department of Agriculture, Animal and Plant Health Inspection Service (2000). "Information Sheet: Highlights of NAHMS Feedlot '99 Part I." (May) http://www.aphis.usda.gov/vs/ceah/cahm/Beef_Feedlot/fd99des1.pdf

U.S. Department of Agriculture, Food Safety and Inspection Service (2003). "Importing to the United States." Accessed January 28, 2004 < http://www.fsis.usda.gov/OPPDE/IPS/Importing.htm > (October 23).

U.S. Department of Agriculture, Food Safety and Inspection Service. "FSIS Security and Guidelines for Food Processors." Accessed February 5, 2003 < http://www.fsis.usda.gov/oa/topics/SecurityGuide.pdf >.

U.S. Department of Agriculture, Grain Inspection Packers and Stockyards Administration (2001). *Packers and Stockyards Statistical Report*, Tables 27,28, 29, GIPSA GR-03-1 (September).

U.S. Department of Agriculture, Office of the Chief Economist (2003). *World Agricultural Supply and Demand Estimates*. WASDE-397. Accessed April 10, 2003 < http://jan.mannlib.cornell.edu/usda/reports/waobr/wasde-bb/2003/wasde397.pdf >.

U.S. Department of Agriculture, National Agricultural Statistics Service (2003). *2002 Livestock Slaughter*. (March).

U.S. Department of Agriculture, National Agricultural Statistics Service. "Agricultural Statistics Data Base, U.S. and State Level Data/Hogs." Accessed January 29, 2004 < http://www.nass.usda.gov:81/ipedb/ >.

U.S. Department of Agriculture, National Agricultural Statistics Service. "Agricultural Statistics Data Base U.S. and State Level Data for Cattle & Calves." Accessed December 4, 2003 < http://www.nass.usda.gov:81/ipedb/ >.

U.S. Department of Commerce, Bureau of the Census (2000). "1997 Economic Census, Wholesale Trade, Commodity Line Sales." (August) Accessed November 25, 2003 < http://www.census.gov/prod/ec97/97w42-ls.pdf >.

U.S. Department of Commerce, Bureau of Economic Analysis. "Private Inventories and Domestic Final Sales of Business by Industry Group," Table 5.12A and "Private Inventories and Domestic Final Sales by Industry" Table 5.12B, National Income and Product Accounts, Accessed May 2, 2003 < http://www.bea.gov/ >.

The U.S. House Committee on Agriculture. "Federal Meat Inspection Act of 1906 – The U.S. House Committee on Agriculture Glossary." Accessed Feb. 28, 2003 < http://agriculture.house.gov/glossary/federal meat_inspection_act_of_1906.htm >.

Wiemers, John (2003). "The U.S. Animal Identification Plan." Presented at the Farm Foundation workshop on Traceability and Quality Assurance, Kansas City, Missouri, November 19, 2003.

Wohler, Shelly. Website Manager, American Institute of Baking, personal discussion with Elise Golan, May 29, 2003.

Index

A

I

J

K

L

M

N

O

P

S

T

U

V

W

Y

Z